Methods in Molecular Biology

For further volumes:
http://www.springer.com/series/7651

Long Non-Coding RNAs

Methods and Protocols

Edited by

Yi Feng

School of Medicine, University of Pennsylvania, Philadelphia, PA, USA

Lin Zhang

School of Medicine, University of Pennsylvania, Philadelphia, PA, USA

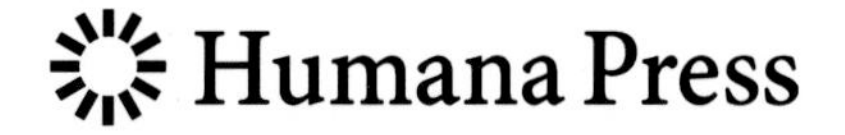

Editors
Yi Feng
School of Medicine
University of Pennsylvania
Philadelphia, PA, USA

Lin Zhang
School of Medicine
University of Pennsylvania
Philadelphia, PA, USA

ISSN 1064-3745 ISSN 1940-6029 (electronic)
Methods in Molecular Biology
ISBN 978-1-4939-3376-1 ISBN 978-1-4939-3378-5 (eBook)
DOI 10.1007/978-1-4939-3378-5

Library of Congress Control Number: 2015957247

Springer New York Heidelberg Dordrecht London

Printed on acid-free paper

Humana Press is a brand of Springer
Springer Science+Business Media LLC New York is part of Springer Science+Business Media (www.springer.com)

Preface

The advent of next-generation sequencing revealed that the majority of the human genome is transcribed to noncoding RNAs. Recent studies on long noncoding RNAs (lncRNAs) suggest that they may represent another class of important, nonprotein regulators of various biological processes, including cell proliferation, differentiation, migration, apoptosis, and transformation. In this column of the Methods in Molecular Biology series, we assemble a broad spectrum of methods used in lncRNA research, ranging from computational annotation of lncRNA genes to molecular and cellular analyses of the function of individual lncRNA. Methods used to study circular RNAs and RNA splicing also are included. Given that many well-characterized lncRNAs exhibit important pathological functions, we also include a chapter to describe the most influential finding on lncRNA in human diseases. We would like to make this book a must-have for anyone who conducts lncRNA research. The intended audience includes molecular biologists, cell and developmental biologists, specialists who conduct disease-oriented research, and bioinformatics experts who seek a better understanding on lncRNA expression and function by computational analysis of the massive sequencing data that are rapidly accumulating in recent years. It is our hope that Long Noncoding RNA Protocol will stimulate the reader to explore diverse ways to understand the mechanisms by which lncRNAs facilitate the molecular aspects of biomedical research. We would like to acknowledge and thank all authors for their valuable contributions, particularly for sharing with readers all hints, tips, and observations that one learns from using a method regularly. We also thank Dr. John Walker for his technical guidance and support.

Philadelphia, PA, USA

Yi Feng, Ph.D.
Lin Zhang, M.D.

Contents

Contributors

SUSAN BOERNER • *Department of Biological Science, Florida State University, Tallahassee, FL, USA*
VENERA D. BOURIAKOV • *Roche NimbleGen Inc., Madison, WI, USA*
DANIEL BURGESS • *Roche NimbleGen, Inc., Madison, WI, USA*
GEORGE A. CALIN • *Center for RNA Interference and Non-Coding RNAs, The University of Texas MD Anderson Cancer Center, Houston, TX, USA; Department of Experimental Therapeutics, The University of Texas MD Anderson Cancer Center, Houston, TX, USA*
LING-LING CHEN • *State Key Laboratory of Molecular Biology, Shanghai Key Laboratory of Molecular Andrology, Institute of Biochemistry and Cell Biology, Shanghai Institutes for Biological Sciences, Chinese Academy of Sciences, Shanghai, China; School of Life Science and Technology, ShanghaiTech University, Shanghai, China*
RUNSHENG CHEN • *Laboratory of Noncoding RNA, Institute of Biophysics, Chinese Academy of Sciences, Beijing, China*
CHONGHUI CHENG • *Division of Hematology/Oncology, Department of Medicine, Robert H. Lurie Comprehensive Cancer Center, Northwestern University Feinberg School of Medicine, Chicago, IL, USA*
NAVROOP K. DHALIWAL • *Department of Cell and Systems Biology, University of Toronto, Toronto, ON, Canada*
YI FAN • *Department of Radiation Oncology, Perelman School of Medicine, University of Pennsylvania, Philadelphia, PA, USA*
LIANG FENG • *Roche NimbleGen Inc., Madison, WI, USA*
YI FENG • *School of Medicine, University of Pennsylvania, Philadelphia, PA, USA*
SRIMONTA GAYEN • *Department of Human Genetics, University of Michigan Medical School, Ann Arbor, MI, USA*
MYRIAM GOROSPE • *Laboratory of Genetics and Genomics, National Institute on Aging-Intramural Research Program, National Institutes of Health, Baltimore, MD, USA*
KIRANMAI GUMIREDDY • *The Wistar Institute, Philadelphia, PA, USA*
CLAIR HARRIS • *Department of Human Genetics, University of Michigan Medical School, Ann Arbor, MI, USA*
SAMUEL E. HARVEY • *Division of Hematology/Oncology, Department of Medicine, Robert H. Lurie Comprehensive Cancer Center, Northwestern University Feinberg School of Medicine, Chicago, IL, USA*
HIROTO HATAKEYAMA • *Department of Gynecologic Oncology and Reproductive Medicine, The University of Texas MD Anderson Cancer Center (MDACC), Houston, TX, USA*
ARTEMIS G. HATZIGEORGIOU • *DIANA-Lab, Department of Electrical and Computer Engineering, University of Thessaly, Volos, Greece*
MICHAEL HINTEN • *Department of Human Genetics, University of Michigan Medical School, Ann Arbor, MI, USA*
XIAOWEN HU • *Center for Research on Reproduction and Women's Health, Perelman School of Medicine, University of Pennsylvania, Philadelphia, PA, USA*

ZHONGYI HU • *Center for Research on Reproduction and Women's Health, Perelman School of Medicine, University of Pennsylvania, Philadelphia, PA, USA*
QIHONG HUANG • *The Wistar Institute, Philadelphia, PA, USA*
HANS E. JOHANSSON • *Research and Development, LGC Biosearch Technologies, Petaluma, CA, USA*
SUNDEEP KALANTRY • *Department of Human Genetics, University of Michigan Medical School, Ann Arbor, MI, USA*
YOUNGSOO KIM • *Antisense Drug Discovery, Oncology, Isis Pharmaceuticals Inc., Carlsbad, CA, USA*
LAN-TIAN LAI • *School of Biological Sciences, Nanyang Technological University, Singapore, Singapore*
CHUNRU LIN • *Department of Molecular and Cellular Oncology, The University of Texas MD Anderson Cancer Center, Houston, TX, USA; Cancer Biology Program, The University of Texas Graduate School of Biomedical Sciences, Houston, TX, USA*
GABRIEL LOPEZ-BERESTEIN • *Center for RNA Interference and Non-Coding RNAs, The University of Texas MD Anderson Cancer Center (MDACC), Houston, TX, USA; Department of Cancer Biology, The University of Texas MD Anderson Cancer Center (MDACC), Houston, TX, USA; Department of Experimental Therapeutics, The University of Texas MD Anderson Cancer Center (MDACC), Houston, TX, USA*
EMILY MACLARY • *Department of Human Genetics, University of Michigan Medical School, Ann Arbor, MI, USA*
A. ROBERT MACLEOD • *Antisense Drug Discovery, Oncology, Isis Pharmaceuticals Inc., Carlsbad, CA, USA*
LINGEGOWDA S. MANGALA • *Department of Gynecologic Oncology and Reproductive Medicine, The University of Texas MD Anderson Cancer Center (MDACC), Houston, TX, USA; Center for RNA Interference and Non-Coding RNA, The University of Texas MD Anderson Cancer Center (MDACC), Houston, TX, USA*
BOTOUL MAQSODI • *Affymetrix, Santa Clara, CA, USA*
KAREN M. MCGINNIS • *Department of Biological Science, Florida State University, Tallahassee, FL, USA*
ZHENYU MENG • *Division of Chemistry and Biological Chemistry, School of Physical and Mathematical Sciences, Nanyang Technological University, Singapore, Singapore*
JENNIFER A. MITCHELL • *Department of Cell and Systems Biology, University of Toronto, Toronto, ON, Canada*
CORINA NIKOLOFF • *Affymetrix, Santa Clara, CA, USA*
ARTURO V. ORJALO JR. • *Research and Development, LGC Biosearch Technologies, Petaluma, CA, USA; Biological Technologies, Analytical Development and Quality Control, Genentech Inc., South San Francisco, CA, USA*
MARIA D. PARASKEVOPOULOU • *DIANA-Lab, Department of Electrical and Computer Engineering, University of Thessaly, Volos, Greece*
TODD A. RICHMOND • *Roche NimbleGen Inc., Madison, WI, USA*
JONATHAN SCHUG • *Department of Genetics, and Institute for Diabetes, Obesity and Metabolism, Perelman School of Medicine, University of Pennsylvania, Philadelphia, PA, USA*
JINDONG SHANG • *Arraystar Inc., Rockville, MD, USA*
FANGWEI SHAO • *Division of Chemistry and Biological Chemistry, School of Physical and Mathematical Sciences, Nanyang Technological University, Singapore, Singapore*
KARYN L. SHEAFFER • *Department of Genetics, and Institute for Diabetes, Obesity and Metabolism, Perelman School of Medicine, University of Pennsylvania, Philadelphia, PA, USA*

YANGGU SHI • *Arraystar Inc., Rockville, MD, USA*
ANIL K. SOOD • *Department of Gynecologic Oncology and Reproductive Medicine, The University of Texas MD Anderson Cancer Center (MDACC), Houston, TX, USA; Center for RNA Interference and Non-Coding RNAs, The University of Texas MD Anderson Cancer Center (MDACC), Houston, TX, USA; Department of Cancer Biology, The University of Texas MD Anderson Cancer Center (MDACC), Houston, TX, USA*
JOHN C. TAN • *Roche NimbleGen Inc., Madison, WI, USA*
SHERRY Y. WU • *Department of Gynecologic Oncology and Reproductive Medicine, The University of Texas MD Anderson Cancer Center (MDACC), Houston, TX, USA*
PENG XIANG • *Department of Biology, Zhongshan School of Medicine, Center for Stem Cell Biology and Tissue Engineering, Sun Yat-Sen University, Guangzhou, People's Republic of China*
ZHEN XING • *Department of Molecular and Cellular Oncology, The University of Texas MD Anderson Cancer Center, Houston, TX, USA*
MINMIN XIONG • *Department of Biology, Zhongshan School of Medicine, Center for Stem Cell Biology and Tissue Engineering, Sun Yat-Sen University, Guangzhou, People's Republic of China*
CONGJIAN XU • *Shanghai Key Laboratory of Female Reproductive Endocrine Related Diseases, Obstetrics and Gynecology Hospital of Fudan University, Shanghai, People's Republic of China*
XIAOWEI XU • *Department of Pathology and Laboratory Medicine, Perelman School of Medicine, University of Pennsylvania, Philadelphia, PA, USA*
JINCHUN YAN • *University of Washington Medical Center, Seattle, WA, USA*
LI YANG • *Key Laboratory of Computational Biology, CAS-MPG Partner Institute for Computational Biology, CAS Center for Excellence in Brain Science, Shanghai Institutes for Biological Sciences, Chinese Academy of Sciences, Shanghai, China*
LIUQING YANG • *Department of Molecular and Cellular Oncology, The University of Texas MD Anderson Cancer Center, Houston, TX, USA; Cancer Biology Program, The University of Texas Graduate School of Biomedical Sciences, Houston, TX, USA; The Center for RNA Interference and Non-Coding RNAs, The University of Texas MD Anderson Cancer Center, Houston, TX, USA*
JE-HYUN YOON • *Department of Biochemistry and Molecular Biology, Medical University of South Carolina, Charleston, SC, USA*
CHAO-XING YUAN • *Department of Pharmacology, Perelman School of Medicine, University of Pennsylvania, Philadelphia, PA, USA*
JIAO YUAN • *Laboratory of Noncoding RNA, Institute of Biophysics, Chinese Academy of Sciences, Beijing, China*
DONGMEI ZHANG • *State Key Laboratory of Biotherapy/Collaborative Innovation Center of Biotherapy, West China Hospital, Sichuan University, Chengdu, People's Republic of China*
LI-FENG ZHANG • *School of Biological Sciences, Nanyang Technological University, Singapore, Singapore*
LIN ZHANG • *School of Medicine, University of Pennsylvania, Philadelphia, PA, USA*
XINNA ZHANG • *Center for RNA Interference and Non-Coding RNAs, The University of Texas MD Anderson Cancer Center, Houston, TX, USA; Department of Gynecologic Oncology, The University of Texas MD Anderson Cancer Center, Houston, TX, USA*

YANG ZHANG • *State Key Laboratory of Molecular Biology, Shanghai Key Laboratory of Molecular Andrology, Institute of Biochemistry and Cell Biology, Shanghai Institutes for Biological Sciences, Chinese Academy of Sciences, Shanghai, China*

YOUYOU ZHANG • *Center for Research on Reproduction and Women's Health, Perelman School of Medicine, University of Pennsylvania, Philadelphia, PA, USA*

YI ZHAO • *Key Laboratory of Intelligent Information Processing, Institute of Computing Technology, Chinese Academy of Sciences, Beijing, China*

XIAOMIN ZHONG • *Department of Biology, Zhongshan School of Medicine, Center for Stem Cell Biology and Tissue Engineering, Sun Yat-Sen University, Guangzhou, People's Republic of China; Shanghai Key Laboratory of Female Reproductive Endocrine Related Diseases, Obstetrics and Gynecology Hospital of Fudan University, Shanghai, People's Republic of China*

TIANYUAN ZHOU • *Antisense Drug Discovery, Oncology, Isis Pharmaceuticals, Inc., Carlsbad, CA, USA*

Chapter 1

LncRNA Pulldown Combined with Mass Spectrometry to Identify the Novel LncRNA-Associated Proteins

Zhen Xing, Chunru Lin, and Liuqing Yang

Abstract

Long noncoding RNAs (LncRNAs) are nonprotein-coding transcripts longer than 200 nucleotides in length. The recent studies have revealed that at least nearly 80 % transcripts in human cells are lncRNA species. Based on their genomic location, most lncRNAs can be characterized as large intergenic noncoding RNAs, natural antisense transcripts, pseudogenes, long intronic ncRNAs, as well as other divergent transcripts. However, despite mounting evidences suggesting that many lncRNAs are likely to be functional, only a small proportion has been demonstrated to be biologically and physiologically relevant due to their lower expression levels and current technique limitations. Thus, there is a greater need to design and develop new assays to investigate the real function of lncRNAs in depth in various systems. Indeed, several methods such as genome-wide chromatin immunoprecipitation-sequencing (ChIP-seq), RNA immunoprecipitation followed by sequencing (RIP-seq) have been developed to examine the genome localization of lncRNAs and their interacting proteins in cells. Here we describe an open-ended method, LncRNA pulldown assay, which has been frequently used to identify its interacting protein partners in the cellular context. Here we provide a detailed protocol for this assay with hands-on tips based on our own experience in working in the lncRNA fields.

Key words Long noncoding RNAs, RNA-protein interaction, Mass spectrometry

1 Introduction

Increasing evidence has indicated that long noncoding RNA (lncRNA) is a new class of players associated with the development and progression of cancer [1], but knowledge of the mechanisms by which they act is still unclear. Several well-studied lncRNAs have provided us important clues about the biology and human disease relevance of these molecules and a few key functional and mechanistic themes have begun to be understood [2, 3]. It is believed that lncRNAs interact with proteins, such as epigenetic modifiers, transcriptional factors/coactivators, and RNP complexes, to regulate the related biological processes [3]. Of note, the specific lncRNA-protein interactions could also be mediated by increasingly identified RNA-binding domains [4, 5]. Here, we

Yi Feng and Lin Zhang (eds.), *Long Non-Coding RNAs: Methods and Protocols*, Methods in Molecular Biology, vol. 1402,
DOI 10.1007/978-1-4939-3378-5_1, © Springer Science+Business Media New York 2016

summarize an unbiased and open-ended lncRNA pulldown assay for better understanding of the lncRNA-associated protein partners. We believe that this methodology will drive the advances of studying lncRNA functions and the related mechanisms.

We used lncRNA pulldown assays to selectively extract a protein-RNA complex from a cell lysate based on principles of physical forces between proteins and RNA. These forces, including dipolar interactions, electrostatic interaction, entropic effects, and dispersion forces, form different degrees during the process of protein binding in a sequence-specific manner (tight).

Also, RNA secondary and tertiary structures play critical roles in protein recognition and binding to certain nucleic acid sequences (loose). Typically, the pulldown assay uses a RNA probe labeled with a high-affinity biotin tag which allows the probe to be recovered. After incubation, the biotinylated RNA probe can bind with a protein/protein complex in a cell lysate and then the complex is purified using magnetic streptavidin beads. The proteins are then eluted from the RNA and detected by Western blot or mass spectrometry. This assay has the advantages of enrichment of low-abundance protein targets, isolation of intact protein complexes, and being compatible with immunoblotting and mass spectrometry analysis.

With this technology, lncRNAs have been shown to regulate chromatin states through recruiting chromatin-modifying complexes PRC2 [6, 7]; regulate transcription programs through binding to androgen receptors [8]; and direct cooperative epigenetic regulation downstream of chemokine signals [9]. In this chapter, we describe a hands-on protocol for conducting a lncRNA pulldown assay with some tips gained from our own direct experiences [8, 9].

2 Materials and Reagents

Owing to widespread contaminating RNase in the experimental environment, we recommend that throughout this assay, all standard precautions should be taken to minimize RNase contamination. All reagents, glassware, and plasticware should be purchased RNase free or be treated with 0.1 % DEPC to make sure that they are RNase free (*see* **Note 1**).

3 Reagents

3.1 For Preparation of In Vitro-Transcribed RNA (IVT)

1. pGEM-3Z Vector (Promega, P2151).
2. Biotin RNA labeling mix.
3. T7 and SP6 RNA polymerase.

4. TURBO DNase.
5. Anti-RNase.
6. RNA Clean and Concentrator (Zymo Research, R1015).
7. Nuclease-free H_2O.
8. UltraPure Agarose (Life Technology, 16500–100).
9. Formaldehyde.
10. 10× MOPS Buffer (Lonza, 50876).
11. Ethidium bromide.
12. RNA loading dye.
13. RNA marker.

3.2 For Preparation of Cell Lysate

1. ProteaPrep Zwitterionic Cell Lysis Kit, Mass Spec Grade (Protea, SP-816).
2. Protease/phosphatase inhibitor cocktail.
3. Panobinostat.
4. Methylstat.
5. 1× Phosphate-buffered saline (PBS) buffer.

3.3 For lncRNA Pulldown

1. BcMag™ Monomer Avidin Magnetic Beads (Bioclone, MMI101).
2. RNA structure buffer
   Tris–HCl (pH 7.0) 10 mM, KCl 0.1 M, $MgCl_2$ 10 mM.
3. RNA capture buffer
   Tris–HCl (pH 7.5) 20 mM, NaCl 1 M, EDTA 1 mM.
4. NT2 buffer
   Tris–HCl (pH 7.4) 50 mM, NaCl 150 mM, MgCl2 1 mM, NP-40 0.05 %.
5. Wash buffer
   NT2 buffer supplemented with sequentially increased NaCl (500 mM to 1 M) or KSCN (750 mM).

3.4 For In-Solution Tryptic Digestion

1. Digestion buffer, ammonium bicarbonate (50 mM).
2. Reducing buffer, DTT (100 mM).
3. Alkylation buffer, iodoacetamide in NH_4HCO_3 (100 mM).
4. Trifluoroacetic acid (TFA), 10 %.
5. Acetonitrole.
6. Peptide recovery buffer, 1 ml.
   400 μl Acetonitrile, 20 μl 10 % TFA, 580 μl 50 mM NH_4HCO_3.
7. Immobilized trypsin.

4 Methods

4.1 IVT of Biotin-Labeled lncRNA

The following protocol is based on the previously method with some modifications according to our own experience (Fig. 1).

4.1.1 Cloning the lncRNA Gene Sequence into pGEM-3Z Vector

The gene sequence should be ligated into the multiple cloning sites locating between the T7 promoter and SP6 promoter.

T7 promoter: 5′ TAA TAC GAC TCA CTA TAG GG 3′
Sp6 promoter: 5′ AAT TTA GGT GAC ACT ATA GAA 3′

4.1.2 Lineation of the pGEM-3Z-lncRNA Vector

Add the restriction endonuclease (RE) to the plasmid.

20 μg of midi-prepared plasmid.
5 μl 10× RE buffer.
5 Unit RE enzyme.
ddH_2O.

Bring the total volume to 50 μl with H_2O and incubate at 37 °C from 2 h to overnight. Purify the linearized plasmid using the QIAquick Gel Purification Kit following the manufacturer's protocol (*see* **Note 2**).

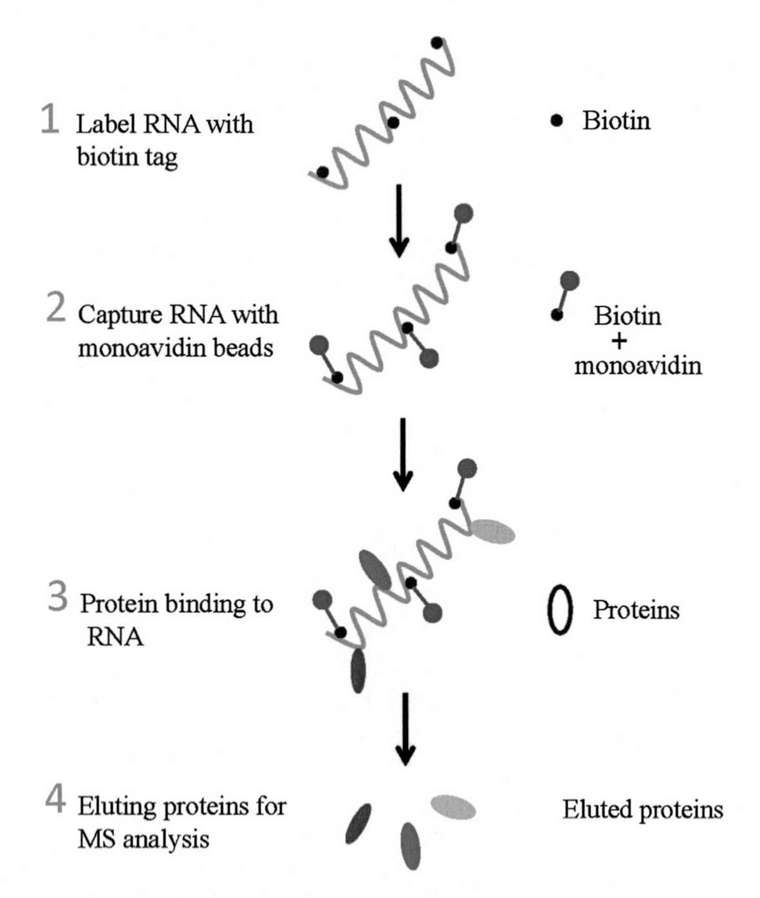

Fig. 1 Schematic diagram showing lncRNA pulldown

4.1.3 IVT Reaction

This reaction is used for the synthesis of "run-off" transcript.

Reaction components.
Nuclease-free H_2O.
Linearized plasmid 1 μg.
Biotin RNA labeling mix, 10×2 μl.
10× transcription buffer 2 μl.
T7/SP6 RNA polymerase 2 μl.

Add the above to a microfuge tube on ice and bring the total volume to 20 μl with the nuclease-free H_2O; mix and centrifuge briefly.

Incubate for 2 h at 37 °C.

Add 2 μl DNase to remove template DNA by incubation for 15 min at 37 °C.

Stop the reaction by adding 2 μl 0.2 M EDTA (pH 8.0) (*see* **Note 3**).

4.1.4 Purification and Size/Yield Check of IVT RNA

Following the DNase treatment and termination of the above reaction. Typically, up to 10 μg IVT RNA eluted into 10–20 μl RNase-free water can be obtained.

RNA Clean and Concentrator Kit Can Be Used to Recover IVT RNA Fragments

Preparing Denatured RNA Gel to Confirm if the RNAs Are Transcribed at the Right Size

Making 100 ml denatured RNA gel.

1 g agarose in 72 ml water heated until dissolved and then cooled to 60 °C.

10 ml 10× MOPS running buffer.

18 ml 37 % formaldehyde.

Prepare the RNA sample and electrophoresis.

Take 1 μl RNA sample, add 5 μl formaldehyde loading dye, 1 μl ethidium bromide (10 μg/ml).

Heat denatured samples at 65–70 °C for 5–15 min.

Load the gel and electrophorese at 5–6 V/cm as far as 2/3 the length of the gel.

Check the RNA size under the UV light. One sharp band with the correct size is a good indication that the RNA is efficiently transcribed. A smeared appearance may be resulted from the enzyme that failed to bind the promoter efficiently or dropped off the template in the half way (*see* **Note 4**).

4.2 Preparation of Cell Lysate

Culture and collect sufficient cells (at least 10×10 cm plate cells for each pulldown); suspend the cell pellet with 1 ml of ProteaPrep Zwitterionic cell lysis buffer supplemented with protease and phosphatases inhibitor cocktails (1:100), anti-RNase (1U/μl), panobinostat (1:100), and methylstat (1:100); incubate the tube on ice for 40 min, and vortex every 10 min; centrifuge the mixture at 14,000 ×*g* for 15 min at 4 °C; transfer supernatant to a new tube for further use (*see* **Note 5**).

4.3 Activation of BcMag™ Monomer Avidin Magnetic Beads

Prior to use, all the reagents in the kit should be equilibrated to room temperature and make 1× working solutions with ddH_2O. Follow the manufacturer's instructions to prepare the activated beads briefly summarized as follows:

1. Gently shake the bottle containing beads until completely suspended. Transfer 50 μl beads to a fresh tube.
2. Place the tube on a magnetic separator for 1 min, and remove the supernatant.

   Note: While the tube remains on the separator remove the tube from the separator for 1 min to prevent bead loss.
3. Wash beads with 200 μl of ddH_2O.
4. Wash the beads with 200 μl of 1× PBS buffer.
5. Block the beads with 150 μl of 1× blocking/elution buffer at room temperature for 5 min. Remove the buffer.
6. Regenerate the beads by adding in 300 μl of 1× regeneration buffer. Remove the buffer.
7. Wash the beads with 200 μl of 1× PBS buffer. The beads are ready to use (*see* **Note 6**).

4.4 lncRNA Pulldown

4.4.1 Pre-clearing the Lysate with Activated Beads to Reduce Nonspecific Binding

1. Remove the PBS from the activated avidin beads.
2. Add 1 ml of cell lysate to the beads and mix well.
3. Incubate for 1 h at 4 °C with end-to-end gentle rotation.
4. Place the tube on a magnetic separator and keep the supernatant for further use. Discard bead pellet.

4.4.2 Folding of IVT RNA

1. 20 μg of biotinylated RNA was heated to 90 °C for 2 min.
2. Chill on ice for 2 min and briefly centrifuge.
3. Bring the total volume to 100 μl with RNA structure buffer.
4. Incubate at room temperature for 20 min to allow proper secondary structure formation.

4.4.3 LncRNA Pulldown

1. Prepare the 50 μl activated avidin magnetic beads.
2. The beads were immediately subject to RNA (20 μg) capture in RNA capture buffer for 30 min at room temperature with gentle agitation.
3. Remove the RNA capture buffer and the RNA-captured beads were washed once with NT2 buffer and incubated with 30 mg pre-cleared cell lysate for 2 h at 4 °C with gentle rotation.

4. Briefly centrifuge the tube and place the tube on a magnetic separator. Remove the supernatant from the beads and discard. The complex of interest should now bind to the beads. The following steps are used to remove the nonspecific binding to the beads.
5. Wash the RNA-binding protein complexes with NT2 buffer twice.
6. Wash the RNA-binding protein complexes with NT2 high-salt buffer (500 mM NaCl) twice.
7. Wash the RNA-binding protein complexes with NT2 high-salt buffer (1 M NaCl) once.
8. Wash the RNA-binding protein complexes with NT2 high-salt buffer (750 mM KSCN) once.
9. Wash the RNA-binding protein complexes with PBS twice (*see* **Note 7**).
10. Elute the beads by incubating the sample with elution buffer for 20 min at 4 °C with frequent agitation. Collect the elute.
11. Repeat elution and pool the elute together.

4.5 MS Analysis

Mass spectrometry (MS) has become the powerful and frequently used choice for protein detection, identification, and quantitation. Due to the complexity of whole workflow, there is no fixed method from sample preparation, instrumentation, and software analysis. Here we summarize the detailed protocol for in-solution digestion with hands-on tips because we think the quality of digested samples also significantly impacts MS results. Next, the resultant peptides can be subject to MS analysis. Here, we recommend the regular liquid chromatography-mass spectrometry (LC-MS) which can be handled by most proteomics facilities in different institutions (Fig. 2).

Procedure for In-Solution Digestion

1. Concentrate and dry the eluted samples with Speedy Vacuum system.
2. Dissolve the dry elute in 10 μl ddH_2O and prepare the reduction reaction.

   15 μl Digestion buffer.
   10 μl Dissolved eluted samples.
   1.5 μl Reducing buffer.
   Bring the total volume to 27 μl with ddH_2O and incubate the reaction at 95 °C for 5 min. Allow samples to cool to room temperature.

3. Perform alkylation reaction by adding 3 μl alkylation buffer to the above tube and incubate in the dark at room temperature for 20 min.

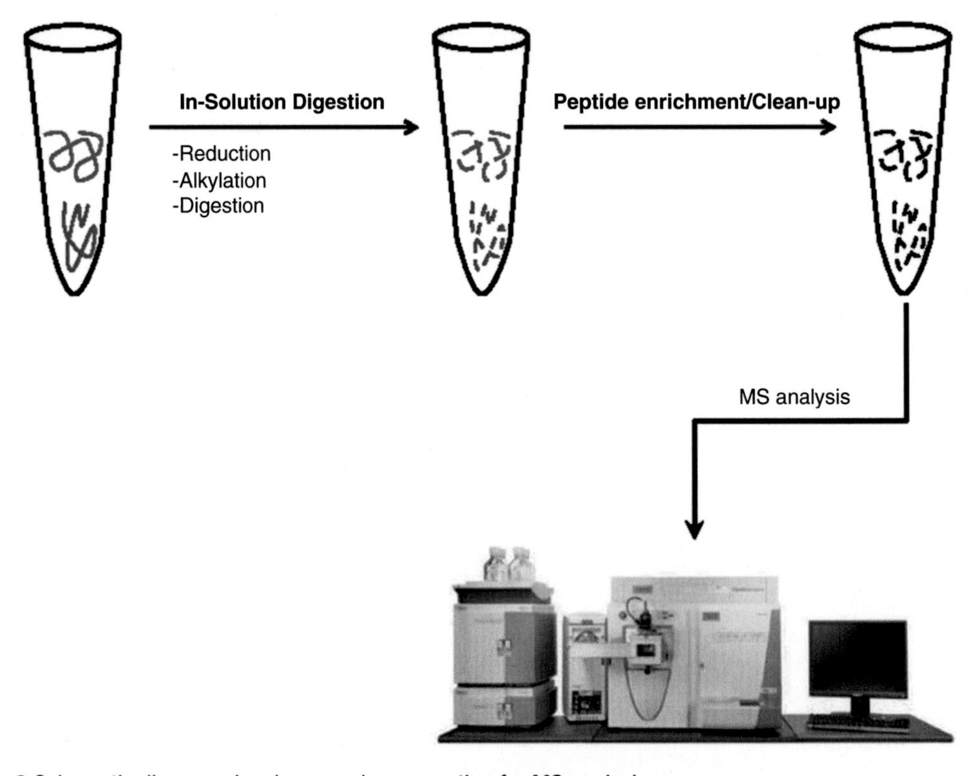

Fig. 2 Schematic diagram showing sample preparation for MS analysis

4. Perform the trypsin digestion overnight with the immobilized trypsin kit following the manufacturer's instructions.
5. The recovered digested peptides were subject to the LS-MS analysis.

5 Notes

1. Because RNA is more susceptible to degradation and RNases are found in all cell types and also widespread in the air, most surfaces, and dust, we strongly recommend all the materials, reagents, and solutions be prepared carefully or purchased commercially certified for RNA work [10].
2. Start with at least 10 μg of plasmid, since most commercial gel purification kit has not more than 50 % recover efficiency, and too low concentration of template will significantly affect the transcription efficiency. We recommend to add the BSA in the restriction enzyme digestion reaction since this will avoid the star activity.
3. Make sure that the insert direction of lncRNA gene fragment into the vector and the use of T7 or SP6 RNA polymerase allow for transcription of both strands of the cloned DNA

(sense and antisense transcripts). Antisense transcripts and beads only can be used as two negative controls for the pulldown experiment, which will be critical for the elucidation of final MS results.

4. The quality and yield of IVT RNA are critical factors for a successful lncRNA pulldown experiment. A sharp and pure band without any smear is recommended for the further experiment.
5. The protease/phosphatase inhibitor cocktail protects protein from degradation by endogenous proteases and phosphatases released during protein extraction and purification. Panobinostat is a deacetylase inhibitor and methylstat is a selective inhibitor of the Jumonji C domain-containing histone demethylases. The supplement of these inhibitors provides the higher chances of identification of novel modification sites related to lncRNA-associated protein/protein complex.
6. These beads are 1 μm, silica-based superparamagnetic beads coated with high density of ultrapure (97 %) avidin subunit monomer on the surface. The bound biotinylated molecules can be easily eluted from the beads by mild elution conditions such as 2 mM biotin-containing buffer. It is very critical that the beads must be used immediately after activation or binding capacity will be dramatically reduced. Some other forms of avidin beads may also be used and the final elution condition needs to be titrated.
7. For each wash, mix the beads gently with wash buffer and then subject to gentle rotation at 4 °C for 5 min. Make sure to carefully remove as much wash buffer as possible from the beads and prevent bead loss.

References

1. Fatica A, Bozzoni I (2014) Long non-coding RNAs: new players in cell differentiation and development. Nat Rev Genet 15:7–21
2. Kung JT, Colognori D, Lee JT (2013) Long noncoding RNAs: past, present, and future. Genetics 193:651–669
3. Rinn JL, Chang HY (2012) Genome regulation by long noncoding RNAs. Annu Rev Biochem 81:145–166
4. Castello A, Fischer B, Eichelbaum K, Horos R, Beckmann BM, Strein C et al (2012) Insights into RNA biology from an atlas of mammalian mRNA-binding proteins. Cell 149: 1393–1406
5. Lunde BM, Moore C, Varani G (2007) RNA-binding proteins: modular design for efficient function. Nat Rev Mol Cell Biol 8:479–490
6. Gupta RA, Shah N, Wang KC, Kim J, Horlings HM, Wong DJ et al (2010) Long non-coding RNA HOTAIR reprograms chromatin state to promote cancer metastasis. Nature 464: 1071–1076
7. Tsai M-C, Manor O, Wan Y, Mosammaparast N, Wang JK, Lan F et al (2010) Long noncoding RNA as modular scaffold of histone modification complexes. Science 329:689–693
8. Yang L, Lin C, Jin C, Yang JC, Tanasa B, Li W et al (2013) lncRNA-dependent mechanisms of androgen-receptor-regulated gene activation programs. Nature 500:598–602
9. Xing Z, Lin A, Li C, Liang K, Wang S, Liu Y et al (2014) lncRNA directs cooperative epigenetic regulation downstream of chemokine signals. Cell 159:1110–1125
10. Sambrook J, Fritsch EF, Maniatis T (1989) Molecular cloning: a laboratory manual. Cold Spring Harbor Laboratory Press, Cold Spring Harbor

Chapter 2

Cross-Linking Immunoprecipitation and qPCR (CLIP-qPCR) Analysis to Map Interactions Between Long Noncoding RNAs and RNA-Binding Proteins

Je-Hyun Yoon and Myriam Gorospe

Abstract

Mammalian cells express a wide range of transcripts, some protein-coding RNAs (mRNA) and many noncoding (nc) RNAs. Long (l)ncRNAscan modulates protein expression patterns by regulating gene transcription, pre-mRNA splicing, mRNA export, mRNA degradation, protein translation, and protein ubiquitination. Given the growing recognition that lncRNAs have a robust impact upon gene expression, there is rising interest in elucidating the levels and regulation of lncRNAs. A number of high-throughput methods have been developed recently to map the interaction of lncRNAs and RNA-binding proteins (RBPs). However, few of these approaches are suitable for mapping and quantifying RBP-lncRNA interactions. Here, we describe the recently developed method CLIP-qPCR (cross-linking and immunoprecipitation followed by reverse transcription and quantitative PCR) for mapping and quantifying RBP-lncRNA interactions.

Key words CLIP, lncRNA, RBP, Ribonucleoprotein complexes, qPCR

1 Introduction

Gene expression programs are influenced transcriptionally via processes such as chromatin remodeling and transcription factor mobilization, and posttranscriptionally through processes like pre-mRNA splicing, 5′ capping, 3′ polyadenylation, editing, mRNA export, localization, and translation. These processes are governed by DNA-binding proteins, RNA-binding proteins (RBPs), and noncoding RNAs [1–4]. In recent years, lncRNAs have emerged as major regulators of transcription, mRNA fate, and protein stability [4–6]. Although they can perform these gene regulatory functions directly via sequence complementarity with target DNA and RNA sequences, most lncRNA functions identified to date involve their interaction with partner DNA-binding proteins and RBPs [4, 6].

RBP-RNA complexes have been studied using a variety of methods [7]. Interactions of specific RBPs and specific RNAs have been examined using traditional procedures such as the analysis of

Yi Feng and Lin Zhang (eds.), *Long Non-Coding RNAs: Methods and Protocols*, Methods in Molecular Biology, vol. 1402,
DOI 10.1007/978-1-4939-3378-5_2, © Springer Science+Business Media New York 2016

ribonucleoprotein immunoprecipitation (RIP) complexes, biotinylated RNA pulldown complexes, and RNA electrophoretic mobility shift assays [8]. Global analyses of mRNAs interacting with RBP were developed using RIP followed by cDNA microarray analysis [9] or by RIP followed by RNA sequencing (RNA-seq [10]). Subsequently, RBP cross-linking using UVC light (254 nm) and immunoprecipitation (CLIP) was developed, initially combined with Sanger Sequencing method [11] and later with high-throughput sequencing (RNA-seq) of the CLIP cDNA library [12, 13]. Further advances were achieved by photoactivatable ribonucleoside-enhanced (PAR)-CLIP analysis, using nucleotide analogs such as 4-thiouridine (4-SU) or 6-thioguanosine (6-SG) following exposure to UVA light (365 nm) [14]. Alternative CLIP methods are also known as individual nucleotide resolution CLIP [15] using reverse transcriptase stall and cross-linking, ligation, and sequencing of hybrids (CLASH), which monitor inter-RNA interactions within tripartite complexes [16].

Despite such enormous advances in RBP-RNA detection, high-throughput analyses require the generation of a small RNA library, and thus a significant amount of effort, cost, time, and optimization. In addition, a substantial amount of bioinformatic analysis is required to map the sequence reads and identify the binding sites. Therefore, we have developed a protocol that combines PAR-CLIP with qPCR analysis to map RBP-binding sites at 100-nt intervals within an lncRNA of interest. This method is based on the novel introduction of partial RNase digestion and the detection of serial, progressive scanning of the target RNA by reverse transcription and qPCR (Fig. 1).

2 Materials

2.1 Mammalian Cell Culture and RNA Labeling

1. DMEM, 10 % FBS, 2 mM L-glutamine, 100 U/ml penicillin/streptomycin.
2. 1 M 4-Thiouridine (4-SU) in DMSO.

2.2 UVA Cross-Linking and Cell Harvesting

1. UV cross-linker.
2. Stratalinker 1800, 365 nm bulb.
3. Ice-cold phosphate-buffered saline(PBS).

2.3 RNase Digestion and Immunoprecipitation

1. RNase T1.
2. Protein A or G Sepharose beads.
3. Antibodies to detect an endogenous protein of interest.
4. Normalized IgG.
5. Proteinase K.

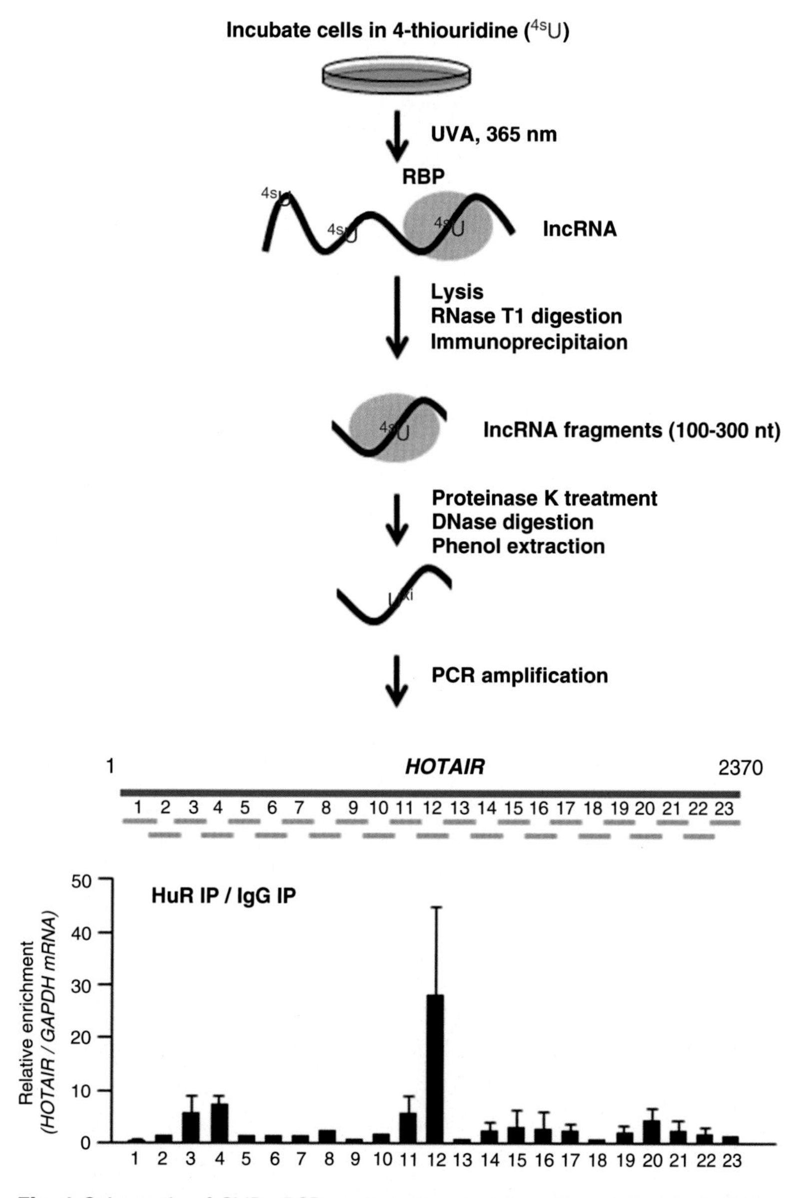

Fig. 1 Schematic of CLIP-qPCR analysis to map interactions between lncRNAs and RBPs. Following the initial PAR-CLIP steps (labeling, cross-linking, and lysis), partial digestion aimed at generating RNA fragments ranging 100–300 nt was performed using RNase T1. After IP using an antibody that recognized a specific RBP (HuR in this case, with IgG in control IP reactions), samples were digested with Proteinase K and DNase I. Following extraction of RNA, PCR amplification was used to quantify the relative abundance of overlapping segments spanning the RNA of interest, in this case the 2370-nt-long lncRNA*HOTAIR*, in order to identify the RNA regions preferentially bound to HuR, and to map the regions of greatest association, identified as those corresponding to fragments 11–12, with minor binding detected around fragments 3–4

6. DNase I.
7. Acidic phenol.
8. Glycoblue.
9. NP-40 lysis buffer:20 mM Tris–HCl at pH 7.5, 100 mM KCl, 5 mM $MgCl_2$, and 0.5 % NP-40.
10. NT2 buffer: 50 mM Tris–HCl at pH 7.5, 150 mM NaCl, 1 mM $MgCl_2$, and 0.05 % NP-40.

2.4 Reverse Transcription and qPCR

1. dNTP mix (10 mM).
2. Random hexamer(150 ng/μl).
3. Reverse transcriptase.
4. Gene-specific primer sets.
5. SYBR mix.

3 Methods

3.1 Mammalian Cell Culture and RNA Labeling

1. Expand human embryonic kidney (HEK) or human cervical cancer (HeLa) cells in 150 mm culture plates, cultured in DMEM supplemented with 10 % (v/v) fetal bovine serum and antibiotics. Grow cells up to 80 % confluence.
2. Sixteen hours before UVA exposure, add 4SU (1 M stock solution in DMSO) to a final concentration of 100 μ min culture medium. Alternatively, 100 μM 6SG can be used, but the cross-linking efficiency is somewhat lower.

3.2 UVA Cross-Linking and Harvesting Cells

1. Aspirate the culture medium, wash cells once with 15 ml ice-cold PBS, and aspirate PBS completely.
2. Set up a tray containing ice and place the plate on the ice.
3. Uncover the plate and irradiate it with 150 mJ/cm² of UVA (365 nm) in Stratalinker or similar device (*see* **Notes 1** and **2**).
4. Scrape cells in 5 ml PBS, transfer to 50 ml conical tube, centrifuge at 2000×*g* at 4 °C for 5 min, and aspirate the supernatant.
5. Cell pellets can be used immediately for lysis or stored at −80 °C for later use.

3.3 RNase Digestion and Immunoprecipitation

1. Resuspend cell pellets by adding three volumes of NP-40 lysis buffer supplemented with protease inhibitors, and 1 mM DTT.
2. Incubate on ice for 10 min and centrifuge at 10,000×*g* for 15 min at 4 °C.
3. Collect supernatants, add RNase T1 to 1U/μl final concentration, and incubate at 22 °C for 2, 4, 6, 8, 10, and 15 min (*see* **Note 3**).

4. Spare 100 μl lysate, add 400 μl water and 500 μl acidic phenol, vortex for 1 min, and centrifuge at 10,000 ×*g* for 20 min at 4 °C.
5. Collect 400 μl of supernatant, add 800 μl 100 % ethanol, 40 μl 3 M sodium acetate, and 1 μl glycoblue. Incubate it at −80 °C for 1 h (or overnight).
6. Centrifuge at 10,000 ×*g* for 20 min at 4 °C, aspirate the supernatant, and add 500 μl of 70 % ethanol.
7. Centrifuge at 10,000 ×*g* for 10 min at 4 °C, aspirate the supernatant, dry pellet at room temperature, and dissolve the pellet with RNase-free water.
8. Run the RNA samples in 1.5 % formaldehyde agarose gel to verify that RNAs are digested in 100- to 300-nt range.
9. Select the samples having RNAs partially digested in the 100- to 300-nt range.
10. To prepare sepharose beads, wash beads with ice-cold PBS three times and resuspend them with equal volume of ice-cold PBS to create a 50 % slurry.
11. Incubate 40 μl of the bead slurry with 10 μg of normalized IgG or antibody of interest for 2 h at 4 °C in NT2 buffer.
12. Centrifuge the beads at 2000 ×*g* for 1 min at 4 °C, and wash three times with NT2 buffer.
13. Add 1 ml of cell lysates to the sepharose beads coated with antibody and incubate them for 3 h at 4 °C.
14. After centrifugation at 2000 ×*g* for 1 min at 4 °C, wash beads three times with NP-40 lysis buffer.
15. Incubate the pellets with 20 units of RNase-free DNase I in 100 μl NP-40 lysis buffer for 15 min at 37 °C.
16. Add 700 μl of NP-40 lysis buffer and centrifuge at 2000 ×*g* for 1 min at 4 °C.
17. Incubate the pellets with 0.1 % SDS and 0.5 mg/ml Proteinase K for 15 min at 55 °C.
18. Collect the supernatant after centrifugation at 10,000 ×*g* at 4 °C for 5 min.
19. Add 500 μl of RNase-free water and 500 μl of acidic phenol, and then vortex for 5 min.
20. Centrifuge at 10,000 ×*g*, 4 °C for 20 min; collect 400 μl supernatant; add 800 μl 100 % ethanol, 40 μl 3 M sodium acetate, and 1 μl glycoblue; and incubate it at −80 °C for 1 h (or overnight).
21. Centrifuge at 10,000 ×*g* for 20 min at 4 °C, remove the supernatant, and add 500 μl of 70 % ethanol.
22. Centrifuge at 10,000 ×*g* for 10 min at 4 °C, remove the supernatant, dry pellet at room temperature, and dissolve the pellet with 12 μl RNase-free water.

3.4 Reverse Transcription and qPCR

1. Mix 1 μl dNTP mix (10 mM) and 1 μl random hexamer (150 ng/μl) with 12 μl purified RNAs.
2. Incubate them at 65 °C for 5 min and 4 °C for 5 min using a thermo cycler.
3. Add 1 μl reverse transcriptase (200U/μl), 1 μl RNase inhibitor (40U/μl), and 4 μl 5× reaction buffer.
4. Incubate samples at 25 °C for 10 min, at 50 °C for 30 min, and at 85 °C for 5 min using a thermo cycler.
5. Mix 2.5 μl of cDNAs, 2.5 μl forward and reverse gene-specific primers (2.5–10 μM) designed to amplify PCR products in 200-nt intervals, and 5 μl of SYBR green master mix (*see* **Notes 4** and **5**).
6. After completion of qPCR, calculate Ct values of IgG IP and specific antibody IP normalized with Ct values of mRNAs encoding housekeeping proteins like GAPDH, ACTB, UBC, and SDHA (*see* **Notes 6** and 7).

4 Notes

1. If a method without RNA labeling is preferred, UVC at 254 nm can be utilized instead of UVA at 365 nm and preincubation with 4-SU or 4-SG can be omitted.
2. If noncanonical RBPs are of interest, RBP and RNA cross-linking may not be successful upon UV exposure. In this case, formaldehyde cross-linking can be utilized instead.
3. It is critical to titrate the amount of RNase T1 (or other RNase) and the time of incubation. Optimization of these parameters in order to obtain 100- to 300-nt RNA fragments (mainly from 18S and 28S ribosomal RNAs) is a key step in CLIP-qPCR analysis. For highly abundant RNAs (e.g., *MALAT1* and *NEAT1*), higher amount of RNase and longer incubation can be utilized.
4. For primer design, divide the RNA of interest in 200-nt overlapping intervals (e.g., spanning positions 1–200, 101–300, 201–400, 301–500). This way, each gene-specific primer will cover all of the full-length transcripts after qPCR. If the full-length target RNA is not in a public database, primer extension or rapid amplification of cDNA ends (RACE) can be done first.
5. Depending on the RBP and RNAs localized in specific cellular compartments, NP-40 lysis buffer can be modified by increasing the concentration of NP-40 (e.g., from 0.5 to 2 %) or by adding more stringent detergents such as Triton X-100 and SDS.
6. If DNA contamination is suspected, RT-minus PCR amplification reactions can be performed. If there are amplifications,

genomic DNA could be contaminated during immunoprecipitation. In this case, increased amount of DNase can be utilized with longer incubation time.

7. This method can be used to map RBPs binding to mRNAs. Some attention should be paid to the fact that mRNAs may be extensively engaged with polyribosomes and hence the coding region may not be available to RBPs and PCR amplification of certain regions may be challenging.

Acknowledgments

JHY and MG were supported by the National Institute on Aging Intramural Research Program, National Institutes of Health and startup fund from Medical University of South Carolina.

References

1. Berger SL (2002) Histone modifications in transcriptional regulation. Curr Opin Genet Dev 12:142–148
2. Glisovic T, Bachorik JL, Yong J, Dreyfuss G (2008) RNA-binding proteins and post-transcriptional gene regulation. FEBS Lett 582:1977–1986
3. Filipowicz W, Bhattacharyya SN, Sonenberg N (2008) Mechanisms of post-transcriptional regulation by microRNAs: are the answers in sight? Nat Rev Genet 9:102–114
4. Yoon JH, Abdelmohsen K, Gorospe M (2013) Posttranscriptional gene regulation by long noncoding RNA. J Mol Biol 425:3723–3730
5. Yoon JH, Abdelmohsen K, Gorospe M (2014) Functional interactions among microRNAs and long noncoding RNAs. Semin Cell Dev Biol 34:9–14
6. Rinn JL, Chang HY (2012) Genome regulation by long noncoding RNAs. Annu Rev Biochem 81:145–166
7. Riley KJ, Steitz JA (2013) The "Observer Effect" in genome-wide surveys of protein-RNA interactions. Mol Cell 49:601–604
8. Lin RJ (2008) RNA -protein interaction protocols. Methods Mol Biol 488:v–vii
9. Tenenbaum SA, Carson CC, Lager PJ, Keene JD (2000) Identifying mRNA subsets in messenger ribonucleoprotein complexes by using cDNA arrays. Proc Natl Acad Sci U S A 97:14085–14090
10. Mukherjee N, Corcoran DL, Nusbaum JD, Reid DW, Georgiev S, Hafner M, Ascano M Jr, Tuschl T, Ohler U, Keene JD (2011) Integrative regulatory mapping indicates that the RNA-binding protein HuR couples pre-mRNA processing and mRNA stability. Mol Cell 43: 327–339
11. Ule J, Jensen KB, Ruggiu M, Mele A, Ule A, Darnell RB (2003) CLIP identifies Nova-regulated RNA networks in the brain. Science 302:1212–1215
12. Licatalosi DD, Mele A, Fak JJ, Ule J, Kayikci M, Chi SW, Clark TA, Schweitzer AC, Blume JE, Wang X, Darnell JC, Darnell RB (2009) HITS-CLIP yields genome-wide insights into brain alternative RNA processing. Nature 456:464–469
13. Chi SW, Zang JB, Mele A, Darnell RB (2009) Argonaute HITS-CLIP decodes microRNA-mRNA interaction maps. Nature 460: 479–486
14. Hafner M, Landthaler M, Burger L, Khorshid M, Hausser J, Berninger P, Rothballer A, Ascano M Jr, Jungkamp AC, Munschauer M, Ulrich A, Wardle GS, Dewell S, Zavolan M, Tuschl T (2010) Transcriptome-wide identification of RNA-binding protein and microRNA target sites by PAR-CLIP. Cell 141:129–141
15. König J, Zarnack K, Rot G, Curk T, Kayikci M, Zupan B, Turner DJ, Luscombe NM, Ule J (2010) iCLIP reveals the function of hnRNP particles in splicing at individual nucleotide resolution. Nat Struct Mol Biol 17:909–915
16. Kudla G, Granneman S, Hahn D, Beggs JD, Tollervey D (2011) Cross-linking, ligation, and sequencing of hybrids reveals RNA-RNA interactions in yeast. Proc Natl Acad Sci U S A 108:10010–10015

Chapter 3

Characterization of Long Noncoding RNA-Associated Proteins by RNA-Immunoprecipitation

Youyou Zhang, Yi Feng, Zhongyi Hu, Xiaowen Hu, Chao-Xing Yuan, Yi Fan, and Lin Zhang

Abstract

With the advances in sequencing technology and transcriptome analysis, it is estimated that up to 75 % of the human genome is transcribed into RNAs. This finding prompted intensive investigations on the biological functions of noncoding RNAs and led to very exciting discoveries of microRNAs as important players in disease pathogenesis and therapeutic applications. Research on long noncoding RNAs (lncRNAs) is in its infancy; yet a broad spectrum of biological regulations has been attributed to lncRNAs. RNA-immunoprecipitation (RNA-IP) is a technique of detecting the association of individual proteins with specific RNA molecules in vivo. It can be used to investigate lncRNA-protein interaction and identify lncRNAs that bind to a protein of interest. Here we describe the protocol of this assay with detailed materials and methods.

Key words Long noncoding RNA, RNA-binding protein, RNA-immunoprecipitation

1 Introduction

lncRNAs are operationally defined as RNA transcripts larger than 200 nt that do not appear to have coding potential [1–5]. Given that up to 75 % of the human genome is transcribed to RNA, while only a small portion of the transcripts encodes proteins [6], the number of lncRNA genes can be large. After the initial cloning of functional lncRNAs such as H19 [7, 8] and XIST [9] from cDNA libraries, two independent studies using high-density tiling array reported that the number of lncRNA genes is at least comparable to that of protein-coding genes [10, 11]. Recent advances in tiling array [10–13], chromatin signature [14, 15], computational analysis of cDNA libraries [16, 17], and next-generation sequencing (RNA-seq) [18–21] have revealed that thousands of lncRNA genes are abundantly expressed with exquisite cell type and tissue specificity in human. In fact, the GENCODE consortium within the framework of the ENCODE project recently reported 14,880

Yi Feng and Lin Zhang (eds.), *Long Non-Coding RNAs: Methods and Protocols*, Methods in Molecular Biology, vol. 1402,
DOI 10.1007/978-1-4939-3378-5_3, © Springer Science+Business Media New York 2016

manually annotated and evidence-based lncRNA transcripts originating from 9277 gene loci in human [6, 21], including 9518 intergenic lncRNAs (also called lincRNAs) and 5362 genic lncRNAs [14, 15, 20]. These studies indicate that (1) lncRNAs are independent transcriptional units; (2) lncRNAs are spliced with fewer exons than protein-coding transcripts and utilize the canonical splice sites; (3) lncRNAs are under weaker selective constraints during evolution and many are primate specific; (4) lncRNA transcripts are subjected to typical histone modifications as protein-coding mRNAs; and (5) the expression of lncRNAs is relatively low and strikingly cell type or tissue specific.

The discovery of lncRNA has provided an important new perspective on the centrality of RNA in gene expression regulation. lncRNAs can regulate the transcriptional activity of a chromosomal region or a particular gene by recruiting epigenetic modification complexes in either *cis*- or *trans*-regulatory manner. For example, Xist, a 17 kb X-chromosome-specific noncoding transcript, initiates X chromosome inactivation by targeting and tethering Polycomb-repressive complexes (PRC) to X chromosome in *cis* [22–24]. HOTAIR regulates the HoxD cluster genes in *trans* by serving as a scaffold which enables RNA-mediated assembly of PRC2 and LSD1 and coordinates the binding of PRC2 and LSD1 to chromatin [12, 25]. Based on the knowledge obtained from studies on a limited number of lncRNAs, at least two working models have been proposed. First, lncRNAs can function as scaffolds. lncRNAs contain discrete protein-interacting domains that can bring specific protein components into the proximity of each other, resulting in the formation of unique functional complexes [25–27]. These RNA-mediated complexes can also extend to RNA-DNA and RNA-RNA interactions. Second, lncRNAs can act as guides to recruit proteins [24, 28, 29], such as chromatin modification complexes, to chromosome [24, 29]. This may occur through RNA-DNA interactions [29] or through RNA interaction with a DNA-binding protein [24]. In addition, lncRNAs have been proposed to serve as decoys that bind to DNA-binding proteins [30], transcriptional factors [31], splicing factors [32–34], or miRNAs [35]. Some studies have also identified lncRNAs transcribed from the enhancer regions [36–38] or a neighbor loci [18, 39] of certain genes. Given that their expressions correlated with the activities of the corresponding enhancers, it was proposed that these RNAs (termed enhancer RNA/eRNA [36–38] or ncRNA-activating/ncRNA-a [18, 39]) may regulate gene transcription.

RNA-immunoprecipitation (RNA-IP) is a technique of detecting the association of individual proteins with specific RNA molecules in vivo. It can be used to investigate lncRNA-protein interaction and identify lncRNAs that bind to a protein of interest. Here we describe the protocol of this assay with detailed materials and methods.

2 Materials

Prepare all solutions using ultra-pure RNase-free water and analytical grade reagents. Contamination of the *solutions* with RNase can result in RNA degradation. Use filtration and/or autoclave sterilization to ensure that all reagents and supplies used in this section are RNase free. Use RNase ZAP to clean all equipment and work surface.

1. Sucrose.
2. 1 M Tris–HCl (pH 7.4).
3. 1 M $MgCl_2$.
4. Triton X-100.
5. 1 M KCl.
6. 0.5 M EDTA.
7. NP-40.
8. 1 M Dithiothreitol (DTT).
9. 10× Phosphate-buffered saline (PBS, Invitrogen, AM9625): To make 1× PBS, mix one part of 10× PBS with nine parts RNase-free water. Store at 4 °C.
10. Protein A/G beads (Sigma, P9424).
11. RNase inhibitor (Invitrogen, 10777-019).
12. Protease inhibitor cocktail (Sigma, P8340).
13. TRIzol RNA extraction reagent (Invitrogen).
14. 1 mL Dounce homogenizer (Fish Scientific, FB56687).
15. Nuclear isolation buffer: 1.28 M Sucrose, 40 mM Tris–HCl (pH 7.4), 20 mM $MgCl_2$, 4 % Triton X-100. Put 40 mL RNase-free water in a beaker with a stir bar and dissolve 21.9 g sucrose in the beaker. Add 2 mL 1 M Tris–HCl (pH 7.5), 1 mL 1 M $MgCl_2$, and 2 mL Triton X-100 and mix well. Make up to a final volume of 50 mL with RNase-free water, and store at 4 °C.
16. RNA immunoprecipitation (RIP) buffer: 150 mM KCl, 25 mM Tris (pH 7.4), 5 mM EDTA, 0.5 % NP-40. Mix 7.5 mL 1 M KCl, 1.25 mL 1 M Tris–HCl (pH 7.4), 500 μL 0.5 M EDTA, and 250 μL NP-40 and make up to a final volume of 48 mL with RNase-free water. Store at 4 °C. Right before use, add DTT (0.5 mM final concentration), RNase inhibitor (100 U/mL final concentration), and protease inhibitor cocktail (1× final concentration).

3 Methods

The procedures must be performed in an RNase-free environment. Use filtered tips and RNase-free tubes and clean all equipment and work surface with RNase ZAP before staring the experiment. lncRNA-IP aims to identify lncRNA species that bind to a protein of interest. The protocol includes two parts: (1) preparing protein lysate from target cells and (2) immunoprecipitating the protein of interest and extract protein-bound RNAs. It is up to the readers to decide the subsequent analysis on the isolated RNAs. Before harvesting cells, precool 1× PBS, RNase-free water, nuclear isolation buffer, and RIP buffer on ice; estimate the amount of RIP buffers needed and add RNase inhibitor and protease inhibitor cocktail to the buffer accordingly (*see* **Note 1**).

3.1 Whole-Cell Lysate Preparation (See Note 2)

If nuclear RNA-protein interaction is the focus of the research, skip this step and go directly to Subheading 3.1.2 for nuclear lysate preparation.

1. Harvest cells using regular trypsinization technique and count the cell number.
2. Wash cells in ice-cold 1× PBS once and resuspend the cell pellet (1.0×10^7 cells) in 1 mL ice-cold RIP buffer containing RNase and protease inhibitors.
3. Shear the cells on ice using a Dounce homogenizer with 15–20 strokes.
4. Centrifuge at 15,000 × *g* for 15 min at 4 °C and transfer the supernatant into a clean tube. This supernatant is the whole-cell lysate.

3.1.1 Cell Harvest and Nuclei Lysate Preparation

1. Harvest cells using regular trypsinization technique and count the cell number.
2. Wash cells in ice-cold 1× PBS three times and resuspend 1.0×10^7 cells in 2 mL ice-cold PBS (*see* **Note 3**).
3. Put cell suspension in 1× PBS on ice, add 2 mL ice-cold nuclear isolation buffer and 6 mL ice-cold RNase-free water into the tube, mix well, and incubate the cells on ice for 20 min with intermittent mixing (four to five times).
4. Harvest nuclei by spinning the tube at 2500 × *g* for 15 min at 4 °C. The pellet contains the purified nuclei.
5. Resuspend nuclei pellet in 1 mL freshly prepared ice-cold RIP buffer containing DTT, RNase inhibitor, and protease inhibitors.
6. Shear the nucleus on ice with 15–20 strokes using a Dounce homogenizer.

7. Pellet nuclear membrane and debris by centrifugation at 16,000 × *g* for 10 min at 4 °C.
8. Carefully transfer the clear supernatant (nuclear lysate) into a new tube. The supernatant is nuclear lysate.

3.1.2 RNA Immune-Precipitation and Purification

1. Wash 40 μL protein A/G beads with 500 μL ice-cold RIP buffer three times. After the wash, spin down the beads at 600 g for 30 s at 4 °C, take off the RIP buffer, and add 40 μL RIP buffer to resuspend the beads.
2. Add the prewashed beads and 5–10 ug IgG into the whole-cell lysate from Subheading 3.1.1 or nuclear lysate from Subheading 3.1.2.
3. Incubate the lysate with IgG and beads at 4 °C with gentle rotation for 1 h. Pellet the IgG with beads by centrifugation at 16,000 × *g* for 5 min.
4. Carefully transfer the supernatant (pre-cleared nuclear lysate) into a new tube. At this point, the lysate can be divided into multiple portions of equal volume for different antibodies and corresponding controls. Take 50 μL lysate and set aside on ice as input control.
5. Add antibody of interest into nuclear lysate (*see* **Note 4**), and incubate the lysate and antibody overnight at 4 °C with gentle rotation.
6. The next day, add 40 μL prewashed protein A/G beads and incubate at 4 °C for 1 h with gentle rotation.
7. Pellet the beads by spinning at 600 × *g* for 30 s at 4 °C, and remove supernatant.
8. Wash the beads with 500 μL ice-cold RIP buffer three times, invert five to ten times during each wash, and pellet the beads by spinning at 600 × *g* for 30 s at 4 °C.
9. Wash the beads with 500 μL ice-cold PBS, pellet the beads by spinning at 600 × *g* for 30 s at 4 °C, and use a fine needle or tip to remove as much PBS as possible without disturbing the beads.
10. Resuspend beads in 1 mL TRIzol RNA extraction reagent and isolate coprecipitated RNA according to the manufacturer's instructions.
11. Dissolve RNA in nuclease-free water and store the RNA at −80 °C for further application (*see* **Note 5**).

4 Notes

1. It is utterly important that the experiment described above is conducted with extra precaution to avoid RNA degradation. All materials have to be RNase free and the buffers need to be precooled on ice.

2. To ensure result reproducibility, the cells need to be maintained consistently.
3. The abundances of different target protein and lncRNAs may vary from cell line to cell line; therefore the amount of lysate input needs to be empirically determined for each assay. We found that 1.0×10^7 cells is a good starting point. In cases more cells are needed, scale up the amount of buffer used to ensure high nuclear lysing efficiency.
4. The amount of antibody used for each experiment needs to be empirically determined. Our suggestion is to start at around 1–2 μg antibody per million cells.
5. The amount of nuclease-free water used to dissolve the RNAs is determined by several factors, including the type of downstream analysis, the amount of lncRNA bound to the target protein, and the cell type. We recommend the researchers start at 20 μL and adjust according to their specific situations.

Acknowledgments

This work was supported, in whole or in part, by the Basser Research Center for BRCA, the NIH (R01CA142776, R01CA190415, P50CA083638, P50CA174523), the Ovarian Cancer Research Fund (XH), the Breast Cancer Alliance, Foundation for Women's Cancer (XH), and the Marsha Rivkin Center for Ovarian Cancer Research.

References

1. Rinn JL, Chang HY (2012) Genome regulation by long noncoding RNAs. Annu Rev Biochem 81:145–166
2. Lee JT (2012) Epigenetic regulation by long noncoding RNAs. Science 338: 1435–1439
3. Prensner JR, Chinnaiyan AM (2011) The emergence of lncRNAs in cancer biology. Cancer Discov 1:391–407
4. Guttman M, Rinn JL (2012) Modular regulatory principles of large non-coding RNAs. Nature 482:339–346
5. Spizzo R, Almeida MI, Colombatti A, Calin GA (2012) Long non-coding RNAs and cancer: a new frontier of translational research? Oncogene 31:4577–4587
6. Djebali S, Davis CA, Merkel A, Dobin A, Lassmann T, Mortazavi A et al (2012) Landscape of transcription in human cells. Nature 489:101–108
7. Pachnis V, Brannan CI, Tilghman SM (1988) The structure and expression of a novel gene activated in early mouse embryogenesis. EMBO J 7:673–681
8. Bartolomei MS, Zemel S, Tilghman SM (1991) Parental imprinting of the mouse H19 gene. Nature 351:153–155
9. Brown CJ, Ballabio A, Rupert JL, Lafreniere RG, Grompe M, Tonlorenzi R et al (1991) A gene from the region of the human X inactivation centre is expressed exclusively from the inactive X chromosome. Nature 349:38–44
10. Kapranov P, Cawley SE, Drenkow J, Bekiranov S, Strausberg RL, Fodor SP et al (2002) Large-scale transcriptional activity in chromosomes 21 and 22. Science 296:916–919
11. Rinn JL, Euskirchen G, Bertone P, Martone R, Luscombe NM, Hartman S et al (2003) The transcriptional activity of human Chromosome 22. Genes Dev 17:529–540
12. Rinn JL, Kertesz M, Wang JK, Squazzo SL, Xu X, Brugmann SA et al (2007) Functional demarcation of active and silent chromatin

domains in human HOX loci by noncoding RNAs. Cell 129:1311–1323
13. Gupta RA, Shah N, Wang KC, Kim J, Horlings HM, Wong DJ et al (2010) Long non-coding RNA HOTAIR reprograms chromatin state to promote cancer metastasis. Nature 464:1071–1076
14. Guttman M, Amit I, Garber M, French C, Lin MF, Feldser D et al (2009) Chromatin signature reveals over a thousand highly conserved large non-coding RNAs in mammals. Nature 458:223–227
15. Khalil AM, Guttman M, Huarte M, Garber M, Raj A, Rivea Morales D et al (2009) Many human large intergenic noncoding RNAs associate with chromatin-modifying complexes and affect gene expression. Proc Natl Acad Sci U S A 106:11667–11672
16. Maeda N, Kasukawa T, Oyama R, Gough J, Frith M, Engstrom PG et al (2006) Transcript annotation in FANTOM3: mouse gene catalog based on physical cDNAs. PLoS Genet 2:e62
17. Jia H, Osak M, Bogu GK, Stanton LW, Johnson R, Lipovich L (2010) Genome-wide computational identification and manual annotation of human long noncoding RNA genes. RNA 16:1478–1487
18. Orom UA, Derrien T, Beringer M, Gumireddy K, Gardini A, Bussotti G et al (2010) Long noncoding RNAs with enhancer-like function in human cells. Cell 143:46–58
19. Prensner JR, Iyer MK, Balbin OA, Dhanasekaran SM, Cao Q, Brenner JC et al (2011) Transcriptome sequencing across a prostate cancer cohort identifies PCAT-1, an unannotated lincRNA implicated in disease progression. Nat Biotechnol 29: 742–749
20. Cabili MN, Trapnell C, Goff L, Koziol M, Tazon-Vega B, Regev A et al (2011) Integrative annotation of human large intergenic noncoding RNAs reveals global properties and specific subclasses. Genes Dev 25:1915–1927
21. Derrien T, Johnson R, Bussotti G, Tanzer A, Djebali S, Tilgner H et al (2012) The GENCODE v7 catalog of human long noncoding RNAs: analysis of their gene structure, evolution, and expression. Genome Res 22:1775–1789
22. Brown CJ, Hendrich BD, Rupert JL, Lafreniere RG, Xing Y, Lawrence J et al (1992) The human XIST gene: analysis of a 17 kb inactive X-specific RNA that contains conserved repeats and is highly localized within the nucleus. Cell 71:527–542
23. Zhao J, Sun BK, Erwin JA, Song JJ, Lee JT (2008) Polycomb proteins targeted by a short repeat RNA to the mouse X chromosome. Science 322:750–756
24. Jeon Y, Lee JT (2011) YY1 tethers Xist RNA to the inactive X nucleation center. Cell 146:119–133
25. Tsai MC, Manor O, Wan Y, Mosammaparast N, Wang JK, Lan F et al (2010) Long noncoding RNA as modular scaffold of histone modification complexes. Science 329:689–693
26. Yap KL, Li S, Munoz-Cabello AM, Raguz S, Zeng L, Mujtaba S et al (2010) Molecular interplay of the noncoding RNA ANRIL and methylated histone H3 lysine 27 by polycomb CBX7 in transcriptional silencing of INK4a. Mol Cell 38:662–674
27. Kotake Y, Nakagawa T, Kitagawa K, Suzuki S, Liu N, Kitagawa M et al (2011) Long noncoding RNA ANRIL is required for the PRC2 recruitment to and silencing of p15(INK4B) tumor suppressor gene. Oncogene 30:1956–1962
28. Huarte M, Guttman M, Feldser D, Garber M, Koziol MJ, Kenzelmann-Broz D et al (2010) A large intergenic noncoding RNA induced by p53 mediates global gene repression in the p53 response. Cell 142:409–419
29. Grote P, Wittler L, Hendrix D, Koch F, Wahrisch S, Beisaw A et al (2013) The tissue-specific lncRNA Fendrr is an essential regulator of heart and body wall development in the mouse. Dev Cell 24:206–214
30. Kino T, Hurt DE, Ichijo T, Nader N, Chrousos GP (2010) Noncoding RNA gas5 is a growth arrest- and starvation-associated repressor of the glucocorticoid receptor. Sci Signal 3:ra8
31. Hung T, Wang Y, Lin MF, Koegel AK, Kotake Y, Grant GD et al (2011) Extensive and coordinated transcription of noncoding RNAs within cell-cycle promoters. Nat Genet 43:621–629
32. Tripathi V, Ellis JD, Shen Z, Song DY, Pan Q, Watt AT et al (2010) The nuclear-retained noncoding RNA MALAT1 regulates alternative splicing by modulating SR splicing factor phosphorylation. Mol Cell 39:925–938
33. Bernard D, Prasanth KV, Tripathi V, Colasse S, Nakamura T, Xuan Z et al (2010) A long nuclear-retained non-coding RNA regulates synaptogenesis by modulating gene expression. EMBO J 29:3082–3093
34. Tripathi V, Shen Z, Chakraborty A, Giri S, Freier SM, Wu X et al (2013) Long noncoding RNA MALAT1 controls cell cycle progression by regulating the expression of oncogenic transcription factor B-MYB. PLoS Genet 9:e1003368

35. Salmena L, Poliseno L, Tay Y, Kats L, Pandolfi PP (2011) A ceRNA hypothesis: the Rosetta Stone of a hidden RNA language? Cell 146:353–358
36. Kim TK, Hemberg M, Gray JM, Costa AM, Bear DM, Wu J et al (2010) Widespread transcription at neuronal activity-regulated enhancers. Nature 465:182–187
37. Wang D, Garcia-Bassets I, Benner C, Li W, Su X, Zhou Y et al (2011) Reprogramming transcription by distinct classes of enhancers functionally defined by eRNA. Nature 474:390–394
38. Melo CA, Drost J, Wijchers PJ, van de Werken H, de Wit E, Oude Vrielink JA et al (2013) eRNAs are required for p53-dependent enhancer activity and gene transcription. Mol Cell 49:524–535
39. Lai F, Orom UA, Cesaroni M, Beringer M, Taatjes DJ, Blobel GA et al (2013) Activating RNAs associate with Mediator to enhance chromatin architecture and transcription. Nature 494:497–501

Chapter 4

Isolation of Protein Complexes Associated with Long Noncoding RNAs

Kiranmai Gumireddy, Jinchun Yan, and Qihong Huang

Abstract

Long noncoding RNAs (lncRNAs) are a new class of regulatory genes that play critical roles in various processes ranging from normal development to human diseases. Recent studies have shown that protein complexes are required for the functions of lncRNAs. The identification of these proteins which are associated with lncRNAs is critical for the understanding of molecular mechanisms of lncRNAs in gene regulation and their functions. In this chapter, we describe a method to isolate proteins associated with lncRNAs. This procedure involves fusion protein maltose-binding protein (MBP) fused to MS2-binding protein to pull down the proteins associated with lncRNA and the identification of these proteins by mass spectrometry.

Key words Long noncoding RNA, Protein complex, MS2

1 Introduction

Human genome sequencing studies have shown that only 2 % of the human genome consists of protein-coding sequences [1]. They have also shown that up to 90 % of the human genome can be transcribed [2]. The transcripts from nonprotein-coding regions include noncoding RNAs. Long noncoding RNAs (lncRNAs) are a novel class of noncoding RNAs. lncRNAs have been shown to have important functions in development, as well as differentiation, X-chromosome inactivation, genomic imprinting, and cellular processes including cell cycle and apoptosis [3–7]. They have also been implicated in human diseases ranging from cancer and amyotrophic lateral sclerosis (ALS) to Alzheimer's disease [8–17]. Recent studies have shown that lncRNAs function through gene regulation in almost all steps of gene expression including chromatin remodeling, transcription, splicing, RNA decay, translation, enhancer function, and epigenetic regulation [18–36]. Gene regulation by lncRNAs requires protein complexes associated with lncRNAs. Thus, identification of proteins associated with lncRNAs

Yi Feng and Lin Zhang (eds.), *Long Non-Coding RNAs: Methods and Protocols*, Methods in Molecular Biology, vol. 1402,
DOI 10.1007/978-1-4939-3378-5_4, © Springer Science+Business Media New York 2016

is critical for the understanding of molecular mechanisms and functions of lncRNAs. Immunoprecipitation is commonly used for the isolation of protein complexes associated with a protein of interest. However, this method is not available for lncRNAs because antibodies generally do not recognize RNAs. This chapter describes a method to isolate protein complexes associated with lncRNAs using fusion protein maltose-binding protein (MBP) fused to MS2-binding protein.

MS2-binding protein binds to RNA sequence MS2. The MS2 sequence needs to be added to either the 5′ or 3′ end of the lncRNA of interest. Sometimes adding the MS2 sequence to the end of lncRNA may change the structure of lncRNA and cause the loss of its functions. Functions of lncRNA-MS2 need to be confirmed by assays to ensure that the fusion lncRNA-MS2 RNA has the same function as the wild-type lncRNA. MBP-MS2 pull-down assay consists of the following steps. The first step involves the expression of MBP fused to MS2-binding protein (MBP-MS2) and immobilization of fusion protein to amylose beads. This fusion protein is ~59 kDa with MBP N-terminal to MS2-binding protein which carries a double mutation (V75Q and A81G) that prevents oligomerization and increases its affinity for RNA. The second step consists of the construction of plasmids and their expression in mammalian cells. A plasmid with MS2 sequence alone without any noncoding RNA is used as a control. Finally, transduce the cell line of interest with MS2 control or lncRNA-MS2 plasmid and lyse the cells. Incubate cytoplasmic or nuclear lysate with MBP-MS2-bound amylose beads, wash, and use for further analysis. PAGE and mass spectroscopy are used to identify the bound proteins.

2 Materials

All the solutions are prepared using ultrapure water from Millipore water system and analytical grade reagents.

1. *Bacterial lysis buffer*: 20 mM HEPES pH7.9, 200 mM KCL, 1 mM EDTA, 0.05 % NP-40, 1 mM PMSF. PMSF protease inhibitor in the lysis buffer is added fresh before each use (*see* **Note 1**).
2. *MBP column buffer*: 200 mM NaCl, 20 mM Tris–HCl, 1 mM EDTA, 1 mM DTT pH7.4. DTT is added fresh before each use (*see* **Note 2**).
3. *Wash buffer 1*: 20 mM HEPES pH 7.9, 200 mM KCl, and 1 mM EDTA.
4. *Wash buffer 2*: 20 mM HEPES pH 7.9, 20 mM KCl, and 1 mM EDTA.

3 Methods

3.1 MS2-MBP Fusion Protein Expression and Immobilization

1. Inoculate 5 ml Luria Broth containing 100 μg/ml ampicillin with a single bacterial colony of BL21 cells transformed with a plasmid expressing MS2-MBP and grow overnight at 37 °C, 220 RPM. Next morning inoculate 1 l of Terrific Broth containing 100 μg/ml ampicillin with 5 ml of overnight culture and grow the cells at 37 °C, 220 RPM for 1 h. Check OD at 600 nm. OD_{600} should be around 0.6–0.8: if lower than 0.6, check OD every 20 min.
2. Once the OD >0.6, induce expression of protein by adding 100 μl of 1 M IPTG (to final concentration of 0.1 mM), and continue to grow cells at 16 °C, 220 RPM overnight.
3. Harvest the cells by centrifugation at 600 RPM for 10 min. All steps should be performed on ice or at 4 °C from this point on.
4. Discard the supernatant and either begin lysis or store the pellet at 80 °C.
5. Suspend the bacterial pellet in 5 ml chilled lysis buffer (20 mM HEPES pH 7.9, 200 mM KCL, 1 mM EDTA, 0.05 % NP-40) containing 1 mM PMSF for every 100 ml culture (for 1 l culture, add 50 ml lysis buffer). Break open the cells by sonication at an output of 15 % for 10 s (1 s on/1 s off), and put tube back on ice. Repeat sonication one more time.
6. Centrifuge the lysate at 15,000 RPM (21000 rcf) at 4 °C for 30 min. Collect the supernatant.
7. Tale 100 μl of amylose magnetic bead suspension and add 500 μl of MBP column buffer (200 mM NaCl, 20 mM Tris–HCl, 1 mM EDTA, 1 mM DTT pH 7.4) and vortex. Place the tube on magnet for 1 min to pull the beads to the side of the tube. Discard the supernatant and repeat wash one more time.
8. Remove the tube from the magnet, add 2 ml of lysate from **step 6** to the washed amylase beads, mix thoroughly, and incubate at 4 °C for 1 h with gentle rotation.
9. Place the tube on a magnet for 2–3 min and discard the supernatant.
10. Wash the coated beads three times with wash buffer.
11. MS2-MBP fusion protein beads are ready to use in next step.

3.2 Transfection and Cytoplasmic Protein Extraction

1. Seed 1×10^6 HeLa cells per 60 mm culture dish in DMEM supplemented with 10 % fetal bovine serum.
2. In a sterile 1.5 ml tube, add 3 μl Lipofectamine™ 2000 for 1 μg of DNA in 250 μl of Opti-MEM® I Medium without serum. Mix gently and incubate for 5 min at room temperature.

3. After 5 min of incubation, add 1 μg of lincRNA-MS2 (*see* **Note 3**) or control MS2 plasmid and incubate at room temperature for 20 min.
4. Add transfection mix dropwise to each plate of cells. Mix gently by rocking the plate back and forth. Incubate the cells at 37 °C in a CO_2 incubator.
5. After 6 h, remove the medium containing the transfection mix and replace with DMEM containing 10 % FBS.
6. 48 h later, wash cells twice with cold PBS, harvest with trypsin-EDTA, and centrifuge at 1000 ×*g* for 5 min.
7. Carefully discard the supernatant, leaving the cell pellet as dry as possible and extract cytoplasmic protein.
8. Add 100 μl ice-cold CER I per 10 μl of packed cell volume to the cell pellet. Vortex the tube vigorously on the highest setting for 15 s to fully suspend the cell pellet. Incubate on ice for 10 min.
9. Add 5.5 μl of ice-cold CERII to the tube, vortex the tube, and incubate for 1 min on ice.
10. Centrifuge at 16,000 ×*g* for 5 min and collect the supernatant (cytoplasmic extract) to a pre-chilled tube. Place the tube on ice until use or store at −80 °C.

3.3 Pull-Down Assay and Protein Separation

1. Incubate the supernatant collected from **step 10** of Subheading 3.2 with the MS2-MBP fusion protein beads from **step 11** of Subheading 3.1 for 3 h at 4 °C.
2. Place the tube on a magnet for 3 min and discard the supernatant.
3. Wash the beads three times with wash buffer 1 (20 mM HEPES pH 7.9, 200 mM KCl, and 1 mM EDTA), twice with wash buffer 2 (20 mM HEPES pH 7.9, 20 mM KCl, and 1 mM EDTA) (*see* **Note 4**), and once with ice-cold PBS.
4. Dissociate bound proteins by boiling with 1× SDS sample buffer for 5 min.
5. Resolve the bound proteins on 15 % SDS-polyacrylamide gel electrophoresis (SDS-PAGE) and stain with colloidal blue stain.
6. Analyze the gel by mass spectroscopy.

Proteins that have higher mass spectrometry counts in cells transfected with lncRNA-MS2 than control MS2 plasmid are selected as potential candidates. Knockdown of these protein candidates with short hairpin RNAs (shRNAs) is used to confirm whether any of the candidates are required for the functions of lncRNA of interest. Further validation of the binding of protein candidates to lncRNA of interest is required to confirm their association.

4 Notes

1. PMFS is from a 1 M stock solution.
2. Pre-measured, dried DTT is purchased from Pierce.
3. MS2 tag can be constructed at the 5′ or 3′ end of lincRNAs. Determining the function(s) of MS2-tagged lincRNA of interest before pull-down assay is critical to ensure that MS2 tag does not affect the function(s) of lincRNA and possibly the interactions with protein complex(es).
4. We find it useful to wash the beads with stringent buffer first and then wash with less stringent buffer. This procedure makes sure that the beads are washed thoroughly but protein interactions are less perturbed.

Acknowledgement

This work was supported by NIH/NCI grant R01CA148759, W. W. Smith Foundation, and Edward Mallinckrodt Jr. Foundation.

References

1. International Human Genome Sequencing Consortium (2004) Finishing the euchromatic sequence of the human genome. Nature 431:931–945
2. Kapranov P, Cheng J, Dike S, Nix DA, Duttagupta R, Willingham AT et al (2007) RNA maps reveal new RNA classes and a possible function for pervasive transcription. Science 316:1484–1488
3. Guttman M, Donaghey J, Carey BW, Garber M, Grenier JK, Munson G et al (2011) lincRNAs act in the circuitry controlling pluripotency and differentiation. Nature 477:295–300
4. Huarte M, Guttman M, Feldser D, Garber M, Koziol MJ, Kenzelmann-Broz D et al (2010) A large intergenic noncoding RNA induced by p53 mediates global gene repression in the p53 response. Cell 142:409–419
5. Kino T, Hurt DE, Ichijo T, Nader N, Chrousos GP (2010) Noncoding RNA gas5 is a growth arrest- and starvation-associated repressor of the glucocorticoid receptor. Sci Signal 3:ra8
6. Mancini-Dinardo D, Steele SJ, Levorse JM, Ingram RS, Tilghman SM (2006) Elongation of the Kcnq1ot1 transcript is required for genomic imprinting of neighboring genes. Genes Dev 20:1268–1282
7. Tian D, Sun S, Lee JT (2010) The long noncoding RNA, Jpx, is a molecular switch for X chromosome inactivation. Cell 143:390–403
8. Beltran M, Puig I, Pena C, Garcia JM, Alvarez AB, Pena R et al (2008) A natural antisense transcript regulates Zeb2/Sip1 gene expression during Snail1-induced epithelial-mesenchymal transition. Genes Dev 22:756–769
9. Benetatos L, Hatzimichael E, Dasoula A, Dranitsaris G, Tsiara S, Syrrou M et al (2010) CpG methylation analysis of the MEG3 and SNRPN imprinted genes in acute myeloid leukemia and myelodysplastic syndromes. Leuk Res 34:148–153
10. Calin GA, Liu CG, Ferracin M, Hyslop T, Spizzo R, Sevignani C et al (2007) Ultraconserved regions encoding ncRNAs are altered in human leukemias and carcinomas. Cancer Cell 12:215–229
11. Faghihi MA, Modarresi F, Khalil AM, Wood DE, Sahagan BG, Morgan TE et al (2008) Expression of a noncoding RNA is elevated in Alzheimer's disease and drives rapid feed-forward regulation of beta-secretase. Nat Med 14:723–730
12. Harismendy O, Notani D, Song X, Rahim NG, Tanasa B, Heintzman N et al (2011) 9p21 DNA variants associated with coronary artery disease impair interferon-gamma signalling response. Nature 470:264–268
13. Li L, Feng T, Lian Y, Zhang G, Garen A, Song X (2009) Role of human noncoding RNAs in the control of tumorigenesis. Proc Natl Acad Sci U S A 106:12956–12961

14. Mourtada-Maarabouni M, Pickard MR, Hedge VL, Farzaneh F, Williams GT (2009) GAS5, a non-protein-coding RNA, controls apoptosis and is downregulated in breast cancer. Oncogene 28:195–208
15. Perez DS, Hoage TR, Pritchett JR, Ducharme-Smith AL, Halling ML, Ganapathiraju SC et al (2008) Long, abundantly expressed noncoding transcripts are altered in cancer. Hum Mol Genet 17:642–655
16. Petrovics G, Zhang W, Makarem M, Street JP, Connelly R, Sun L et al (2004) Elevated expression of PCGEM1, a prostate-specific gene with cell growth-promoting function, is associated with high-risk prostate cancer patients. Oncogene 23:605–611
17. Wang X, Arai S, Song X, Reichart D, Du K, Pascual G et al (2008) Induced ncRNAs allosterically modify RNA-binding proteins in cis to inhibit transcription. Nature 454:126–130
18. Azzalin CM, Reichenbach P, Khoriauli L, Giulotto E, Lingner J (2007) Telomeric repeat containing RNA and RNA surveillance factors at mammalian chromosome ends. Science 318:798–801
19. Bernard D, Prasanth KV, Tripathi V, Colasse S, Nakamura T, Xuan Z et al (2010) A long nuclear-retained non-coding RNA regulates synaptogenesis by modulating gene expression. EMBO J 29:3082–3093
20. Flynn RL, Centore RC, O'Sullivan RJ, Rai R, Tse A, Songyang Z et al (2011) TERRA and hnRNPA1 orchestrate an RPA-to-POT1 switch on telomeric single-stranded DNA. Nature 471:532–536
21. Gong C, Maquat LE (2011) lncRNAs transactivate STAU1-mediated mRNA decay by duplexing with 3′ UTRs via Alu elements. Nature 470:284–288
22. Gumireddy K, Li A, Yan J, Setoyama T, Johannes GJ, Orom UA et al (2013) Identification of a long non-coding RNA-associated RNP complex regulating metastasis at the translational step. EMBO J 32:2672–2684
23. Heo JB, Sung S (2011) Vernalization-mediated epigenetic silencing by a long intronic noncoding RNA. Science 331:76–79
24. Kim TK, Hemberg M, Gray JM, Costa AM, Bear DM, Wu J et al (2010) Widespread transcription at neuronal activity-regulated enhancers. Nature 465:182–187
25. Kotake Y, Nakagawa T, Kitagawa K, Suzuki S, Liu N, Kitagawa M et al (2011) Long non-coding RNA ANRIL is required for the PRC2 recruitment to and silencing of p15(INK4B) tumor suppressor gene. Oncogene 30:1956–1962
26. Liu F, Marquardt S, Lister C, Swiezewski S, Dean C (2010) Targeted 3′ processing of antisense transcripts triggers Arabidopsis FLC chromatin silencing. Science 327:94–97
27. Maison C, Bailly D, Roche D, Montes de Oca R, Probst AV, Vassias I et al (2011) SUMOylation promotes de novo targeting of HP1alpha to pericentric heterochromatin. Nat Genet 43:220–227
28. Martianov I, Ramadass A, Serra Barros A, Chow N, Akoulitchev A (2007) Repression of the human dihydrofolate reductase gene by a non-coding interfering transcript. Nature 445:666–670
29. Nagano T, Mitchell JA, Sanz LA, Pauler FM, Ferguson-Smith AC, Feil R et al (2008) The air noncoding RNA epigenetically silences transcription by targeting G9a to chromatin. Science 322:1717–1720
30. Orom UA, Derrien T, Beringer M, Gumireddy K, Gardini A, Bussotti G et al (2010) Long noncoding RNAs with enhancer-like function in human cells. Cell 143:46–58
31. Rinn JL, Kertesz M, Wang JK, Squazzo SL, Xu X, Brugmann SA et al (2007) Functional demarcation of active and silent chromatin domains in human HOX loci by noncoding RNAs. Cell 129:1311–1323
32. Schmitz KM, Mayer C, Postepska A, Grummt I (2010) Interaction of noncoding RNA with the rDNA promoter mediates recruitment of DNMT3b and silencing of rRNA genes. Genes Dev 24:2264–2269
33. Tripathi V, Ellis JD, Shen Z, Song DY, Pan Q, Watt AT et al (2010) The nuclear-retained noncoding RNA MALAT1 regulates alternative splicing by modulating SR splicing factor phosphorylation. Mol Cell 39:925–938
34. Tsai MC, Manor O, Wan Y, Mosammaparast N, Wang JK, Lan F et al (2010) Long noncoding RNA as modular scaffold of histone modification complexes. Science 329:689–693
35. Wang KC, Chang HY (2011) Molecular mechanisms of long noncoding RNAs. Mol Cell 43:904–914
36. Yap KL, Li S, Munoz-Cabello AM, Raguz S, Zeng L, Mujtaba S et al (2010) Molecular interplay of the noncoding RNA ANRIL and methylated histone H3 lysine 27 by polycomb CBX7 in transcriptional silencing of INK4a. Mol Cell 38:662–674

Chapter 5

Profiling Long Noncoding RNA Expression Using Custom-Designed Microarray

Xinna Zhang, Gabriel Lopez-Berestein, Anil K. Sood, and George A. Calin

Abstract

Long noncoding RNAs (lncRNAs) are an important class of pervasive genes involved in a variety of biological functions. The abnormal expression of lncRNAs has been implicated in a range of many human diseases, including cancer. But only a small number of functional lncRNAs have been well characterized to date. lncRNA expression profiling may help to identify useful molecular biomarkers and targets for novel therapeutic approaches in the future. In this chapter, we describe a highly efficient lncRNA expression profiling method using a custom-designed microarray.

Key words lncRNA expression, Microarray profiling, cDNA synthesis, cDNA labeling

1 Introduction

The most well-studied sequences in the human genome are those of protein-coding genes. However, the coding exons of these genes account for only 1.5 % of the genome [1]. In recent years, it has become increasingly apparent that the nonprotein-coding portion of the genome is of crucial functional importance in the regulation of multiple biological processes including development, differentiation, and metabolism [2]. MicroRNAs (miRNA) are the most widely studied class of ncRNAs. It has been shown that epigenetic and genetic defects in miRNAs and their processing machinery are a common hallmark of disease [3, 4]. However, other ncRNAs, such as transcribed ultraconserved regions (T-UCRs), small nucleolar RNAs (snoRNAs), PIWI-interacting RNAs (piRNAs), large intergenic noncoding RNAs (lincRNAs), and the heterogeneous group of lncRNAs, might also contribute to the development of many different human disorders [2, 5].

Noncoding RNAs are grouped into two major classes based on transcription size: small ncRNAs and long ncRNAs. Small ncRNAs include the well-documented miRNAs, siRNAs, and piRNAs. lncRNAs are a heterogeneous group of noncoding transcripts

Yi Feng and Lin Zhang (eds.), *Long Non-Coding RNAs: Methods and Protocols*, Methods in Molecular Biology, vol. 1402,
DOI 10.1007/978-1-4939-3378-5_5, © Springer Science+Business Media New York 2016

ranging in length from 200 nt to ~100 kilobases and lack significant open reading frames. This class of ncRNAs makes up the largest portion of the mammalian noncoding transcriptome. LncRNAs' expression levels appear to be lower than protein-coding genes, and some lncRNAs are preferentially expressed in specific tissues [6]. Various mechanisms of transcriptional regulation of gene expression by lncRNAs have been proposed. Early discoveries support a paradigm in which lncRNAs regulate transcription via chromatin modulation. They can bind directly to RNA or DNA, provide scaffolds for multimodular complexes controlling gene expression, or exert guide functions to target proteins to specific genomic locations [7]. Recent evidence indicates that lncRNAs function in various cellular contexts, including posttranscriptional regulation, posttranslational regulation of protein activity, organization of protein complexes, cell-cell signaling, as well as recombination [8].

LncRNAs are developmental and tissue specific, and have been associated with a spectrum of biological processes. The abnormal expression of lncRNAs has been implicated in a range of many human diseases, including cancer, ischemic heart disease, and Alzheimer's disease. The altered expression of lncRNAs is a feature of many types of cancers and has been shown to promote the development, invasion, and metastasis of tumors by a variety of mechanisms. Recent findings suggest that levels of T-UCR transcription are altered in human tumorigenesis and that the specific T-UCR expression profiles can be used to distinguish different types of human cancer. T-UCR expression signatures have been described for chronic lymphocytic leukemia (CLL), colorectal cancer (CRC), and hepatocellular carcinoma (HCC). Both downregulation and upregulation of different T-UCRs are seen when comparing expression in tumors with normal tissues [9]. HOTAIR was also found to be involved in human neoplasia [10]. In epithelial cancer cells, HOTAIR overexpression causes polycomb to be retargeted across the genome. The invasive capacity of these cells and propensity to metastasize are also increased in these cells, mediated by the polycomb protein PRC2. By contrast, cancer invasiveness is decreased when HOTAIR expression is lost; as a result these cells showed higher than usual levels of PRC2 activity. As such, HOTAIR might have an active role in modulating the cancer epigenome and mediating cell transformation. A similar function has been postulated for some other lincRNAs, such as lincRNA-p21, which function as a repressor in p53-dependent transcriptional response [11].

An investigation of the differentially expressed lncRNAs in human diseases may help identify useful biomarkers for diagnosis and prognosis of human diseases. In addition, there is currently great interest in the emerging opportunities for targeting lncRNAs using novel therapeutic approaches. Numerous techniques are available for gaining information about lncRNA signatures. Frequently used strategies include northern blot analysis [12],

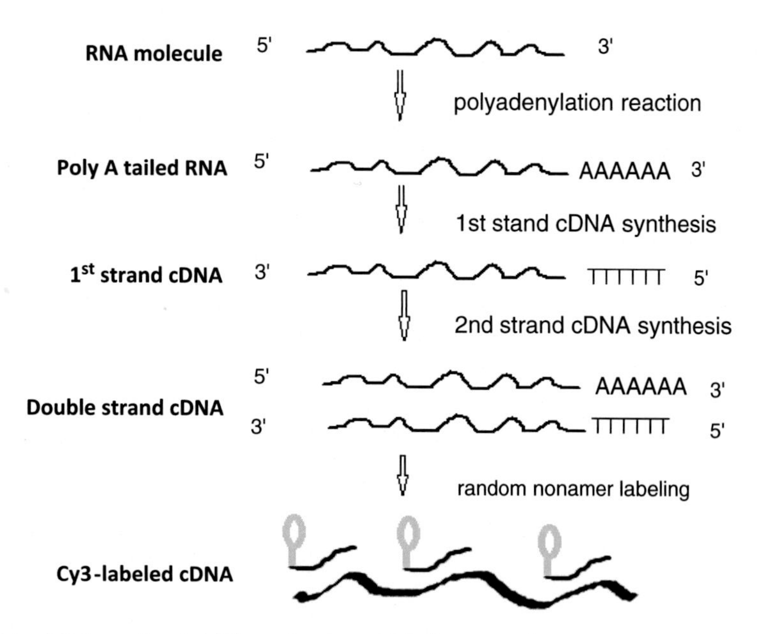

Fig. 1 Schematic of cDNA synthesis and labeling procedure

microarray analysis [13], quantitative real-time polymerase chain reaction (qRT-PCR) [12], deep sequencing [14], ChIP-seq [15], and in situ hybridization [16]. RNA microarray is a commonly used high-throughput technology and is based on nucleic acid hybridization between a mixture of labeled RNA identified as a target and their corresponding complementary probes. In order to profile lncRNA expression in human cancer and identify novel biomarkers involving cancer initiation and progression, we have developed a highly efficient lncRNA microarray profiling method (Fig. 1) using a custom microarray platform containing 22 K probes for pyknons, UCRs, and other lncRNAs.

2 Materials

2.1 RNA Purification

1. DNA-free kit (Life Technologies, AM1906).

2.2 RNA PolyA Tailing and cDNA Synthesis

1. Poly(A) Tailing Kit (Life Technologies, AM1350).
2. cDNA Synthesis System (Roche, 11117831001).
3. 5 mg/ml Glycogen.
4. 7.5 M Ammonium acetate.
5. Absolute ethanol.
6. Agilent RNA 6000 Nano Kit (Agilent, 5067-1511).
7. Isopropanol.
8. Phenol:chloroform:isoamyl alcohol (25:24:1) (Ambion, 9730).

2.3 cDNA Labeling

1. Cy3 Random Nonamers (Trilink, N46-0001).
2. Klenow Fragment (NEB, M0212M).
3. 1 M Tris–HCl pH 7.4 (Sigma, T-2663).
4. 1 M MgCl2 (Sigma, M-1028).
5. β-Mercaptoethanol (Sigma, M3148).
6. 5 M NaCl (Sigma, 71386).
7. 100 mM dNTPs (Life Technologies, 10297-018).
8. 0.5 M EDTA (Sigma, E-7889).

2.4 DNA Hybridization

1. Gene Expression Hybridization Kit (Agilent, 5188-5242).

2.5 Equipment

1. A thermo cycler, such as Surecycler 8800 (Agilent Technologies).
2. A desktop refrigerated centrifuge such as the Eppendorf 5430 (Eppendorf).
3. A spectrophotometer such as the NanoDrop 2000 (Thermo Scientific).
4. A Bioanalyzer 2100 (Agilent).
5. A heat block.
6. A Microarray Scanner such as SureScan Microarray Scanner (Agilent).
7. Hybridization Oven (Agilent).

3 Methods

3.1 DNase Treatment

1. In most cases, 2–5 μg of total RNA is suitable as starting material (*see* **Note 1**). To remove the contamination of DNA, incubate 3 μg of RNA with 1 μl rDNase I (2 U) in a 15 μl reaction for 30 min at 37 °C (*see* **Note 2**).
2. Add 2 μl of DNase inactivation reagent to stop the reaction, and incubate for 2 min at room temperature, mixing occasionally.
3. Centrifuge at 10,000 × *g* for 1.5 min and transfer the RNA to a fresh tube (*see* **Notes 3** and **4**).

3.2 Poly (A) Tailing

1. Add 4 μl 5× reaction buffer, 2 μl 25 mM MnCl2, 1 μl 1 mM ATP, 1 μl poly A polymerase to the RNA (12 μl from last step), adjust the final volume to 20 μl, mix well, and then incubate the tube at 37 °C for 30 min (*see* **Notes 5–7**).

3.3 First Strand cDNA Synthesis

1. Add 2 μl oligo dT primer to the poly (A)-tailed RNA from last step, incubate at 70 °C for 5 min (*see* **Note 8**), and then place the tube immediately on ice.

2. Add the following components and mix well:

Component	Volume	Final concentration
RT buffer, 5× conc	8 μl	1×
DTT, 0.1 M	4 μl	10 mM
AMV, 25 U/μl	2 μl	50 U
Protector RNase inhibitor, 25 U/μl	1 μl	25 U
dNTP, 10 mM	3 μl	

Incubate the samples at 42 °C for 60 min. Then place the tube on ice.

3.4 Second-Strand Synthesis

1. Add the following components to the first-strand reaction(s) in the indicated order on ice:

   Incubate at +16 °C for 2 h.

Component	Volume
cDNA from RT reaction	40 μl
Second-strand synthesis buffer, 5× conc.	30 μl
dNTP mix, 10 mM	2.5 μl
Second-strand enzyme blend	6.5 μl
Water, PCR grade	71 μl
Total volume	150 μl

2. Add 20 μl of 5 U/μl T4 DNA polymerase to each reaction. Incubate at +16 °C for an additional 5 min. Do not allow the reaction temperature to exceed +16 °C during this step.
3. Stop reaction by adding 17 μl EDTA, 0.2 M.

3.5 Digestion of RNA

1. Add 1.5 μl of 15 U RNase I to the tubes from the step above, and digest for 30 min at 37 °C.
2. Add 190 μl of phenol:chloroform:isoamyl alcohol to stop the reaction. Vortex well, and then centrifuge at 12,000 ×*g* for 5 min. Transfer the upper, aqueous layer to a clean, labeled 1.5 ml tube.
3. Add 19 μl of 7.5 M ammonium acetate, then add 8 μl of 5 mg/ml glycogen to the samples, and mix by repeated inversion at each step. Add 380 μl of ice-cold absolute ethanol to the samples. Centrifuge at 12,000 ×*g* for 20 min.
4. Wash the supernatant with 500 μl of ice-cold 80 % ethanol (v/v) once, and dissolve the DNA pellet in 20 μl of VWR water (*see* **Note 9**).

3.6 QC of cDNA

1. Measure the concentration of cDNA by Nanodrop, and verify that all samples meet the following requirements: concentration ≥ 100 ng/μl; A260/A280 ≥ 1.8; A260/A230 ≥ 1.8. Analyze the samples using the Agilent Bioanalyzer and RNA 6000 Nano Kit, and verify that all samples meet the following requirement for acceptance: median size ≥ 400 bp when compared to a DNA ladder.

3.7 Labeling

1. Prepare Random 9mer Buffer (*see* **Note 10**) as follows:

VWR deionized water	8.6 ml
1 M Tris–HCl	1.25 ml
1 M $MgCl_2$	125 μl
β-Mercaptoethanol	17.5 μl
Total	*10 ml*

2. Dilute Cy3 dye-labeled 9mers to 1 O.D./42 μl Random 9mer Buffer. Aliquot 40 μl in 0.2 ml thin-walled PCR tubes and store at −20 °C (*see* **Note 10**).
3. Assemble the following components in separate 0.2 ml thin-walled PCR tubes:

Component	Volume
cDNA	1 μg
Cy3-9mer primers	40 μl
VWR water	To volume
Total	*80 μl*

4. Heat-denature samples at 98 °C for 10 min. Quick-chill in an ice-water bath for 2 min (*see* **Note 11**).
5. Prepare the following dNTP/Klenow Master Mix:

Component	Volume
dNTP mix	10 μl
VWR deionized water	8 μl
Klenow (50 U/μl)	2 μl
Total	*20 μl*

6. Add 20 μl of dNTP/Klenow Master Mix to the denatured samples from **step 4**. Mix well, and then incubate at 37 °C for 2 h.
7. Stop the reaction by addition of 10 μl 0.5 M EDTA.
8. To precipitate the cDNA, add 11.5 μl 5 M NaCl and 110 μl of isopropanol and then incubate for 10 min on ice.

9. Centrifuge at maximum speed of 12,000 × *g* for 10 min. Discard supernatant. Wash the pellet with 500 μl 80 % ice-cold ethanol.
10. Air-dry the pellet, and then rehydrate pellets in 25 μl VWR deionized water.
11. Determine the concentration of each sample (*see* **Note 12**).

3.8 Hybridization

1. Prepare the hybridization mix as follows (*see* **Notes 13** and **14**): Incubate at 100 °C for 5 min. Immediately transfer to an ice-water bath for 5 min.

Component	Volume
Labeled cDNA sample	5 μg
10 GE blocking agent	11 μl
2× Hi-RPM hybridization buffer	55 μl
Total volume	110 μl

2. Quickly spin in a centrifuge to collect any condensation at the bottom of the tube. Immediately proceed to hybridization.
3. Hybridize at 55 °C for 18 h (*see* **Note 15**).

4 Notes

1. High-quality RNA is required for optimal cDNA synthesis yield and cDNA labeling for microarray hybridization (free of interfering substances and with high integrity). Total RNAs can be prepared by TRIzol Reagent or other commercially available kits.
2. The reaction can be scaled up, but the RNA concentration should not exceed 0.2 μg/μl.
3. When transferring the RNA-containing supernatant to a fresh tube, avoid introducing the DNase inactivation reagent into solutions because it can sequester divalent cations and change the buffer conditions.
4. Analyze samples using the Agilent Bioanalyzer and RNA 6000 Nano Kit. Degraded samples appear as significantly lower intensity traces with the main peak area shifted to the left and typically exhibit much more noise in the trace. Verify that all samples meet the following requirements: $A_{260}/A_{280} \geq 1.8$; $A_{260}/A_{230} \geq 1.8$.
5. Enzymes (e.g., poly A polymerase) should be mixed gently without generating bubbles. Pipet the enzymes carefully and slowly; otherwise, the viscosity of the 50 % glycerol in the buffer can lead to pipetting errors.

6. If high concentration of RNA is used, the concentration of ATP should be increased too.
7. The *no-PAP* control cDNA can be prepared from a polyadenylation reaction in which the poly A polymerase is omitted.
8. Heat denature the tailed RNA no longer than 5 min, since the divalent cations from the tailing reaction can cause degradation of RNA in high temperature. The polyadenylated RNA can also be purified with a phenol-chloroform extraction and ethanol precipitation before the cDNA synthesis step.
9. cDNA is stable at −20 °C for 1 month, but try to avoid multiple freeze-thaw cycles.
10. Cy3-9mer dilution buffer should be prepared freshly each time when the primers are resuspended. Diluted primers can be stored at −20 °C for 4 months.
11. Snap-chilling after denaturation is critical for high-efficiency labeling.
12. Use a NanoDrop spectrophotometer to measure the concentration of labeled cDNA. Typical yields range from 25 to 50 μg per reaction.
13. To prepare the 10× blocking agent, add 500 μl of nuclease-free water to the vial containing lyophilized 10× gene expression blocking agent supplied with the gene expression hybridization kit, and gently mix on a vortex mixer. If the pellet does not go into solution completely, heat the mix for 4–5 min at 37 °C. It can then be stored at −20 °C for 2 months.
14. Cyanine dyes (Cy) are ozone sensitive. It is important to regularly monitor ozone levels in the lab environment and take the necessary precautions to maintain atmospheric ozone levels below 5 ppb (parts per billion).
15. For the microarray wash and scan, we follow the instruction of Agilent miRNA microarray protocol. But the condition can be further optimized based on the design of the custom arrays.

References

1. Alexander RP, Fang G, Rozowsky J, Snyder M, Gerstein MB (2010) Annotating non-coding regions of the genome. Nat Rev Genet 11:559–571
2. Mercer TR, Dinger ME, Mattick JS (2009) Long non-coding RNAs: insights into functions. Nat Rev Genet 10:155–159
3. Hammond SM (2007) MicroRNAs as tumor suppressors. Nat Genet 39:582–583
4. Croce CM (2009) Causes and consequences of microRNA dysregulation in cancer. Nat Rev Genet 10:704–714
5. Esteller M (2011) Non-coding RNAs in human disease. Nat Rev Genet 12:861–874
6. Ponting CP, Oliver PL, Reik W (2009) Evolution and functions of long noncoding RNAs. Cell 136:629–641
7. Wang KC, Chang HY (2011) Molecular mechanisms of long noncoding RNAs. Mol Cell 43:904–914
8. Geisler S, Coller J (2013) RNA in unexpected places: long non-coding RNA functions in diverse cellular contexts. Nat Rev Mol Cell Biol 14:699–712

9. Calin GA, Liu CG, Ferracin M, Hyslop T, Spizzo R, Sevignani C, Fabbri M, Cimmino A, Lee EJ, Wojcik SE et al (2007) Ultraconserved regions encoding ncRNAs are altered in human leukemias and carcinomas. Cancer Cell 12:215–229
10. Gupta RA, Shah N, Wang KC, Kim J, Horlings HM, Wong DJ, Tsai MC, Hung T, Argani P, Rinn JL et al (2010) Long non-coding RNA HOTAIR reprograms chromatin state to promote cancer metastasis. Nature 464:1071–1076
11. Huarte M, Guttman M, Feldser D, Garber M, Koziol MJ, Kenzelmann-Broz D, Khalil AM, Zuk O, Amit I, Rabani M et al (2010) A large intergenic noncoding RNA induced by p53 mediates global gene repression in the p53 response. Cell 142:409–419
12. Feng Y, Hu X, Zhang Y, Zhang D, Li C, Zhang L (2014) Methods for the study of long noncoding RNA in cancer cell signaling. Methods Mol Biol 1165:115–143
13. Xu G, Chen J, Pan Q, Huang K, Pan J, Zhang W, Chen J, Yu F, Zhou T, Wang Y (2014) Long noncoding RNA expression profiles of lung adenocarcinoma ascertained by microarray analysis. PLoS One 9:e104044
14. Malouf GG, Zhang J, Yuan Y, Comperat E, Roupret M, Cussenot O, Chen Y, Thompson EJ, Tannir NM, Weinstein JN et al (2014) Characterization of long non-coding RNA transcriptome in clear-cell renal cell carcinoma by next-generation deep sequencing. Mol Oncol 9:32–43
15. Chakravarty D, Sboner A, Nair SS, Giannopoulou E, Li R, Hennig S, Mosquera JM, Pauwels J, Park K, Kossai M et al (2014) The oestrogen receptor alpha-regulated lncRNA NEAT1 is a critical modulator of prostate cancer. Nat Commun 5:5383
16. Ling H, Spizzo R, Atlasi Y, Nicoloso M, Shimizu M, Redis RS, Nishida N, Gafa R, Song J, Guo Z et al (2013) CCAT2, a novel noncoding RNA mapping to 8q24, underlies metastatic progression and chromosomal instability in colon cancer. Genome Res 23: 1446–1461

Chapter 6

Long Noncoding RNA Expression Profiling Using Arraystar LncRNA Microarrays

Yanggu Shi and Jindong Shang

Abstract

Arraystar LncRNA microarrays are designed for global gene expression profiling of both LncRNAs and mRNAs on the same array. The array contents feature comprehensive collections of LncRNAs and include entire sets of known coding mRNAs. Each RNA transcript is detected by a splice junction-specific probe or a unique exon sequence, such that the alternatively spliced transcript isoforms or variants are reliably and accurately detected. The highly optimized experimental protocols and efficient workflow ensure sensitive, robust, and accurate microarray data generation. Standard data analyses are provided for microarray raw data processing, data quality control, gene expression clustering and heat map visualization, differentially expressed LncRNAs and mRNAs, LncRNA subcategories, regulatory relationships of LncRNAs with the mRNAs, gene ontology, and pathway analysis. The LncRNA microarrays are powerful tools for the study of LncRNAs in biology and disease, with broad applications in gene expression profiling, gene regulatory mechanism research, LncRNA functional discovery, and biomarker development.

Key words Long noncoding RNA, LncRNA, lincRNA, Microarray, Gene expression profiling

1 Introduction

LncRNAs have diverse regulatory roles in gene expression [1–7] and are involved in many biological processes and diseases [3, 8–21]. Expression profiling of LncRNAs has become increasingly essential to unravel how the genes are expressed and regulated. Also, the LncRNA expression patterns during biological development or pathogenesis are often more specific than those of mRNAs, which may be utilized as a desirable property for biomarker applications [2, 22]. In a sense, past gene expression profiling studies missing the LncRNAs may benefit from revisiting the LncRNA expression to gain new insights.

For gene expression profiling that includes LncRNAs, microarray is a preferred platform compared to RNA-seq [23]. Typically, LncRNAs are expressed at much lower abundance (~1/10 of the median mRNA level) [1, 2, 24–28], which can be overwhelmed by

Yi Feng and Lin Zhang (eds.), *Long Non-Coding RNAs: Methods and Protocols*, Methods in Molecular Biology, vol. 1402, DOI 10.1007/978-1-4939-3378-5_6,

the highly abundant mRNAs and impeded for accurate quantification in RNA-sequencing [29–31]. On the other hand, different RNA sequences are independently hybridized to the probes on the microarray. LncRNAs at lower abundance levels are relatively unaffected by the presence of highly abundant but unrelated RNA molecules [23, 32]. Also, a significant population of LncRNAs (50–70 %) are transcribed as natural antisense RNAs (NAT), the strand directions of which can be readily and natively determined by the complementarity of the array probes. Like mRNAs, LncRNAs also undergo alternative splicing, which may, without the protein-coding constraints, produce functionally diverse or even opposite transcripts. Due to the difficulty of transcript reconstruction from short RNA sequencing reads, less than 35 % of the RNA transcript isoforms can be correctly reconstructed by the state-of-the-art RNA-seq algorithms [33]. With Arraystar LncRNA microarrays, the splice junction-specific or unique exon-specific probes unambiguously and reliably detect RNAs at transcript level. At a practical level, the microarray procedures are robust, efficient, and simpler. Microarray RNA targets are generated by T7 promoter-driven linear amplification, which is better at preserving the fidelity of native RNA abundance levels and avoids distortions introduced by exponential PCR amplifications during the RNA-sequencing processes [34–37].

Arraystar LncRNA microarrays curate from various data sources and scientific publications for comprehensive LncRNA collection. Subcategories of lincRNAs, enhancer LncRNAs, transcribed HOX clusters, and pseudogenes are included. Importantly, the stringent computational pipeline ensures that the LncRNA collections are both extensive and of high quality. Based on the genomic arrangements and the regulatory potentials on the nearby protein-coding genes, LncRNAs are systematically and concisely classified in five subgroups: intergenic, bidirectional, intronic, antisense, and sense-overlapping. All the known protein-coding mRNAs in the Collaborative Consensus Coding Sequence project (CCDS) are entirely covered. The RNA labeling system is formulated for both LncRNA and mRNA. Cy3-labeled cRNAs are generated along the entire length of the transcript without 3′-end bias. The procedure enhances the detection of RNAs in limited quantities, degraded RNAs, and LncRNAs without a poly(A) tail. The highly optimized protocols and well-established workflow maximize the success of producing sensitive, accurate, and reproducible results. The standard data analyses are furnished with detailed annotation, LncRNA subcategories, noncoding and coding relationships, abundance levels, differential expression, clustering heat maps, Volcano plots, gene ontology, and pathway analysis, to get close to the biology of LncRNAs.

2 Materials

2.1 LncRNA Expression Microarrays

Arraystar LncRNA Expression Microarrays are available for human, mouse, and rat species (Arraystar Inc., Maryland, USA). Due to the low phylogenetic conservation of LncRNA sequences [2], an LncRNA array designed for one species cannot be used for other species. The technical specifications are listed in Table 1.

2.2 Arraystar Flash RNA Labeling Kit, One Color

Arraystar Flash RNA Labeling Kit, One Color is designed to prepare fluorophore-labeled RNA targets from an input RNA amount between 1.5 and 3 μg. The kit reverse transcribes RNA to double-stranded cDNA using oligo(dT) and random primers, both containing a T7 promoter. The target RNAs are then linearly amplified from the T7 promoter by T7 polymerase in an in vitro transcription (IVT) reaction. The reaction simultaneously amplifies cRNA and incorporates Cy3- or Cy5-UTP substrate. The IVT linear amplification preserves the native RNA abundance levels much better than PCR-based exponential amplification.

2.3 Microarray Hybridization

1. Tecan HS Pro Hybridization Station, or
2. Agilent Microarray Hybridization Oven G2545A.
3. Agilent SureHyb Microarray Hybridization Chamber G2534A.
4. Agilent Hybridization Gasket Slide Kit.
5. Rectangular slide staining dishes and slide rack.

2.4 Compatible Microarray Scanners for Standard 1″ × 3″ Slide Format

1. Agilent Microarray Scanner, Model SureScan.
2. Agilent Microarray Scanner, Model G2565B.
3. GenePix® 4000A Microarray Scanner (Molecular Devices).
4. GenePix® 4000B Microarray Scanner (Molecular Devices).

2.5 Agilent 2100 BioAnalyzer

2.6 NanoDrop Spectrophotometer, Equipped with Microarray Module (Thermo Scientific)

2.7 PCR Thermal Cycler

2.8 Software Packages

1. Agilent Feature Extraction Software.
2. Agilent GeneSpring GX Software Package.

Table 1
Arraystar LncRNA Expression Microarray specifications

Microarray	Arraystar Human LncRNA Expression Microarray V3.0	Arraystar Mouse LncRNA Expression Microarray V3.0	Arraystar Rat LncRNA Expression Microarray
GEO platform accession	GPL16956	GPL19286	GPL15690
Species	Human, *Homo sapiens*	Mouse, *Mus musculus*	Rat, *Rattus norvegicus*
Total number of distinct probes	58,944	60,804	24,745
Probe length	60 nt	60 nt	60 nt
Probe region	Unique exon- or splice junction-specific probes along the entire length of the transcript		
Probe specificity	Transcript specific		
Protein-coding mRNAs	26,109	24,881	15,200
LncRNAs	30,586	35,923	9,300
Transcribed pseudogenes	577	3,419	
LncRNAs with open reading frames	709	1,428	
LncRNA sources	NCBI RefSeq, USCS Known Genes 6.0, GENECODE v13, RNAdb 2.0, NRED, lincRNAs (Khalil et al.), lincRNAs (Cabili et al.), LncRNAs with enhancer-lie function (Orom et al.), HOX LncRNAs, T-UCRs	NCBI RefSeq, USCS Known Genes 6.0, Ensembl 38.71, Fantom3, RNAdb 2.0, and NRED, lincRNAs (Guttman et al., Khalil et al., Alexander et al., Sigova et al.), T-UCRs, Evolutionary constrained LncRNAs	NCBI RefSeq, USCS all_mrna, Mouse LncRNA orthologs, UCRs
mRNA sources	The collaborative consensus coding sequence project (CCDS)	The collaborative consensus coding sequence project (CCDS)	NCBI RefSeq
Array format	8 × 60 k	8 × 60 k	4 × 44 k
Technology	In situ oligonucleotide		

2.9 Additional Equipment

1. Water bath/heating block.
2. Powder-free gloves.
3. Clean, blunt forceps.
4. Micropipettors.
5. Sterilized and nuclease-free pipette tips.
6. Sterilized and nuclease-free microcentrifuge tubes.
7. High-speed microcentrifuge.
8. Low-speed tabletop microcentrifuge with slide holder attachment.
9. Vortex mixer.

2.10 Additional Reagents

1. Arraystar Spike-In RNA kit.
2. Agilent Gene Expression Wash Buffer Kit.
3. Deionized nuclease-free water.
4. TRIzol® Reagent (Life Technologies).
5. QIAGEN RNeasy® Mini Kit (Qiagen).

3 Methods

An LncRNA Expression Microarray experiment consists of several major steps in a workflow shown in Fig. 1.

3.1 RNA Samples

Plan the samples in biological replicates and set up comparison groups based on the experimental objective and design (*see* **Note 1**). The amount of total RNA per sample for a microarray experiment is typically 2 μg (Table 2). A lower amount may place significant constraints on reliable handling, storage, recovery, sample QC, experiment repeat, success rate, and data quality. As a good practice, reserve an aliquot of the RNA from the same preparation for future qPCR confirmation of the microarray result.

Total RNA can be extracted and purified in sufficient amount from 2×10^6 cells or 10–25 mg tissue as the starting material. We recommend using TRIzol® Reagent (Life Technologies) or RNeasy® Mini Kit (Qiagen). Please refer to the manufacturer's instructions accordingly.

For RNA extraction from blood or plasma, heparin as anticoagulant should be avoided. Heparin inhibits many enzymatic reactions and is difficult to be removed. EDTA- or citrate-based anticoagulants can be used instead.

3.2 RNA Sample QC

The RNA sample quality and quantity are critical to the success of microarray experiment, which should be evaluated before proceeding with the experiment.

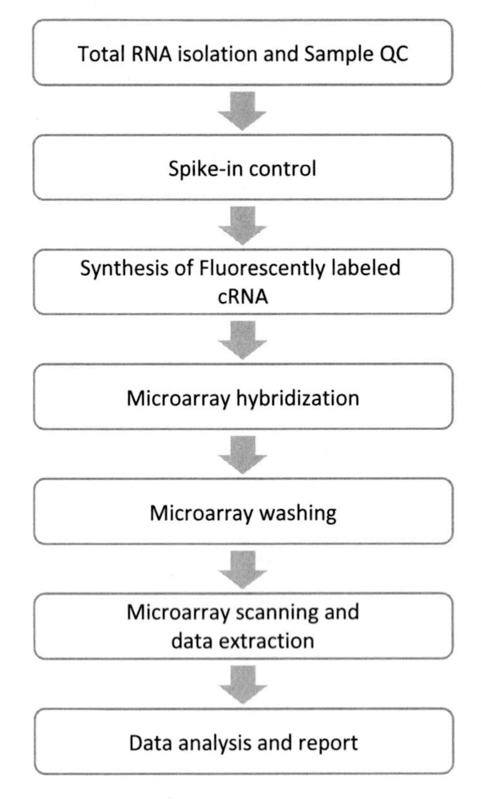

Fig. 1 Overview of lncRNA microarray experiment

Table 2
Input RNA amount per sample per array

Recommendation	Input RNA amount (μg)
Optimal	2
Minimum	1.5
Maximum	3

The concentration of RNA can be measured by UV absorbance at 260 nm with a NanoDrop spectrophotometer. A260/A280 and A260/A230 ratios are indicators for the presence of impurities and the values should be close to 2.0 (1.9–2.1 are acceptable).

The RNA integrity can be evaluated by one of the following methods:

1. Agilent 2100 Bioanalyzer, with an RNA Integrity Number (RIN) score greater than 7.

2. Denaturing gel electrophoresis, showing sharp and intense 28S and 18S rRNA bands, with a 28S:18S rRNA band intensity ratio close to 2:1.

RNAs extracted from formaldehyde-fixed-paraffin-embedded (FFPE) tissues are heavily degraded; it may be difficult to obtain a reliable BioAnalyzer RIN number, and gel electrophoresis is used.

3.3 Spike-In Control

RNA samples are spiked with spike-in RNA mix as an internal control prior to RNA labeling. The spiked-in control RNA is then fluorescently labeled and hybridized identically as the sample RNA during the entire processes. The control serves as the embedded reference in RNA labeling, scanner calibration, data normalization, sensitivity, and variability, which greatly improves the data quality assessment.

1. The spike-in mix is used in equal amount to the sample RNA in labeling reaction. Decide an input RNA amount to use (Table 2). Prepare fresh spike-in control dilution according to Table 3. Diluted spike-in mix is for single use only and should not be saved for reuse.
2. Add the volume of the diluted spike-in mix to the RNA sample (*see* Tables 2 and 3).

3.4 cRNA Synthesis and Labeling

Fluorescently labeled target RNA is synthesized using Arraystar Flash RNA Labeling Kit, specially formulated for use with Arraystar LncRNA Expression Microarrays (Fig. 2). The first-strand cDNA synthesis uses a mixture of poly(dT) and random RT primers tagged with a T7 promoter, allowing copying mRNAs and LncRNAs either with or without polyadenylation. The mixed primers also reduce 3′-biased reverse transcription from the poly(A) tail. After the cDNA synthesis, the cRNA is transcribed in vitro from the strong T7 promoter by T7 RNA polymerase, using fluorescently labeled nucleoside triphosphate substrate.

Table 3
Spike-in mix dilution and amount to use

Amount of sample RNA (μg)	Dilution of spike-in control	Amount of spike-in control (μl)
1	1:100	1
1.5	1:100	1.5
2	1:100	2
3	1:100	3

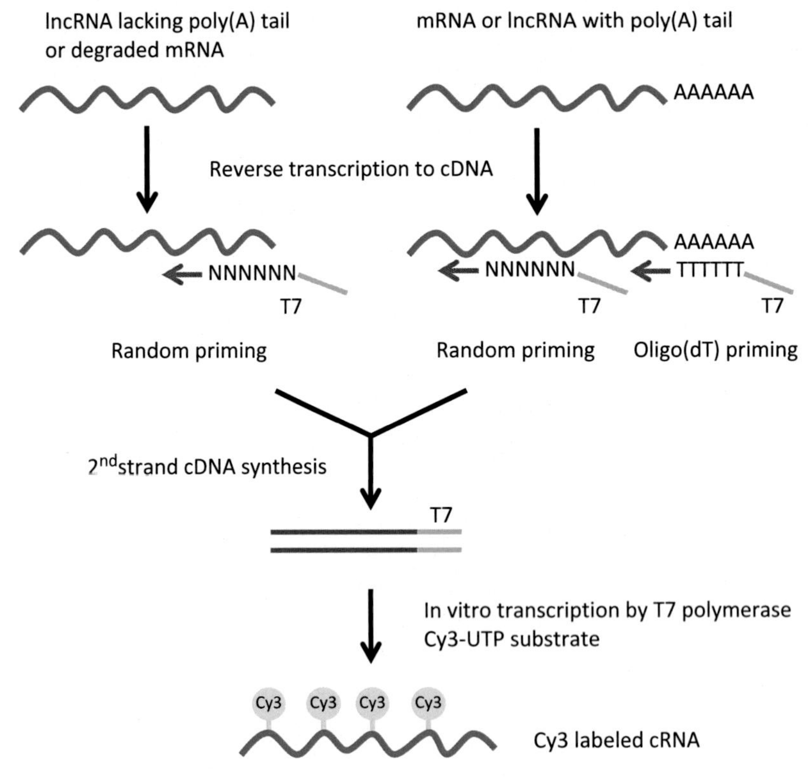

Fig. 2 Linear amplification and Cy-3 labeling of RNA targets

3.4.1 Double-Strand cDNA Synthesis

1. Anneal T7-RT primer mix to the RNA sample in annealing mix. Incubate at 65 °C for 5 min, and then immediately chill on ice for 1–2 min.

Annealing mix	
x μl	RNase-free water
y μl	Total RNA sample (2 μg)
1 μl	T7-RT primer mix
5.2 μl	Total reaction volume

2. Prepare cDNA synthesis master mix:

cDNA synthesis master mix	
2.5 μl	cDNA synthesis buffer
1 μl	DTT
1 μl	MuLV RT
0.3 μl	RNase inhibitor
4.8 μl	Total reaction volume

3. Combine the cDNA synthesis master mix with the T7-RT primer-annealed RNA sample. Incubate at 40 °C for 90 min, and then at 65 °C for 10 min to inactivate the MuLV RT.

3.4.2 In Vitro Transcription of cRNA

1. Prepare In vitro transcription master mix on ice:

In vitro transcription master mix	
8 μl	T7 transcription buffer
0.5 μl	Cy3-UTP
3.8 μl	IVT enzyme mix
12.3 μl	Total reaction volume

2. Add 12.3 μl of the in vitro transcription master mix to each synthesized cDNA reaction. Incubate at 42 °C for 3.5–4 h in the dark.

3.4.3 cRNA Purification

Add 500 μl TRIzol Reagent to the mix. Proceed with the cRNA extraction according to the manufacturer's instructions (Life Technologies).

3.4.4 Quantification of the cRNA

To verify the RNA yield and dye incorporation, we recommend using a Microarray module-equipped NanoDrop ND-1000 Spectrophotometer. Select Cy3 dye setting in the Microarray module. Please consult the NanoDrop manual for operating details.

1. Determine the cRNA yield by A260 reading, and select "RNA" type for an extinction coefficient of 40 in the measurement setting:

$$\text{cRNA amount}\left(\mu\text{g}\right)=\frac{\text{cRNA concentration}\left(\text{ng}/\mu\text{l}\right)\times\text{Volume}\left(\mu\text{l}\right)}{1000}$$

   A typical cRNA yield is >15 μg, which is sufficient for hybridization of at least two slides.

2. Calculate the specific activity of the labeled cRNA (pmoles of dye per μg of cRNA):

$$\text{Specific activity}=\frac{\text{Dye}\left(\text{pmol}/\mu\text{l}\right)}{\text{cRNA}\left(\mu\text{g}/\mu\text{l}\right)}$$

   If the cRNA yield is <1.65 μg or the specific activity is <9.0 pmol/μg (dye/cRNA), do not proceed with the microarray experiment.

3.5 Microarray Hybridization

Microarray hybridization is carried out in Agilent Microarray Hybridization Chambers. Refer to its user manual for detailed instructions of how to load samples, and assemble and disassemble the chambers.

1. Prepare cRNA fragmentation mix as follows:

Component	Amount
Labeled cRNA	1.65 μg
Blocking buffer	5 μl
Nuclease-free water	Bring the volume to 24 μl
25× Fragmentation buffer	1 μl
Total volume	25 μl

2. Incubate the mix at 60 °C for exactly 30 min to fragment the labeled cRNA.
3. Add 25 μl 2× hybridization buffer supplied with the Arraystar LncRNA Microarray Kit to stop the fragmentation reaction.
4. Spin at 13,000 × *g* for 1 min at room temperature to suppress air bubbles.
5. Place the sample on ice and pipette onto the array surface as soon as possible.
6. Place a 24 × 30 mm cover slip on top of the slide; be careful not to trap air bubbles.
7. Clamp the microarray/backing gasket slide sandwich into the SureHyb hybridization chamber. Rotate the slide assembly to wet the surface. Ensure that the air pockets are not in stuck positions and can move freely.
8. Incubate at 65 °C for 17 h in a hybridization oven with a rotation speed of 10 rpm. We recommend Tecan HS Pro hybridization station for microarray hybridization. Refer to its user manual for detailed instructions.

3.6 Microarray Washing

For washing the array, we strongly recommend Agilent Gene Expression Wash Buffer Kits. Please see its manual for detailed instructions.

1. Prepare and pre-warm all washing solutions to their desired temperatures 1 h prior to washing. Disassemble hybridization chamber with the array surface facing up.
2. Immerse the slide in 100 ml of wash solution B1 (pre-warmed to 42 °C) in a Coplin jar until the cover slip moves freely away from the slide.
3. Remove the cover slip with a forceps and decant the wash solution.

4. Move the slide to another jar filled with 100 ml of wash solution B1 (pre-warmed to 42 °C) and shake at room temperature for 3 min.
5. Move the slide to another jar filled with 100 ml of wash solution B2 (pre-warmed to room temperature) and shake at room temperature for 3 min.
6. Repeat **step 5**.
7. Move the slide to another jar filled with 100 ml of wash solution B3 (pre-warmed to room temperature) and shake at room temperature for 3 min.
8. Repeat **step 7**.
9. Place the slide in centrifuge rack and centrifuge at 950 rpm for 5 min for drying.
10. The microarray slide is now ready for scanning.

3.7 Microarray Scanning and Data Extraction

The following microarray scanners are compatible with Arraystar LncRNA Microarray products:

1. Agilent Microarray Scanner, Model SureScan.
2. Agilent Microarray Scanner, Model G2565B.
3. GenePix 4000A Microarray Scanner (Molecular Devices).
4. GenePix 4000B Microarray Scanner (Molecular Devices).

3.7.1 Agilent Scanner Settings

Turn on the scanner to pre-warm the laser for at least 20 min before the scanning.

1. Set the scan settings for one-color scan:

Setting	Value
Dye channel	Green
Scan region	61 × 21.6 mm
Scan resolution (μm)	3
Tiff	20 bit
Green PMT	XDR Hi 100 %; XDR Lo 10 %

2. Scan the microarray at 3 μm resolution.
3. Save the TIFF images.

3.7.2 GenePix Scanner Settings

Only GenePix 4000A and 4000B models are supported for scanning Arraystar LncRNA microarrays. Refer to the manufacturer's user guide for appropriate scanner settings.

3.7.3 Data Extraction

After the microarray image scanning, raw data are extracted from the .tif images using Agilent Feature Extraction Software. Please refer to its Reference Guide for detailed instructions.

1. Launch Agilent Feature Extraction (FE) software.
2. Add the image (.tif) files to be extracted to the FE Project.
3. Select options in the FE Project Properties.
4. Check the Extraction Set Configuration.
   (a) Select the Extraction Set Configuration tab.
   (b) Verify that the correct grid template is assigned to each extraction set in the Grid Name column.
   (c) Verify that the correct protocol is assigned to each extraction set in the Protocol Name column.
5. Save the FE Project (.fep) by selecting File > Save As and browse to a desired file location.
6. Verify that the icons for the image files in the FE Project Window no longer have a red X through them.
7. Select Project > Start Extracting.
8. After the extraction is completed successfully, examine the QC report for each extraction set by double-clicking the QC Report link in the Summary Report tab (Fig. 3). Refer to Agilent Feature Extraction Software Reference Guide for detailed explanations.

3.8 Data Analysis

We recommend using GeneSpring GX (Agilent, USA) software for microarray data analysis, which features extensive, graphic user interface-guided microarray analyses on raw or processed data. Please refer to the User Manual for instructions. Many software are also freely available for microarray data analysis (*see* **Note 2**).

The procedures to import Arraystar LncRNA microarray raw data into GeneSpring are described below.

3.8.1 Set Up Raw Data Import

To read LncRNA microarray raw data into GeneSpring GX, set up the import for the first-time use.

1. Select Annotations/Create Technology/Custom from File menu.
2. Select Technology Type: "Single color," Technology name: desired_technology_name; Organism: select_one; Choose a sample data file: a_rawdatafile; Number of samples in single data file: "One Sample."
3. The raw data will be displayed in a spread sheet in the "Format data file" step.
4. Define the data rows to be imported in the "Select Row Scope for Import." Keep the rows starting with the column headings to the end row.
5. Select Identifier: "ProbeName"; BG Corrected Signal: "gProcessedSignal."

QC Report -Agilent Technologies: 1 Color

Date	Wednesday, August 14, 2013 – 15:11	Grid	045997_D_F_20121218
Image	US10450393_S01 [1_3]	FE Version	11.0.1.1
Protocol	GE1_1100_July11 (Read Only)	Saturation Value	778178 (g)

Spot Finding of the Four Corners of the Array

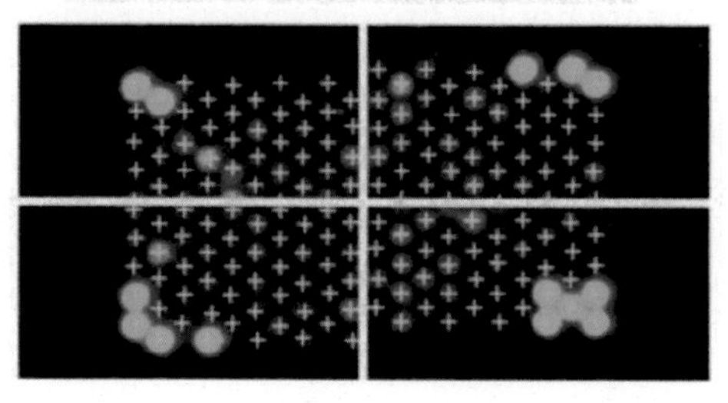

	Feature	Local Background
	Green	Green
Non Uniform	29	53
Population	127	2826

Spatial Distribution of All Outliers on the Array

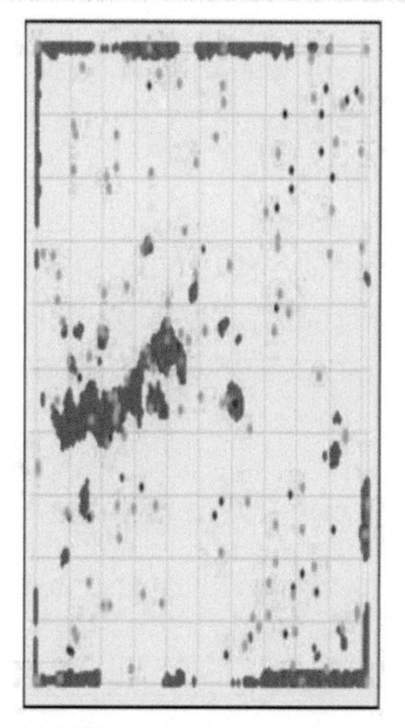

FeatureNonUnif (Green) = 29 (0.05%)

GeneNonUnif (Green) = 29 (0.049%)

•BG NonUniform •BG Population

•Green FeaturePopulation •Green Feature NonUniform

Non-Control probes Net Signal Statistics:

	Green
# Saturated Features	0
99% of Sig. Distrib.	19491
50% of Sig. Distrib.	72
1% of Sig. Distrib.	42

Histogram of Signals Plot

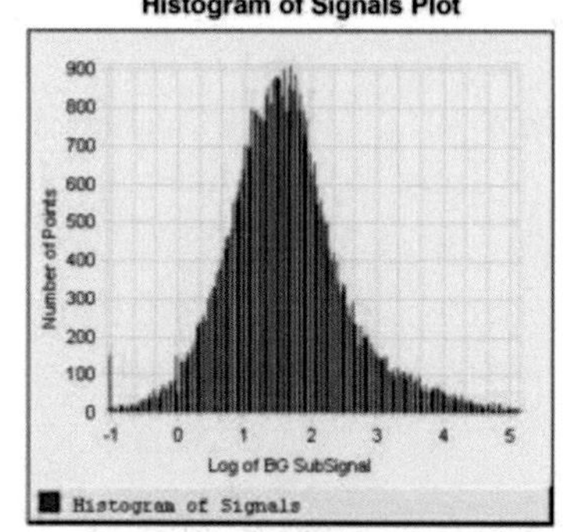

Features (NonCtrl) with BGSubSignal < 0:11251 (Green)

Negative Control Stats

	Green
Average Net Signals	46.07
StdDev Net Signals	4.29
Average BG Sub Signal	-6.79
StdDev BG Sub Signal	4.01

Local Bkg (inliers)

	Green
Number	58358
Avg	32.78
SD	1.71

Reproducibility: Median %CV for Replicated Signal (inliers) Non-Control probes

	Green
BGSubSignal	6.34
ProcessedSignal	6.34

Fig. 3 An example QC Report for a 8×60K microarray, generated by the Feature Extraction Software

3.8.2 Create New Project for Data Analysis

1. Select Project/New experiment menu.
2. Select Experiment type: "Generic Single Color" and Workflow type: "Advanced Analysis."
3. In the Load Data step, select the Technology: desired_technology_name you created previously.
4. Choose the raw data files.
5. Uncheck "Please select if your data is in log scale." as the intensity values in raw data are not log2 transformed. Threshold raw signals to: "5.0." Select Normalization Algorithm: "Quantile."
6. In the Preprocess Baseline Options step, select "Do not perform baseline transformation."
7. Once the data are imported, proceed with microarray data processing, analysis, interpretation, graphics, and result reporting using the functionalities provided by the software suite.

3.9 Standard Analyses

The standard analyses provided by the Arraystar LncRNA Expression Microarray service include common microarray data analyses as well as annotations and analyses specific to LncRNAs.

1. Microarray scan images.
2. Feature extraction and raw intensity values.
3. Array QC metrics.
4. Data quality and low-intensity filtering/flagging (*see* **Note 3**).
5. Data normalization (*see* **Note 4**).
6. Box plots of intensity distribution.
7. Scatterplots of intensities between groups or samples.
8. Differentially expressed LncRNAs (*see* **Note 5**).
9. Differentially expressed mRNAs (*see* **Note 5**).
10. Heat maps and hierarchical clustering (*see* **Note 6**).
11. Volcano plots.
12. LncRNA annotations with coding type, database source, sequence accession, genomic location coordinates, transcription, xhyb, probe name and sequence, relationship with associated coding gene, and associated coding gene genomic information (*see* **Note 7**).
13. LncRNA classification and subgroup as enhancer LncRNAs, antisense LncRNAs, lincRNAs, and the associated protein-coding mRNA gene expression data.
14. mRNA annotations with coding type, database source, sequence accession, genomic location coordinates, transcription, xhyb, probe name and sequence, gene ontology terms, and protein product.

15. Gene ontology analysis of differentially expressed mRNAs (*see* **Note 8**).
16. Pathway analysis of differentially expressed mRNAs (*see* **Note 9**).
17. Gene Expression Omnibus (GEO) repository support. Minimum Information About a Microarray Experiment (MIAME) compliant (*see* **Note 10**).

4 Notes

1. In a basic experiment design to compare gene expression between conditions and to screen differentially expressed genes, one group of samples are designated for each condition for comparison. At least three samples per group are needed as biological replicates. More biological replicates are strongly recommended. The effect of replication on gene expression microarray experiments has been discussed previously [38]. The variabilities in clinical samples are typically much larger than experiments using cell line samples, requiring more biological replicates. Due to the high-quality manufacturing process and excellent reproducibility of the microarrays, technical replications, i.e., the same RNA sample on replicated arrays, are not needed.
2. Open-source or free software are also available for data analysis, for example, GenePattern (www.broadinstitute.org/cancer/software/genepattern), MultiExperiment Viewer MeV (www.tm4.org/mev.html), Chipster (chipster.csc.fi), and R/Bioconductor packages (www.bioconductor.org).
3. Each microarray feature (spot) is evaluated and flagged as "Present" or "Absent" based on whether the signal is positive and significant, uniform, above the background, oversaturated, or population outlier. Additionally, a feature may be flagged as "Marginal" if the background is not uniform or the background reading is a population outlier.
4. The raw intensities need to be "normalized" first to be comparable. Quantile normalization is commonly used in microarray data normalization [39]. The method does not rely on housekeeping genes as the reference. The normalized intensity values are log2 transformed.
5. When comparing differential gene expression, for example, Group2 vs. Group1, the magnitude of change is expressed as *fold change* (*FC*), which can be obtained by

$$FC = 2^{|\text{Norm2}-\text{Norm1}|}$$

where Norm2 and Norm1 are the arithmetic means of the log2-transformed normalized intensities for samples in Group2 and Group1, respectively. Conventionally, a decrease in gene expression (downregulation) is indicated by a negative sign of the *FC*:

$FC = -|FC|$ if norm2 < norm1, for down regulation.

For example, a gene is downregulated by 13.93-fold when comparing Group2 vs. Group1:

Group1				Group2				*FC*
Sample1	Sample2	Sample3	Norm1 (average)	Sample4	Sample5	Sample6	Norm2 (average)	$-2^{\|3.3729-7.1731\|}$
6.4413	7.2712	7.8067	7.1731	3.6496	3.2023	3.2670	3.3729	-13.93

The *p*-value for a gene expression comparison is calculated by *t*-test. As many *t*-tests are performed for the genes on the array, the *p*-values are adjusted for multiple testing correction as false discovery rates (*FDR*), using Benjamini Hochberg procedure [40].

Other statistical analyses have been devised to assess differential gene expression more robustly and accurately. For example, R/Bioconductor *limma* package implements linear models and empirical Bayes methods to perform differential expression analysis [41].

The differentially expressed genes (DEG) are initially selected by using minimally stringent cutoffs of $p \leq 0.05$ and $FC \geq 2$. These DEGs are displayed on Volcano plots along the $-\log_{10}(p)$ and $\log_2(FC)$ axes. The DEGs on the list can be further ranked by *FC* and applied with more stringent *p*-value (or *FDR*) cutoffs using the Excel Data/Filter functionality.

When selecting differentially expressed genes for confirmation, the fold change magnitudes, *p*-values (adjusted), and raw signal intensities should be considered. Genes showing significant *FC* and *p*-values but having very low raw intensities may still not be confirmed by qPCR. Empirically, at least one of the group raw intensities must be greater than 200–500.

To confirm the differentially expressed genes, either SYBR® Green qPCR or Taqman® assay can be used. It is a good practice to perform qPCR on the same RNA prep used for the array and follow all the qPCR design guidelines by the qPCR manufacturer. The qPCR primers should be situated as close as possible to the array probe location. If the LncRNA is a natural antisense LncRNA, a strand-specific primer, instead of oligo(dT) primer, should be used for first-strand cDNA synthesis. qPCR has a typical limit of detection equivalent to 0.5 $\Delta\Delta$Ct (or $FC \approx 1.5$) in differential expression analysis.

There are cases where (1) no or very few individual genes meet the threshold for DEGs, because the relevant biological differences are modest; (2) too many genes show up as DEGs;

and (3) single-gene analysis may miss important effects on pathways. Biological processes often affect sets of genes acting in concert. Consistent changes in all genes in a pathway may have more significant effect by a large change of a single gene. To overcome these challenges, Gene Set Enrichment Analysis (GSEA) or gene ontology or pathway analysis can be considered to prioritize the genes for further studies [42].

6. Correct sample grouping by unsupervised clustering is an indication of successful classification by gene expression patterns. However, if the groups are identical or too similar in gene expression, samples may appear incorrectly clustered with other groups. This is not necessarily an indication of microarray data problem.

   Due to space limitation, gene names are not displayed on the heat maps. To identify individual genes on the heat map, an interactive software is needed. Alternatively, limit the number of the genes on the heat map by including only a small subset of genes of interest, for example, DEGs at high stringency in a pathway, to allow the gene names be printed.

7. The LncRNAs on the array were collected from multiple data sources. To obtain the sequences and to access the genomic records, use the corresponding databases and accession numbers/IDs as indicated. In the report, a probe may have multiple annotation entries for the same LncRNA, due to multiple relationships with the protein-coding gene(s) or transcript(s).
8. Gene ontology analysis is performed for protein-coding genes only. The GO terms are hyperlinked to and described at geneontology.org.
9. Pathway analysis is performed for protein-coding genes only. The default pathway database is KEGG: Kyoto Encyclopedia of Genes and Genomes (http://www.genome.jp/kegg/).
10. GEO is a public functional genomics data repository supporting MIAME-compliant data submissions. Many journals require microarray data to be deposited and the associated GEO accession number obtained prior to publication. The GEO platform records for Arraystar LncRNA Expression Microarrays have been created by Arraystar (Table 1), which can be referenced for the preparation of GEO submissions.

References

1. Cabili MN, Trapnell C, Goff L et al (2011) Integrative annotation of human large intergenic noncoding RNAs reveals global properties and specific subclasses. Genes Dev 25:1915–1927
2. Derrien T, Johnson R, Bussotti G et al (2012) The GENCODE v7 catalog of human long noncoding RNAs: analysis of their gene structure, evolution, and expression. Genome Res 22:1775–1789
3. Batista PJ, Chang HY (2013) Long noncoding RNAs: cellular address codes in development and disease. Cell 152:1298–1307

4. Kung JT, Colognori D, Lee JT (2013) Long noncoding RNAs: past, present, and future. Genetics 193:651–669
5. Mercer TR, Dinger ME, Mattick JS (2009) Long non-coding RNAs: insights into functions. Nat Rev Genet 10:155–159
6. Amaral PP, Clark MB, Gascoigne DK et al (2011) LncRNAdb: a reference database for long noncoding RNAs. Nucleic Acids Res 39:D146–D151
7. Fatica A, Bozzoni I (2014) Long non-coding RNAs: new players in cell differentiation and development. Nat Rev Genet 15:7–21
8. Kornienko AE, Guenzl PM, Barlow DP et al (2013) Gene regulation by the act of long noncoding RNA transcription. BMC Biol 11:59
9. Nie L, Wu HJ, Hsu JM et al (2012) Long non-coding RNAs: versatile master regulators of gene expression and crucial players in cancer. Am J Transl Res 4:127–150
10. Lee JT (2012) Epigenetic regulation by long noncoding RNAs. Science 338:1435–1439
11. Wang KC, Chang HY (2011) Molecular mechanisms of long noncoding RNAs. Mol Cell 43:904–914
12. Wapinski O, Chang HY (2011) Long noncoding RNAs and human disease. Trends Cell Biol 21:354–361
13. Chen G, Wang Z, Wang D et al (2013) LncRNADisease: a database for long-noncoding RNA-associated diseases. Nucleic Acids Res 41:D983–D986
14. Taft RJ, Pang KC, Mercer TR et al (2010) Non-coding RNAs: regulators of disease. J Pathol 220:126–139
15. Broadbent HM, Peden JF, Lorkowski S et al (2008) Susceptibility to coronary artery disease and diabetes is encoded by distinct, tightly linked SNPs in the ANRIL locus on chromosome 9p. Hum Mol Genet 17:806–814
16. Scheuermann JC, Boyer LA (2013) Getting to the heart of the matter: long non-coding RNAs in cardiac development and disease. EMBO J 32:1805–1816
17. Gomez JA, Wapinski OL, Yang YW et al (2013) The NeST long ncRNA controls microbial susceptibility and epigenetic activation of the interferon-gamma locus. Cell 152:743–754
18. Cabianca DS, Casa V, Bodega B et al (2012) A long ncRNA links copy number variation to a polycomb/trithorax epigenetic switch in FSHD muscular dystrophy. Cell 149:819–831
19. Sana J, Faltejskova P, Svoboda M et al (2012) Novel classes of non-coding RNAs and cancer. J Transl Med 10:103
20. Gutschner T, Diederichs S (2012) The hallmarks of cancer: a long non-coding RNA point of view. RNA Biol 9:703–719
21. Han P, Li W, Lin CH et al (2014) A long noncoding RNA protects the heart from pathological hypertrophy. Nature 514:102–106
22. Kumarswamy R, Bauters C, Volkmann I et al (2014) Circulating long noncoding RNA, LIPCAR, predicts survival in patients with heart failure. Circ Res 114:1569–1575
23. Xu W, Seok J, Mindrinos MN et al (2011) Human transcriptome array for high-throughput clinical studies. Proc Natl Acad Sci U S A 108:3707–3712
24. Kampa D, Cheng J, Kapranov P et al (2004) Novel RNAs identified from an in-depth analysis of the transcriptome of human chromosomes 21 and 22. Genome Res 14:331–342
25. Cawley S, Bekiranov S, Ng HH et al (2004) Unbiased mapping of transcription factor binding sites along human chromosomes 21 and 22 points to widespread regulation of noncoding RNAs. Cell 116:499–509
26. Ravasi T, Suzuki H, Pang KC et al (2006) Experimental validation of the regulated expression of large numbers of non-coding RNAs from the mouse genome. Genome Res 16:11–19
27. Guttman M, Garber M, Levin JZ et al (2010) Ab initio reconstruction of cell type-specific transcriptomes in mouse reveals the conserved multi-exonic structure of lincRNAs. Nat Biotechnol 28:503–510
28. Yan L, Yang M, Guo H et al (2013) Single-cell RNA-Seq profiling of human preimplantation embryos and embryonic stem cells. Nat Struct Mol Biol 20(9):1131–1139
29. Jiang L, Schlesinger F, Davis CA et al (2011) Synthetic spike-in standards for RNA-seq experiments. Genome Res 21:1543–1551
30. Labaj PP, Leparc GG, Linggi BE et al (2011) Characterization and improvement of RNA-Seq precision in quantitative transcript expression profiling. Bioinformatics 27:i383–i391
31. Toung JM, Morley M, Li M et al (2011) RNA-sequence analysis of human B-cells. Genome Res 21:991–998
32. Kretz M, Webster DE, Flockhart RJ et al (2012) Suppression of progenitor differentiation requires the long noncoding RNA ANCR. Genes Dev 26:338–343
33. Steijger T, Abril JF, Engstrom PG et al (2013) Assessment of transcript reconstruction methods for RNA-seq. Nat Methods 10:1177–1184
34. King C, Guo N, Frampton GM et al (2005) Reliability and reproducibility of gene expression measurements using amplified RNA from

laser-microdissected primary breast tissue with oligonucleotide arrays. J Mol Diagn 7:57–64
35. Li L, Roden J, Shapiro BE et al (2005) Reproducibility, fidelity, and discriminant validity of mRNA amplification for microarray analysis from primary hematopoietic cells. J Mol Diagn 7:48–56
36. Klur S, Toy K, Williams MP et al (2004) Evaluation of procedures for amplification of small-size samples for hybridization on microarrays. Genomics 83:508–517
37. Wilson CL, Pepper SD, Hey Y et al (2004) Amplification protocols introduce systematic but reproducible errors into gene expression studies. Biotechniques 36:498–506
38. Pavlidis P, Li Q, Noble WS (2003) The effect of replication on gene expression microarray experiments. Bioinformatics 19:1620–1627
39. Bolstad BM, Irizarry RA, Astrand M et al (2003) A comparison of normalization methods for high density oligonucleotide array data based on variance and bias. Bioinformatics 19: 185–193
40. Benjamini Y, Hochberg Y (1995) Controlling the false discovery rate: a practical and powerful approach to multiple testing. J R Stat Soc Ser B 57:289–300
41. Smyth GK (2004) Linear models and empirical bayes methods for assessing differential expression in microarray experiments. Stat Appl Genet Mol Biol 3: Article3
42. Subramanian A, Tamayo P, Mootha VK et al (2005) Gene set enrichment analysis: a knowledge-based approach for interpreting genome-wide expression profiles. Proc Natl Acad Sci U S A 102:15545–15550

Chapter 7

Nuclear RNA Isolation and Sequencing

Navroop K. Dhaliwal and Jennifer A. Mitchell

Abstract

Most transcriptome studies involve sequencing and quantification of steady-state mRNA by isolating and sequencing poly (A) RNA. Although this type of sequencing data is informative to determine steady-state mRNA levels it does not provide information on transcriptional output and thus may not always reflect changes in transcriptional regulation of gene expression. Furthermore, sequencing poly (A) RNA may miss transcribed regions of the genome not usually modified by polyadenylation which includes many long noncoding RNAs. Here, we describe nuclear-RNA sequencing (nucRNA-seq) which investigates the transcriptional landscape through sequencing and quantification of nuclear RNAs which are both unspliced and spliced transcripts for protein-coding genes and nuclear-retained long noncoding RNAs.

Key words Transcriptome profiling, LncRNA, Nuclear RNA, Cellular fractionation, Massively parallel sequencing

1 Introduction

RNA-sequencing (RNA-seq) is a technique that uses next-generation sequencing in order to investigate the dynamic transcriptome of the cell [1]. Most transcriptomic studies are based on RNA-seq that involves generation of poly (A)-positive RNA libraries from total cellular RNA [2]. For mammalian genomes a large proportion of the transcribed genome is noncoding and transcribed along with protein-coding genes, contributing to the complexity of mammalian transcriptomes [3]. Nuclear RNA-sequencing (nucRNA-seq) characterizes genome-wide transcriptional output by performing deep sequencing of a cDNA library generated from nuclear RNA isolated from purified intact nuclei [4].

The quality of nucRNA-seq data depends on isolation of pure intact nuclei from single cells, purification of RNA from the nuclear fraction, and generation of the double-stranded cDNA library. Isolation of intact nuclei is the first critical step and must be optimized for different cell types to obtain pure nuclei. Here, we describe an optimized nuclear isolation protocol for mouse

Yi Feng and Lin Zhang (eds.), *Long Non-Coding RNAs: Methods and Protocols*, Methods in Molecular Biology, vol. 1402,
DOI 10.1007/978-1-4939-3378-5_7,

embryonic stem (ES) cells. The optimal protocol for erythroid cells has been described by Mitchell et al. [4] and was used to develop the mouse ES cell protocol described here.

2 Materials

2.1 Components for Isolation and Testing of Intact Nuclei

Prepare all solutions using molecular biology-grade water and reagents that are certified RNase free. Work on ice for all steps unless otherwise indicated. Chill all buffers completely on ice prior to use.

1. Molecular biology-grade Sigma water: Sigma-Aldrich, W4502.
2. 1 M Sucrose: Weigh 17.115 g sucrose, transfer to a 100 ml bottle, and dissolve in Sigma water to a final volume of 50 ml.
3. 1× PBS: Sigma-Aldrich.
4. 10× RSB: Mix 2.5 ml of 1 M Tris–HCl pH 7.5, 0.5 ml of 5 M NaCl, 750 μl of 1 M $MgCl_2$, and 21.25 ml of Sigma water to make final concentrations of 100 mM Tris–HCl, 100 mM NaCl, and 30 mM $MgCl_2$.
5. 1× RSB: 1/10 Dilution of 10× RSB with Sigma water.
6. Buffer A: Mix 0.5 ml of 10× RSB, 1.0 ml of 1 % Triton X-100, 2.5 μl of 1 M DTT, and 0.5 ml of 1 M sucrose stock and adjust final volume to 5 ml with Sigma water generating final concentrations of 1× RSB, 0.2 % Triton X-100 (*see* **Note 1**), 0.5 mM DTT, and 0.1 M sucrose.
7. Buffer B: Mix 0.5 ml of 10× RSB, 0.5 ml of 1 % Triton X-100, 2.5 μl of 1 M DTT, and 1.25 ml of 1 M sucrose and adjust final volume to 5 ml with Sigma water generating working concentrations as 1× RSB, 0.1 % Triton X-100, 0.5 mM DTT, and 0.25 M sucrose.
8. Buffer C: Mix 16.7 ml of 1 M sucrose, 250 μl of 1 M $MgCl_2$, 500 μl of 1 M Tris–HCl (pH 8.0), and 25 μl 1 M DTT and adjust final volume to 50 ml with Sigma water generating working concentrations of 5 mM $MgCl_2$, 10 mM Tris–HCl, 0.5 mM DTT, and 0.33 M sucrose.
9. 10× RIPA (Cell Signalling Technology, 9806), protease inhibitor cocktail (Roche 04693159001).
10. Antibodies used to test the purity of the nuclear fraction by western blotting: Anti-Ubfl rabbit polyclonal (Santa Cruz Biotechnology, sc-9131), anti-Cyclophilin A mouse monoclonal (Abcam, ab58144), Goat anti-rabbit conjugated with HRP (Bio-Rad, 1662408EDU), Goat anti-mouse conjugated with HRP (Bio-Rad, 1721011).

2.2 Components for Nuclear RNA Isolation

1. Trizol LS Reagent: Thermo Fisher Scientific, 10296-028.
2. Chloroform: Sigma-Aldrich 472476.
3. 100 % Isopropanol: Sigma-Aldrich I9516.
4. 75 % Ethanol: Mix 37.5 ml 100 % EtOH with 12.5 ml Sigma water.
5. DNase I, RNase free: MBI Fermentas, EN0525, 1 U/μl.
6. Acid phenol/chloroform: Ambion, AM9720.
7. Ribolock: Thermo Fisher Scientific, E00382, 40 U/μl.

2.3 Required Equipment

1. Polypropylene round-bottom tubes: 14.5 ml, Sarstedt, 62.515.006.
2. Dounce homogenizer: B-type (tight) 7 ml, VWR, 71000-518.
3. RNase Zap: For cleaning the Dounce homogenizers between each use, Thermo Fisher Scientific, AM9780.
4. Centrifuge: To pellet the nuclei the centrifuge should ideally be equipped with a swinging bucket rotor able to accommodate the 14.5 ml tubes.

3 Methods

3.1 Isolating Intact Nuclei and Testing the Purity of the Nuclear Fraction

In order to perform optimal nuclear RNA sequencing, it is critical to isolate pure intact nuclei.

3.1.1 Isolation of Intact Nuclei

Note: Before starting make fresh 10× and 1× RSB, 1 M sucrose, and buffers A, B, and C. Place all buffers on ice. Chill centrifuge to 4 °C.

1. Pour an 8 ml sucrose cushion of buffer C in the 14.5 ml round-bottom tube for each sample and chill on ice.
2. Collect $1–5 \times 10^7$ dissociated single cells in cold 1× PBS. Centrifuge cells at $120 \times g$ for 4 min at 4 °C.
3. Resuspend single cells in 1 ml ice-cold buffer A (*see* **Note 2**).
4. Dounce gently with 10–12 strokes in a chilled B-type (tight) 7 ml Dounce homogenizer (*see* **Note 3**).
5. Dilute with equal volume, 1 ml, of ice-cold buffer B. Wash walls of Dounce as you add buffer B (*see* **Note 4**). Pipette gently to mix the two buffers.
6. Pipette the 2 ml of buffer A/B suspension gently onto the wall of the tube immediately above the sucrose cushion provided by buffer C. There should be minimal mixing of the cell suspension with the sucrose cushion (Fig. 1).
7. Centrifuge at $300 \times g$ for 5 min at 4 °C.

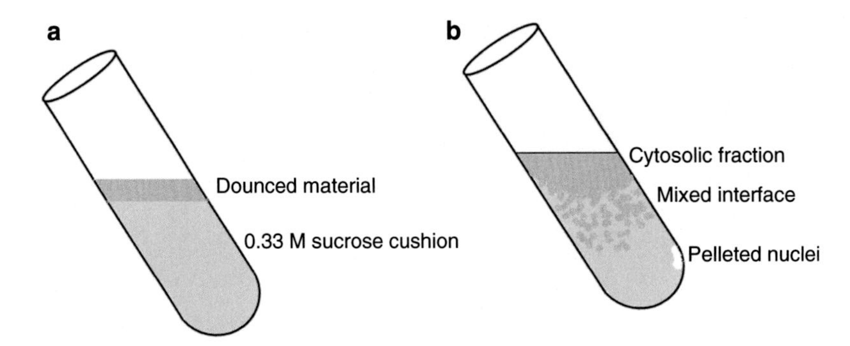

Fig. 1 (**a**) This cartoon demonstrates that dounced material should be placed over the sucrose cushion by tilting the round-bottom tube to an angle and adding the dounced material along the walls of the tube close to the interface with the sucrose cushion. (**b**) This demonstrates the position for cytosolic and nuclear fraction. Nuclei will be pelleted at the bottom of the tube whereas the top 1 ml fraction will be the cytosolic fraction

8. If the cytoplasmic fraction is to be retained remove 1 ml of the cytoplasmic fraction from the top layer avoiding the interface with the sucrose cushion and place on ice in a fresh 14.5 ml tube. Pipette off and dispose of the mixed interface of remaining cytoplasm-contaminated solution and about half of the sucrose cushion. Pour off remaining sucrose cushion and resuspend the nuclear pellet in 1 ml ice-cold 1× RSB.

3.1.2 Testing the Purity of Nuclear Fraction

1. Remove 50 μl of the suspended nuclei and cytoplasmic fraction, dilute each with 50 μl 2× RIPA buffer containing protease inhibitors, mix by pipetting, and incubate on ice for 5 min. Centrifuge at 10,000 × *g* for 10 min at 4 °C.
2. Nuclear and cytoplasmic fractions can be tested using a standard Western blotting protocol. After transfer the membrane can be incubated with both rabbit polyclonal anti-Ubf1 antibody (nuclear protein) at a 1:1000 dilution and mouse monoclonal anti-cyclophilin A (cytoplasmic protein) antibody at a 1:1000 dilution for 2 h at room temperature, washed and incubated with secondary antibodies: goat anti-rabbit and goat anti-mouse, HRP-conjugated both at 1:1000 dilution, and washed and developed using Bio-Rad Clarity Western ECL kit (Fig. 2).

3.2 Trizol Extraction of Nuclear RNA

1. Working in the 14.5 ml round-bottom tubes, add 3 ml Trizol LS to each 1 ml nuclear or cytoplasmic fraction, mix well by pipetting to homogenize the sample; if necessary, snap freeze on dry ice, and store at −80 °C for later processing while checking the purity of the nuclear and cytoplasmic fractions.
2. When ready thaw samples completely on ice.
3. Incubate at room temperature for 5 min.

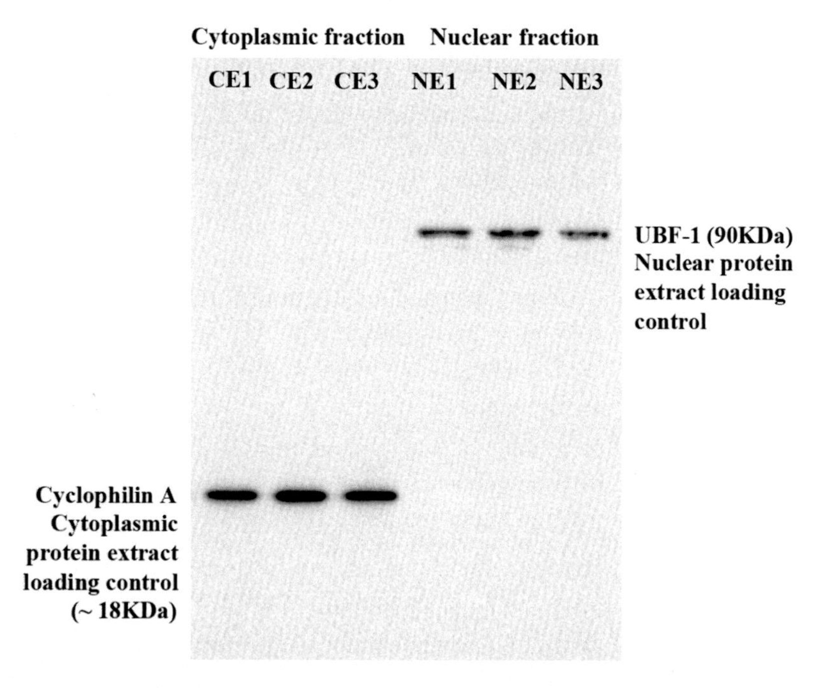

Fig. 2 Immunoblot showing the purity of three different ES cell cytoplasmic (CE1, CE2, CE3) and nuclear extractions (NE1, NE2, NE3). Cyclophilin A (cytoplasmic protein loading control) and UBF-1 (nuclear protein loading control) were detected simultaneously in all fractions on one immunoblot. Only UBF-1 is detected in the nuclear fractions and only Cyclophilin A is detected in the cytoplasmic fraction indicating that the fractions are pure

4. Add 0.8 ml chloroform, cap tightly, and mix by inverting ten times.
5. Incubate at room temperature for 2–15 min.
6. Centrifuge at 12,000 × *g* for 15 min at 4 °C.
7. Remove the upper aqueous phase to a new tube avoiding the interphase which contains genomic DNA.
8. Add 2 ml of 100 % isopropanol to the aqueous phase. Cap tightly, mix, and incubate for 10 min at room temperature.
9. Centrifuge at 12,000 × *g* for 10 min at 4 °C.
10. Remove the supernatant from the tube, leaving the RNA pellet.
11. Wash pellet with 4 ml of 75 % EtOH.
12. Centrifuge at 7500 × *g* for 5 min at 4 °C.
13. Mark the location of the RNA pellet on the outside of the tube with an EtOH resistant marker and then carefully pour off the supernatant.
14. Centrifuge briefly at 7500 × *g* at 4 °C to pellet any remaining 75 % EtOH which should be carefully removed with a pipette, avoiding touching the RNA pellet.

15. Air-dry the pellet on ice for about 5 min but do not allow it to dry completely or the RNA will become less soluble.
16. Resuspend the pellet in 50 μl RNase-free Sigma water and transfer to a clear 1.5 ml tube. Quantitate samples to determine yield and purity.
17. Working on ice, for each sample take ≤10 μg of RNA and add Sigma water to a final volume of 79 μl.
18. To each sample add 10 μl 10× DNase I buffer containing $MgCl_2$, 1 μl Ribolock, and 10 μl DNase I. Mix by flicking gently and give the tubes a quick spin.
19. Incubate for 30 min at 37 °C.
20. Add 100 μl acid phenol/chloroform (*see* **Note 5**). Mix by inverting ten times and incubate at room temperature for 2 min.
21. Centrifuge at 12,000 × *g* for 15 min at 4 °C.
22. Remove the upper aqueous phase to a fresh tube avoiding the interphase. Store this tube on ice during **steps 23–25**.
23. Add 150 μl Sigma water to the remaining acid phenol. Mix by inverting ten times and incubate at room temperature for 2 min.
24. Centrifuge at 12,000 × *g* for 15 min at 4 °C.
25. Remove the upper aqueous phase, avoiding the interphase, and combine it with the first aqueous fraction.
26. Add 1/10 volume 3 M sodium acetate (pH 5.2) and 2.5 volumes of ice-cold 100 % ethanol to precipitate the RNA.
27. Centrifuge at 12,000 × *g* for 10 min at 4 °C.
28. Remove the supernatant from the tube leaving the RNA pellet.
29. Wash pellet with 1 ml of 75 % EtOH.
30. Centrifuge at 7500 × *g* for 5 min at 4 °C.
31. Mark the location of the RNA pellet on the outside of the tube with an EtOH-resistant marker and then carefully pour off the supernatant.
32. Centrifuge briefly at 7500 × *g* at 4 °C to pellet any remaining 75 % EtOH which should be carefully removed with a pipette, avoiding touching the RNA pellet.
33. Air-dry the pellet on ice for about 5 min but do not allow it to dry completely or the RNA will become less soluble.
34. Resuspend the pellet in 20–50 μl RNase-free Sigma water. Quantitate samples to determine yield and purity.
35. Run the sample on an Agilent Bioanalyzer to determine RNA integrity and quantity (*see* **Note 6**).
36. RT-qPCR may be used to determine the location of specific transcripts within the cell (*see* **Note 7**).

3.3 Generation of Illumina Paired-End Library for Sequencing

In Mitchell et al. 2012 library preparation was performed according to the Illumina PE protocol, incorporating improvements suggested in Quail et al. 2008, with reactions scaled according to starting DNA quantity [5]. Other methods that preserve RNA strand information could also be used, for example the Illumina TruSeq Stranded Total RNA Sample Preparation Kit.

4 Notes

1. Triton X-100 is a mild nonionic detergent used to solubilize cytoplasmic membrane proteins but the nuclear membrane needs to remain intact; vary the concentration of detergent to optimize the protocol for different cell types. We routinely test 0.1, 0.2, and 0.5 %, although some cells may require more or less detergent for optimal results.
2. In order to obtain a more concentrated cell suspension, less volume of buffers A and B can be used, for example 0.5 ml of each.
3. Gentle douncing is required to prevent bursting the nuclei. Nuclei can be observed using an inverted microscope; nuclei should be round and intact with few cytoplasmic shreds. The Dounce homogenizer should be cleaned between each use with RNase Zap (Thermo Fisher Scientific, AM9780) followed by Sigma water.
4. Use 14.5 ml polypropylene round-bottom (Sarstedt, 62.515.006) tubes for sucrose cushions. Pipette solution gently and slowly onto the wall of the tube immediately above the sucrose cushion. Tilting the tube will reduce mixing of the cell fraction with the sucrose cushion (Fig. 1).
5. Acid phenol/chloroform is used to separate RNA from DNA. The RNA will partition into the upper aqueous phase while the DNA will remain at the interphase with the phenol solution. Avoid contaminating the RNA with excessive DNA by leaving the last 100 μl of the aqueous phase.
6. This step may be done for you by a sequencing facility prior to library construction.
7. Perform reverse transcription using random primers as many nuclear RNA will not be polyadenylated. For protein-coding genes containing intron sequences, introns will be preferentially found in the nuclear fraction as introns are co-transcriptionally spliced [6], whereas exon sequences will usually be equally partitioned between the nuclear and cytoplasmic fractions. Many lncRNAs are retained in the nucleus, for example *Airn*, and will be found almost exclusively in the nuclear fraction [4]. Primers that can be used to detect intron, exon, and noncoding transcripts in the mouse genome are provided in Table 1.

Table 1
Primers used to assess the purity of the nuclear and cytoplasmic RNA fractions

Gene and feature	Fraction	Forward primer	Reverse primer
Slc4a1 exon	Nuclear/cytoplasmic	AGTTGGGAGCTCAGCCAGT	GGTCCTTCGGGAAGTCCT
Slc4a1 intron	Nuclear	GATCTGAGGCCCAGCCAGAA	CCCCACCTCCTTCACCTTCC
Airn (lncRNA)	Nuclear	ACTTTGACAGAACAATCGGCTCAG	GAACATTTGCAAAGGACAGTCGAG
Malat1 (lncRNA)	Nuclear	CACTTGTGGGGAGACCTTGT	GTTACCAGCCCAAACCTCAA

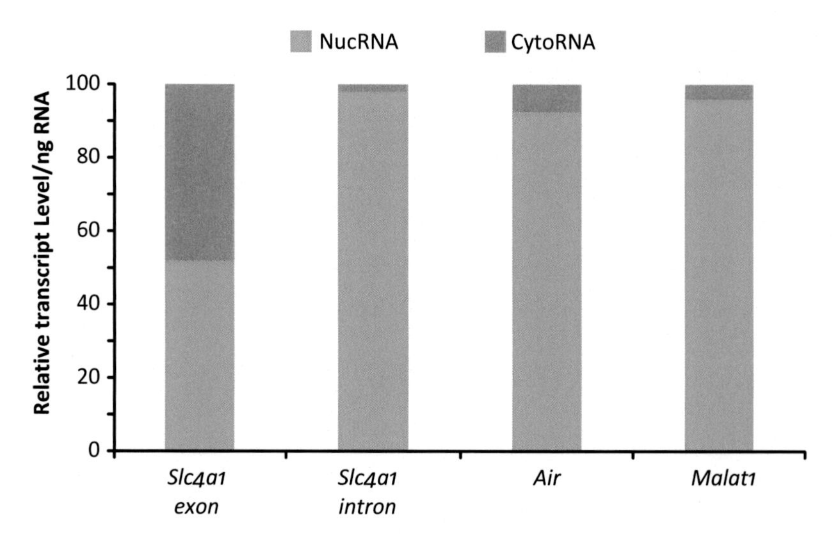

Fig. 3 Relative quantities of each sequence in the nuclear (NucRNA) and cytoplasmic (CytoRNA) fractions

Note that these primers work well for qPCR using iQ™ SYBR Green Supermix (Bio-Rad, 170-8880). Absolute quantification (using the standard curve method) per ng RNA used in the reverse transcription reaction should be applied as a reference gene is not used to normalize expression between the nuclear and cytoplasmic fractions (Fig. 3).

Acknowledgements

This work was supported by the Canadian Institutes of Health Research, the Canada Foundation for Innovation, and the Ontario Ministry of Research and Innovation (operating and infrastructure grants held by J.A.M.). Studentship funding held by N.K.D. was provided by Ontario Graduate Scholarships.

References

1. Chu Y, Corey DR (2012) RNA sequencing: platform selection, experimental design, and data interpretation. Nucleic Acid Ther 22(4): 271–274
2. Morozova O, Hirst M, Marra MA (2009) Applications of new sequencing technologies for transcriptome analysis. Annu Rev Genomics Hum Genet 10:135–151
3. The FANTOM Consortium, RIKEN Genome Exploration Research Group, Genome Science Group (Genome Network Project Core Group) (2005) The transcriptional landscape of the mammalian genome. Science 309(5740):1559–1563
4. Mitchell JA, Clay I, Umlauf D, Chen CY, Moir CA, Eskiw CH, Schoenfelder S, Chakalova L, Nagano T, Fraser P (2012) Nuclear RNA sequencing of the mouse erythroid cell transcriptome. PLoS One 7(11):e49274
5. Quail MA, Kozarewa I, Smith F, Scally A, Stephens PJ, Durbin R, Swerdlow H, Turner DJ (2008) A large genome center's improvements to the Illumina sequencing system. Nat Methods 5(12):1005–1010
6. Tilgner H, Knowles DG, Johnson R, Davis CA, Chakrabortty S, Djebali S, Curado J, Snyder M, Gingeras TR, Guigó R (2012) Deep sequencing of subcellular RNA fractions shows splicing to be predominantly co-transcriptional in the human genome but inefficient for lncRNAs. Genome Res 22(9):1616–1625

Chapter 8

Targeted LncRNA Sequencing with the SeqCap RNA Enrichment System

John C. Tan, Venera D. Bouriakov, Liang Feng, Todd A. Richmond, and Daniel Burgess

Abstract

Sequencing-based whole-transcriptome analysis (i.e., RNA-Seq) can be a powerful tool when used to measure gene expression, detect novel transcripts, characterize transcript isoforms, and identify sequence polymorphisms. However, this method can be inefficient when the goal is to study only one component of the transcriptome, such as long noncoding RNAs (lncRNAs), which constitute only a small fraction of transcripts in a total RNA sample. Here, we describe a target enrichment method where a total RNA sample is converted to a sequencing-ready cDNA library and hybridized to a complex pool of lncRNA-specific biotinylated long oligonucleotide capture probes prior to sequencing. The resulting sequence data are highly enriched for the targets of interest, dramatically increasing the efficiency of next-generation sequencing approaches for the analysis of lncRNAs.

Key words Target enrichment, SeqCap RNA, lncRNA sequencing, RNA-Seq, Rare transcript detection

1 Introduction

Long noncoding RNAs (lncRNAs) have been increasingly appreciated for their role in gene regulation and transcript splicing [1–3], and impact on development [4–6] and disease [7–9]. Next-generation sequencing (NGS)-based methods have greatly advanced the identification and understanding of lncRNAs [10–13]. However, standard whole-transcriptome sequencing-based methods are inefficient when a researcher is interested primarily in lncRNA transcripts, since a large proportion of the data is typically derived from the transcripts of highly expressed protein-coding genes. For example, Cabili et al. [13] observed approximately ten-fold lower median maximal expression level of lncRNAs relative to protein-coding genes in a dataset comprised of nearly 4.2 billion sequencing reads.

Yi Feng and Lin Zhang (eds.), *Long Non-Coding RNAs: Methods and Protocols*, Methods in Molecular Biology, vol. 1402,
DOI 10.1007/978-1-4939-3378-5_8, © Springer Science+Business Media New York 2016

Researchers have demonstrated the use of Roche NimbleGen Sequence Capture technology for hybrid capture of RNA following conversion into cDNA libraries for targeted sequencing [14, 15]. These studies demonstrated that the target enrichment process preserves the relative abundance of the transcripts. By performing target enrichment with capture probes designed specifically for annotated lncRNAs, a researcher can focus their sequencing output on lncRNA transcripts to reduce the number of reads needed per sample to obtain results comparable to standard RNA-Seq methods.

This document describes the Roche NimbleGen SeqCap RNA Enrichment System. The general SeqCap RNA workflow is displayed in Fig. 1. The process, which starts with total RNA and results in a sequencing ready library, occurs over 2 days. On day 1, total RNA is converted into a stranded cDNA library which is hybridized to capture probes in an overnight incubation step. Multiple cDNA libraries can be pooled prior to hybridization if each library contains a unique molecular barcode (i.e., indexed adapter); this is termed "pre-capture multiplexing." On day 2, the hybridized probe-target complexes are captured using streptavidin beads, washed to remove non-targeted cDNA, and amplified by PCR. If the enriched library meets the QC criteria, it is ready to sequence.

The SeqCap lncRNA Enrichment Kit (Roche NimbleGen, Inc., Madison, WI) contains a pool of capture probes targeting an annotated set of lncRNAs and TUCPs (transcripts of uncertain coding potential). The annotation for this design was obtained from GENCODE [16] v19 and the Human Body Map Project [13], and targets approximately 17 Mb of sequence representing 32,808 lncRNA and TUCP transcript isoforms. Target enrichment with this capture design enriches for lncRNA transcripts when compared to sequencing libraries without a target enrichment step (i.e., standard RNA-Seq methods). Figure 2 shows the increased sensitivity for lncRNA isoforms when comparing SeqCap RNA with whole-transcriptome RNA-Seq. Even when comparing with larger RNA-Seq datasets, the efficiency gain with SeqCap RNA is clear; one million reads obtained using SeqCap RNA detected more lncRNA isoforms than 20 million RNA-Seq reads. The capture methodology preserves the relative abundance of transcripts in each sample and allows more sensitive detection of rare transcripts, requiring less sequencing data on a per sample basis. If desired, the capture probes used during the hybridization step can be custom designed to target other transcripts of interest, and are available as part of the SeqCap RNA Enrichment Kits (Roche NimbleGen, Inc., Madison, WI).

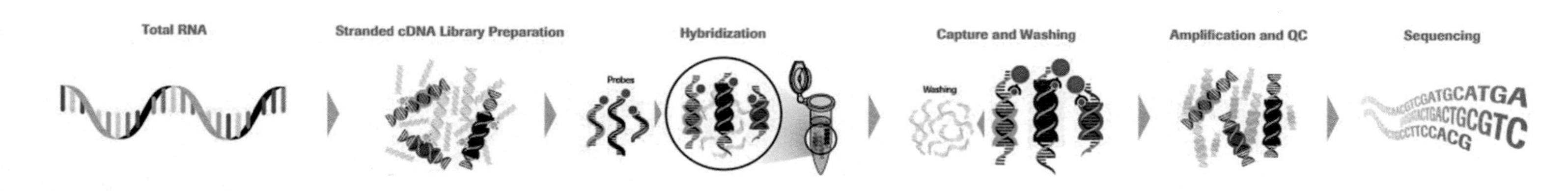

Fig. 1 SeqCap RNA Enrichment System workflow. The workflow requires 100 ng of total RNA as input material. The RNA is converted into a stranded cDNA library using the KAPA Stranded RNA-Seq Library Preparation Kit distributed by Roche. Library preparation generally takes 6–8 h. If desired, several cDNA libraries can be pooled (pre-capture multiplexed) prior to the subsequent hybridization step. Hybridization occurs overnight for approximately 16–20 h. Then, it takes approximately 5 h to wash, recover, and amplify the hybridization reaction. The resulting library is ready for paired-end sequencing on an Illumina sequencing instrument

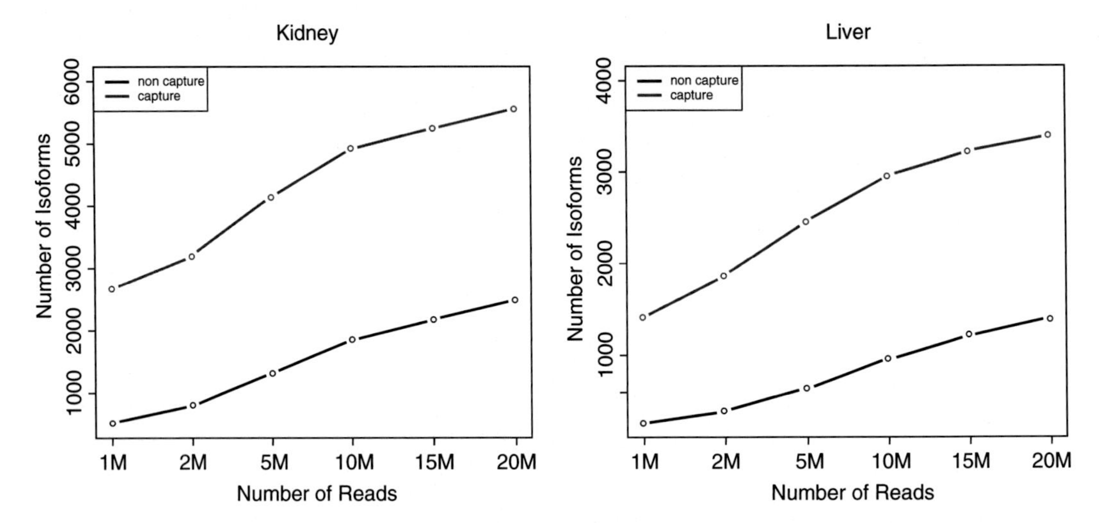

Fig. 2 Comparison of lncRNA isoform detection for captured (SeqCap RNA) and non-captured (RNA-Seq) libraries. SeqCap RNA and RNA-Seq were used to analyze kidney and liver RNA samples. SeqCap RNA data was generated starting with total RNA. RNA-Seq data was generated starting with ribo-depleted RNA. The sequence data was subsampled to various numbers of reads (*x*-axis) from 1 to 20 million reads. SeqCap RNA data (*red*) which includes a capture step detects more lncRNA isoforms than non-captured RNA-Seq data (*black*) even with 20× less sequencing data

2 Materials

1. SeqCap lncRNA Enrichment Kit (4, 48, or 384 reactions: Roche catalog # 07277270001, 07277288001, or 07277296001, respectively; *see* **Note 1**).
2. Consumables available from Roche Diagnostics.
   (a) KAPA Stranded RNA-Seq Library Preparation Kit (24 or 96 reactions: Roche catalog # 07277261001 or 07277253001, respectively).
   (b) SeqCap Adapter Kits A and/or B (96 reactions: Roche catalog # 07141530001 or 07141548001 for Kit A and B, respectively).
   (c) SeqCap Hybridization and Wash Kit (24 or 96 reactions: Roche catalog # 05634261001 or 05634253001, respectively).
   (d) SeqCap EZ Accessory Kit v2 (24 or 96 reactions: Roche catalog # 07145594001 or 06776345001, respectively).
   (e) SeqCap HE-Oligo Kits A and/or B (96 reactions: Roche catalog # 06777287001 or 06777317001 for Kit A and B, respectively).
   (f) SeqCap Pure Capture Bead Kit (24 reactions: Roche catalog # 06977952001).
   (g) Water, PCR grade (4 × 25 ml: Roche catalog # 03315843001).

3. Consumables purchased from other vendors.
   (a) Agilent DNA 1000 Kit (1 kit: Agilent catalog # 5067-1504).
   (b) Agilent RNA 6000 Nano Kit (1 kit: Agilent catalog # 5067-1511).
   (c) Agencourt AMPure XP Beads (5, 60, or 450 ml: Beckman Coulter catalog # A63880, A63881, or A63882, respectively).
   (d) ERCC RNA Spike-In Mix (1 kit: Life Technologies catalog # 4456740).
   (e) Ethanol, 200 proof, for molecular biology.
   (f) TE buffer, 1× solution pH 8.0, low EDTA.
   (g) Elution buffer (10 mM Tris–HCl, pH 8.0) available from multiple vendors.

3 Methods

3.1 Store the SeqCap RNA Enrichment System Reagents

3.1.1 Aliquot the SeqCap RNA Enrichment System Probe Pool (See **Note 2**)

1. If frozen, thaw the tube of SeqCap RNA probe pool on ice.
2. Vortex the SeqCap RNA probe pool for 3 s.
3. Centrifuge the tube of SeqCap RNA probe pool at $10,000 \times g$ for 30 s to ensure that the liquid is at the bottom of the tube before opening the tube.
4. Aliquot the SeqCap RNA probe pool into single-use aliquots (4.5 μl/aliquot) in 0.2 ml PCR tubes (or 96-well plates, if desired) and store at −15 to −25 °C until use. The presence of some residual volume after dispensing all single-use aliquots is normal.
5. When ready to perform the experiment, thaw the required number of single-use SeqCap RNA probe pool aliquots on ice.

   Note: The SeqCap RNA probe pool should not undergo multiple freeze/thaw cycles. To help ensure the highest performance of the SeqCap RNA probe pool, Roche NimbleGen recommends aliquoting the SeqCap RNA probe pool into single-use volumes to prevent damage from successive freeze/thaw cycles.

3.1.2 Store the Frozen Reagents

1. Upon receipt, store the KAPA Stranded RNA-Seq Library Preparation Kit, SeqCap EZ Accessory Kit v2, SeqCap Hybridization and Wash Kit, SeqCap Adapter Kits A and B, and SeqCap HE-Oligo Kits A and B at −15 to −25 °C until use.

3.1.3 Store the Refrigerated Reagents

1. Upon receipt, store the SeqCap Pure Capture Bead Kit at +2 to +8 °C until use.

   Note: The SeqCap Pure Capture Bead Kit must not be frozen (*see* **Note 3**).

3.2 QC the RNA Sample and Prepare the Sample Library

Note: Resuspend the Index Adapters on ice. Care should be taken when opening tubes to avoid loss of the lyophilized pellet.

3.2.1 Resuspend the Index Adapters

1. Spin the lyophilized index adapters, contained in the SeqCap Adapter Kit A and/or B, briefly to allow the contents to pellet at the bottom of the tube.
2. Add 50 μl cold, PCR-grade water to the each of the 12 tubes labeled "SeqCap Index Adapter" in the SeqCap Adapter Kit A and/or B. Keep adapters on ice.
3. Briefly vortex the index adapters plus PCR-grade water and spin down the resuspended index adapter tubes.
4. The resuspended index adapter tubes should be stored at −15 to −25 °C.

3.2.2 QC the RNA Sample

1. Analyze RNA samples using the Agilent Bioanalyzer and RNA 6000 Nano Kit following the manufacturer's instructions.
2. Bioanalyzer traces of total RNA samples derived from human tissue should exhibit two strong peaks from ribosomal RNA with little background signal. Compare your traces to reference traces from RNA for the organism being researched. If the RNA sample appears degraded, then obtain a new RNA sample. Proceeding with a degraded RNA sample may lead to suboptimal results (*see* **Note 4**).

3.2.3 Prepare the Sample Library

1. Spike-in ERCC controls into 100 ng of total RNA following the manufacturer's instructions.
2. Pipet up and down ten times to mix.
3. Adjust the volume of the combined RNA sample and ERCC spike-in control to a total volume of 10 μl in PCR-grade water.
4. Perform RNA fragmentation.
   (a) To each 10 μl spiked sample add 10 μl 2× fragment, prime, and elute buffer, resulting in a total volume of 20 μl (Table 1).
   (b) Keeping the samples on ice, mix thoroughly by gently pipetting up and down several times.
   (c) Fragment and prime sample(s) by incubating for 8 min @ 94 °C.
   (d) Following incubation, place tube(s) on ice and proceed immediately to the next step.
5. Synthesize first-strand cDNA.
   (a) Assemble first-strand synthesis master mix on ice (Table 2; volumes include 20 % excess pipetting volume per sample).
   (b) To each 20 μl fragmented RNA sample, add 10 μl of first-strand synthesis master mix, resulting in a total volume of 30 μl (Table 3).

Table 1
Fragment, prime, and elute reagents

Component	Volume (μl)
RNA sample (100 ng) + ERCC control	10
2× fragment, prime, and elute buffer	10
Total	20

Table 2
First-strand synthesis master mix

First-strand synthesis master mix	Per individual sample library (μl)
First-strand synthesis buffer	11
KAPA script	1
Total	12

Table 3
First-strand synthesis reagents

Component	Volume (μl)
Fragmented, primed RNA	20
First-strand synthesis master mix	10
Total	30

(c) Keeping the sample on ice, mix thoroughly by gently pipetting up and down several times.

(d) Synthesize first-strand cDNA following the first-strand synthesis program (*see* **Note 2**):

- Step 1: 10 min @ +25 °C.
- Step 2: 15 min @ +42 °C.
- Step 3: 15 min @ +70 °C.
- Step 4: Hold @ +4 °C.

(e) Keep the tube(s) on ice and proceed immediately to the next step.

6. Synthesize and mark second-strand cDNA.

(a) Assemble second-strand synthesis and marking master mix on ice (Table 4; volumes include 10 % excess pipetting volume per sample).

Table 4
Second-strand synthesis and marking master mix

Second-strand synthesis and marking master mix	Per individual sample library (μl)
Second-strand marking buffer	31
Second-strand synthesis enzyme mix	2
Total	33

Table 5
Second-strand synthesis reagents

Component	Volume (μl)
First-strand cDNA	30
Second-strand synthesis and marking master mix	30
Total	60

(b) To each 30 μl sample add 30 μl of second-strand synthesis and marking master mix, resulting in a total volume of 60 μl (Table 5).

(c) Mix thoroughly by gently pipetting up and down several times.

(d) Synthesize and mark second-strand cDNA following the second-strand synthesis program (*see* **Note 2**):

- Step 1: 60 min @ +16 °C.
- Step 2: Hold @ +4 °C.

(e) Keep tube(s) on ice and proceed immediately to the next step.

7. Clean up double-stranded cDNA.

(a) To each 60 μl second-strand cDNA synthesis reaction, add 108 μl of Agencourt AMPure XP beads, prewarmed to room temperature, resulting in a total volume of 168 μl (Table 6; *see* **Note 5**).

(b) Mix thoroughly by pipetting up and down multiple times.

(c) Incubate the tube at room temperature for 15 min to allow the cDNA to bind to the beads.

(d) Place the tube on a magnet to capture the beads. Incubate until the liquid is clear.

(e) Carefully remove and discard the supernatant.

Table 6
Second-strand synthesis bead-based cleanup

Component	Volume (μl)
Second-strand synthesis reaction	60
Agencourt AMPure XP beads	108
Total	168

Table 7
A-tailing master mix

A-tailing master mix	Per individual sample library (μl)
PCR-grade water	24
10× KAPA A-tailing buffer	3
KAPA A-tailing enzyme	3
Total	30

(f) Keeping the tube on the magnet, add 200 μl of freshly prepared 80 % ethanol (*see* **Note 6**).

(g) Incubate the tube at room temperature for ≥30 s.

(h) Carefully remove and discard the ethanol.

(i) Keeping the tube on the magnet, add 200 μl of freshly prepared 80 % ethanol.

(j) Incubate the tube at room temperature for ≥30 s.

(k) Carefully remove and discard the ethanol. Try to remove all residual ethanol without disturbing the beads.

(l) Allow the beads to dry at room temperature, sufficiently for all the ethanol to evaporate.
Note: Over-drying the beads may result in dramatic yield loss.

(m) Remove the tube from the magnet.

(n) Proceed immediately with the next step.

8. Perform A-tailing.

(a) Assemble A-Tailing master mix (Table 7).

(b) Thoroughly resuspend the beads with 30 μl of the A-tailing master mix (per reaction) by pipetting up and down multiple times.

Table 8
SeqCap adapter reagents

SeqCap adapter working dilution (700 nM)	Per individual sample library (μl)
PCR-grade water	9.3
Stock concentration SeqCap Adapter (10 μM)	0.7
Total	10

Table 9
Adapter ligation master mix

Adapter ligation master mix	Per individual sample library (μl)
PCR-grade water	16
5× KAPA ligation buffer	14
KAPA T4 DNA ligase	5
Total	35

(c) Perform A-tailing reaction following the cDNA A-tailing program (*see* **Note 2**):

- Step 1: 30 min @ +30 °C.
- Step 2: 30 min @ +60 °C.
- Step 3: Hold @ +4 °C.

(d) Proceed immediately to the next step.

9. Adapter ligation.

(a) Generate a 700 nM adapter working dilution for the adapter ligation reaction (Table 8; volumes include 100 % excess pipetting volume per sample; *see* **Note 7**).

(b) Assemble adapter ligation master mix (Table 9).

(c) To each 30 μl A-tailing reaction add 35 μl of the adapter ligation master mix, resulting in a total volume of 65 μl.

(d) Mix by pipetting up and down multiple times.

(e) To the 65 μl mixture of ligation master mix plus cDNA and beads add 5 μl of the SeqCap Library Adapter working dilution (with the desired index), resulting in a total volume of 70 μl (Table 10).

Note: Ensure that you record the index used for each sample.

Table 10
Adapter ligation reagents

Component	Per individual sample library (μl)
A-tailing reaction	30
Adapter ligation master mix	35
SeqCap library adapter	5
Total	70

Table 11
Adapter ligation, first cleanup reagents

Component	Per individual sample library (μl)
Adapter ligation reaction	70
PEG/NaCl SPRI solution	70
Total	140

(f) Pipette up and down ten times to mix.

(g) Incubate the adapter ligation reaction for 15 min @ +20 °C.

(h) Proceed immediately to the next step.

10. First post-ligation cleanup.

(a) To each 70 μl adapter ligation reaction add 70 μl of thawed PEG/NaCl SPRI solution, resulting in a total volume of 140 μl (Table 11).

(b) Mix thoroughly by pipetting up and down multiple times.

(c) Incubate the tube at room temperature for 15 min to allow the cDNA to bind to the beads.

(d) Place the tube on a magnet to capture the beads. Incubate until the liquid is clear.

(e) Carefully remove and discard 135 μl of supernatant.

(f) Keeping the tube on the magnet, add 200 μl of freshly prepared 80 % ethanol.

(g) Incubate the tube at room temperature for ≥30 s.

(h) Carefully remove and discard the ethanol.

(i) Keeping the tube on the magnet, add 200 μl of freshly prepared 80 % ethanol.

(j) Incubate the tube at room temperature for ≥30 s.

(k) Carefully remove and discard the ethanol. Try to remove all residual ethanol without disturbing the beads.

Table 12
Adapter ligation, second cleanup reagents

Component	Per individual sample library (μl)
First post-ligation cleanup	50
PEG/NaCl SPRI solution	50
Total	100

(l) Allow the beads to dry at room temperature, sufficiently for all the ethanol to evaporate.
Note: Over-drying the beads may result in dramatic yield loss.

(m) Remove the tube from the magnet.

(n) Thoroughly resuspend the beads in 50 μl of elution buffer (10 mM Tris–HCl, pH 8.0).

(o) Incubate the tube at room temperature for 2 min to allow the cDNA to elute off the beads.

(p) Proceed immediately to the next step.

11. Second post-ligation cleanup.

(a) To each 50 μl resuspended cDNA with beads add 50 μl of thawed PEG/NaCl SPRI solution, resulting in a total volume of 100 μl (Table 12).

(b) Mix thoroughly by pipetting up and down multiple times.

(c) Incubate the tube at room temperature for 15 min to allow the cDNA to bind to the beads.

(d) Place the tube on a magnet to capture the beads. Incubate until the liquid is clear.

(e) Carefully remove and discard 95 μl of supernatant.

(f) Keeping the tube on the magnet, add 200 μl of freshly prepared 80 % ethanol.

(g) Incubate the tube at room temperature for ≥30 s.

(h) Carefully remove and discard the ethanol.

(i) Keeping the tube on the magnet, add 200 μl of freshly prepared 80 % ethanol.

(j) Incubate the tube at room temperature for ≥30 s.

(k) Carefully remove and discard the ethanol. Try to remove all residual ethanol without disturbing the beads.

(l) Allow the beads to dry at room temperature, sufficiently for all the ethanol to evaporate.
Note: Over-drying the beads may result in dramatic yield loss.

Table 13
Pre-capture LM-PCR master mix

Pre-capture LM-PCR master mix	Per individual sample library or control (μl)
2× KAPA HiFi HotStart Readymix	25
10× KAPA library amplification primer mix[a]	5
Total	30

[a]Pre-capture LM-PCR oligos are in the 10× KAPA Library Amplification Primer Mix, which is contained within the KAPA Stranded RNA Library Preparation Kit

(m) Remove the tube from the magnet.

(n) Thoroughly resuspend the beads in 22.5 μl of elution buffer (10 mM Tris–HCl, pH 8.0).

(o) Incubate the tube at room temperature for 2 min to allow the cDNA to elute off the beads.

(p) Place the tube on a magnet to capture the beads. Incubate until the liquid is clear.

(q) Transfer 20 μl of the clear supernatant to a new 0.2 ml PCR tube.

(r) Proceed to the next step, or store the solution at +4 °C for up to 1 week, or at −20 °C for up to 1 month.

3.3 Amplify the Sample Library Using LM-PCR

3.3.1 Prepare the Pre-capture LM-PCR Master Mix

1. Assemble the pre-capture LM-PCR master mix on ice (Table 13; *see* **Note 8**).
2. Mix well by pipetting up and down ten times.
3. To each 20 μl sample in a PCR tube/well add 30 μl of pre-capture LM-PCR master mix, resulting in a total volume of 50 μl.
4. Mix well by pipetting up and down several times. Do not vortex.

3.3.2 Perform the Pre-capture PCR Amplification

1. Place the PCR tube (or 96-well PCR plate) in the thermocycler.
2. Amplify the sample library using the following cDNA pre-capture LM-PCR program (*see* **Notes 2** and **9**):

   (a) Step 1: 45 s @ +98 °C.

   (b) Step 2: 15 s @ +98 °C.

   (c) Step 3: 30 s @ +60 °C.

   (d) Step 4: 30 s @ +72 °C.

(e) Step 5: Go to **step 2**, repeat ten times (for a total of 11 cycles).

(f) Step 6: 5 min @ +72 °C.

(g) Step 7: Hold @ +4 °C.

3.3.3 Purify the Amplified Sample Library with Agencourt AMPure XP Beads (See ***Note 10****)*

1. Allow the Agencourt AMPure XP Beads, contained in the SeqCap Pure Capture Bead Kit, to warm to room temperature for at least 30 min before use.
2. Vortex the beads for 10 s before use to ensure a homogenous mixture of beads.
3. Add 50 μl (or 1.0× volume) Agencourt AMPure XP Beads to the 50 μl amplified, sample library.
4. Mix thoroughly by pipetting up and down several times.
5. Incubate at room temperature for 15 min to allow the cDNA to bind to the beads.
6. Place the tube on a magnetic particle collector to capture the beads. Incubate until the liquid is clear.
7. Carefully remove and discard 95 μl of supernatant.
8. Keeping the tube on the magnetic particle collector, add 200 μl of freshly prepared 80 % ethanol.
9. Incubate the tube at room temperature for ≥30 s.
10. Carefully remove and discard the ethanol.
11. Keeping the tube on the magnetic particle collector, add 200 μl of freshly prepared 80 % ethanol.
12. Incubate the tube at room temperature for ≥30 s.
13. Carefully remove and discard the ethanol. Try to remove all residual ethanol without disturbing the beads.
14. Allow the beads to dry at room temperature, sufficiently for all the ethanol to evaporate.
    Note: Over-drying the beads may result in dramatic yield loss.
15. Remove the tube from the magnetic particle collector and resuspend the cDNA using 52 μl of PCR-grade water.

    Note: It is critical that the amplified sample library is eluted with PCR-grade water and not buffer EB or 1× TE.
16. Pipette up and down ten times to mix to ensure that all of the beads are resuspended.
17. Incubate at room temperature for 2 min.
18. Place the tube back in the magnetic particle collector and allow the solution to clear.
19. Remove 50 μl of the supernatant that now contains the amplified sample library and transfer into a new 1.5 ml tube.

3.3.4 Check the Quality of the Amplified Sample Library

1. Measure the A260/A280 ratio of the amplified sample library to quantify the cDNA concentration using a NanoDrop spectrophotometer and determine the cDNA quality.
   Note: When working with samples that will be pooled for hybridization (i.e., pre-capture multiplexing; *see* **Note 11**), accurate quantitation is essential. Alternative quantitation methods, such as those that are fluorometry based, should be used in place of, or in addition to, the NanoDrop spectrophotometer. Slight differences in the molar concentration of each sample pooled to form the "multiplex cDNA sample library pool" will result in variations in the total number of sequencing reads obtained for each library in the pool.
   (a) The A260/A280 ratio should be 1.7–2.0.
   (b) The sample library yield should be >1.0 μg.
   (c) The negative control yield should be negligible. If this is not the case, the measurement may be high due to the presence of unincorporated primers carried over from the LM-PCR reaction and not an indication of possible contamination between amplified sample libraries.
2. Run 1 μl of each amplified sample library (and any negative controls) on an Agilent DNA 1000 chip. Run the chip according to the manufacturer's instructions.
   (a) The Bioanalyzer should indicate that average fragment size falls between 150 and 500 bp (Fig. 3) when fragmenting input RNA to a 100–250 bp size range.

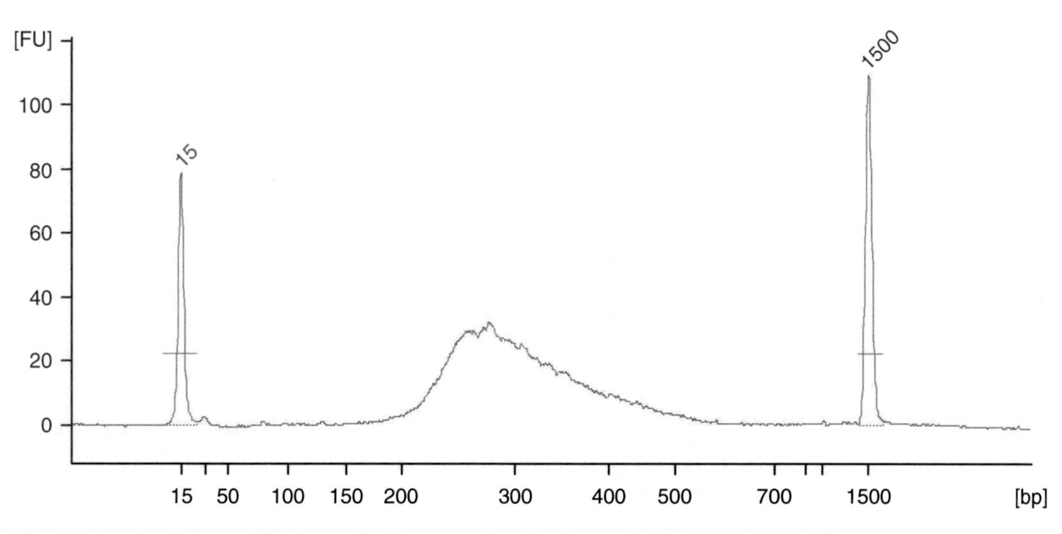

Fig. 3 Example of an amplified sample library analyzed using an Agilent DNA 1000 chip

(b) The negative control should not show any significant signal within this size range, which could indicate contamination between amplified sample libraries. A sharp peak may be visible below 150 bp. If the negative control reaction shows a positive signal by the NanoDrop spectrophotometer but the Bioanalyzer trace indicates only the presence of a sharp peak below 150 bp in size, then the negative control should not be considered contaminated.

3. If the amplified sample library meets these requirements, proceed to the next step. If the amplified sample library does not meet these requirements, reconstruct the library.

3.4 Hybridize the Sample and SeqCap RNA Probe Pool

Note: The hybridization protocol requires a thermocycler capable of maintaining +47 °C for 16–20 h (*see* **Note 12**). A programmable heated lid is required.

3.4.1 Prepare for Hybridization

1. Turn on a heat block to +95 °C and let it equilibrate to the set temperature.
2. Remove the appropriate number of 4.5 μl SeqCap RNA probe pool aliquots (one per sample library) from the −15 to −25 °C freezer and allow them to thaw on ice.

3.4.2 Resuspend the SeqCap HE Universal and SeqCap HE Index Oligos

1. Spin the lyophilized oligo tubes, contained in the SeqCap HE-Oligo Kits A and/or B, briefly to allow the contents to pellet to the bottom of the tube.
2. Add 120 μl PCR-grade water to the SeqCap HE Universal Oligo tube (1000 μM final concentration).
3. Add 10 μl PCR-grade water to each SeqCap HE Index Oligo tube (1000 μM final concentration).
4. Vortex the primers plus PCR-grade water for 5 s and spin down the resuspended oligo tube.
5. The resuspended oligo tube should be stored at −15 to −25 °C.

   Note: To prevent damage to the hybridization-enhancing (HE) oligos due to multiple freeze/thaw cycles, once resuspended the oligos can be aliquoted into smaller volumes to minimize the number of freeze/thaw cycles.

3.4.3 Prepare the Multiplex cDNA Sample Library Pool

1. Thaw on ice (if frozen) each of the uniquely indexed amplified cDNA sample libraries that will be included in the multiplex capture experiment.
2. Mix together equal amounts (by mass) of each of these amplified cDNA sample libraries to obtain a single pool with a combined mass of 1 μg. This mixture will subsequently be referred to as the "multiplex cDNA sample library pool."
   Note: To obtain equal numbers of sequencing reads from each component library in the multiplex cDNA sample library pool upon completion of the experiment, it is very important to

Table 14
Example multiplex cDNA sample library pool

Component	Amount
SeqCap HE Universal Oligo	1000 pmol (1 µl of 1000 µM)
SeqCap HE Index 2 Oligo	250 pmol (0.25 µl of 1000 µM)
SeqCap HE Index 4 Oligo	250 pmol (0.25 µl of 1000 µM)
SeqCap HE Index 6 Oligo	250 pmol (0.25 µl of 1000 µM)
SeqCap HE Index 8 Oligo	250 pmol (0.25 µl of 1000 µM)
Total	2000 pmol (2 µl of 1000 µM)

combine identical amounts of each independently amplified cDNA sample library at this step. Accurate quantification and pipetting are critical.

3.4.4 Prepare the Multiplex Hybridization Enhancing Oligo Pool

1. Thaw on ice the resuspended SeqCap HE Universal Oligo (1000 µM) and each resuspended SeqCap HE Index oligo (1000 µM) that matches a DNA Adapter Index included in the multiplex cDNA sample library pool.
2. Mix together the HE oligos so that the resulting multiplex hybridization-enhancing oligo pool contains, by mass, 50 % SeqCap HE Universal Oligo and 50 % of a mixture of the appropriate SeqCap HE Index oligos. The total combined mass of the multiplex hybridization-enhancing oligo pool should be 2000 pmol, which is the amount required for a single sequence capture experiment.
3. For example, if a multiplex cDNA sample library pool contains four cDNA sample libraries prepared with SeqCap Adapter Indexes 2, 4, 6, and 8, respectively, then the multiplex hybridization-enhancing oligo pool would contain the reagents listed in Table 14:

Note: Due to the difficulty of accurately pipetting small volumes, it is recommended to either prepare a larger volume of the multiplex hybridization-enhancing oligo pool using the 1000 µM stocks or dilute the 1000 µM stocks and then pool. These pools can be dispensed into individual single-use aliquots that can be stored at −15 to −25 °C until needed. Additionally, it is important that the individual SeqCap HE oligos contained in a multiplex hybridization-enhancing oligo pool are precisely matched with the adapter indexes present in the multiplex cDNA sample library pool in a multiplexed sequence capture experiment.

3.4.5 Prepare the Hybridization Sample

1. Add 5 µl of COT human DNA (1 mg/ml), contained in the SeqCap EZ Accessory Kit v2, to a new 1.5 ml tube.

Table 15
Multiplex library with blocker and hybridization-enhancing reagents

Component	Amount	Volume (μl)
COT human DNA	5 μg	5
Multiplex cDNA sample library pool	1 μg	≤50
Multiplex hybridization-enhancing oligo pool	2000 pmol	2
	Total	≤57

2. Add 1 μg of multiplex cDNA sample library to the 1.5 ml tube containing 5 μl of COT human DNA.
3. Add 2000 pmol of Multiplex hybridization enhancing oligo pool (1000 pmol SeqCap HE Universal Oligo and 1000 pmol SeqCap HE Index Oligo pool) to the multiplex cDNA sample library pool plus COT human DNA (Table 15).
4. Close the tube's lid and make a hole in the top of the tube's cap with an 18–20 G or smaller needle.
   Note: The closed lid with a hole in the top of the tube's cap is a precaution to suppress contamination in the DNA vacuum concentrator.
5. Dry the multiplex cDNA sample library pool/COT human DNA/multiplex hybridization-enhancing oligo pool in a DNA vacuum concentrator on high heat (+60 °C).
   Note: Denaturation of the DNA with high heat is not problematic after linker ligation because the hybridization utilizes single-stranded DNA.
6. To each dried-down multiplex cDNA sample library pool/COT human DNA/multiplex hybridization-enhancing oligo pool, add the following reagents contained in the SeqCap hybridization and wash kit:
   (a) 7.5 μl of 2× Hybridization buffer (vial 5).
   (b) 3 μl of Hybridization component A (vial 6).
   The tube with the multiplex cDNA sample library pool/COT human DNA/multiplex hybridization-enhancing oligo pool should now contain the components listed in Table 16:
7. Cover the hole in the tube's cap with a sticker or small piece of laboratory tape.
8. Vortex the multiplex cDNA sample library pool/COT human DNA/multiplex hybridization enhancing oligo pool plus hybridization cocktail (2× hybridization buffer + hybridization component A) for 10 s.
9. Centrifuge at maximum speed for 10 s.

Table 16
Multiplex library with blocker, hybridization-enhancing, and hybridization buffer reagents

Component	Solution capture
COT human DNA	5 μg
Multiplex cDNA sample library pool	1 μg
Multiplex hybridization-enhancing oligo pool	2000 pmol[a]
2× Hybridization buffer (vial 5)	7.5 μl
Hybridization component A (vial 6)	3 μl
Total	10.5 μl

[a]Composed of 50 % (1000 pmol) SeqCap HE Universal Oligo and 50 % (1000 pmol) of a *mixture* of the appropriate SeqCap HE Index oligos

10. Place the multiplex cDNA sample library pool/COT human DNA/multiplex hybridization-enhancing oligo pool/hybridization cocktail in a +95 °C heat block for 10 min to denature the cDNA.
11. Centrifuge the multiplex cDNA sample library pool/COT human DNA/multiplex hybridization-enhancing oligo pool/hybridization cocktail at maximum speed for 10 s at room temperature.
12. Transfer the multiplex cDNA sample library pool/COT human DNA/multiplex hybridization-enhancing oligo pool/hybridization cocktail to a 4.5 μl aliquot of SeqCap RNA probe pool in a 0.2 ml PCR tube (the entire volume can also be transferred to one well of a 96-well PCR plate).
13. Vortex for 3 s.
14. Centrifuge at maximum speed for 10 s.
    The hybridization sample should now contain the components listed in Table 17:
15. Incubate in a thermocycler at +47 °C for 16–20 h. The thermocycler's heated lid should be turned on and set to maintain +57 °C (10 °C above the hybridization temperature).

3.5 Wash and Recover the Captured Multiplex cDNA Sample

3.5.1 Prepare Sequence Capture and Bead Wash Buffers

Note: If a sample has noticeable volume loss after the hybridization step (e.g., due to sample evaporation), do not proceed to sample processing. Perform a new hybridization with remaining or newly generated material. It is extremely important that the water bath temperature be closely monitored and remains at +47 °C. Because the displayed temperatures on many water baths are often

Table 17
Multiplex library with all SeqCap RNA hybridization reagents

Component	Solution capture
COT human DNA	5 μg
Multiplex cDNA sample library pool	1 μg
Multiplex hybridization-enhancing oligo pool	2000 pmol[a]
2× Hybridization buffer (vial 5)	7.5 μl
Hybridization component A (vial 6)	3 μl
SeqCap RNA probe pool	4.5 μl
Total	15 μl

[a]Composed of 50 % (1000 pmol) SeqCap HE Universal Oligo and 50 % (1000 pmol) of a *mixture* of the appropriate SeqCap HE Index oligos

Table 18
Wash buffer reagents

Concentrated buffer	Volume of concentrated buffer (μl)	Volume of PCR-grade water (μl)	Total volume of 1× buffer[a] (μl)
10× Stringent wash buffer (vial 4)	40	360	400
10× Wash buffer I (vial 1)	30	270	300
10× Wash buffer II (vial 2)	20	180	200
10× Wash buffer III (vial 3)	20	180	200
2.5× Bead wash buffer (vial 7)	200	300	500

[a]Store working solutions at room temperature (+15 to +25 °C) for up to 2 weeks. The volumes in this table are calculated for a single experiment; scale up accordingly if multiple samples will be processed

imprecise, Roche NimbleGen recommends that you place an external, calibrated thermometer in the water bath.

Equilibrate 1× stringent wash buffer and 1× wash buffer I at +47 °C for at least 2 h before washing the captured multiplex cDNA sample.

Volumes for an individual capture are shown here. When preparing 1× buffers for processing multiple reactions, prepare an excess volume of ~5 % to allow for complete pipetting (liquid-handling systems may require an excess of ~20 %).

1. Dilute 10× wash buffers (I, II, III, and Stringent) and 2.5× bead wash buffer to create 1× working solutions. All buffers are contained in the SeqCap hybridization and wash kit (Table 18).

2. Preheat the following wash buffers to +47 °C in a water bath:
   (a) 400 μl of 1× stringent wash buffer.
   (b) 100 μl of 1× wash buffer I.

3.5.2 Prepare the Capture Beads

1. Allow the capture beads, contained in the SeqCap Pure Capture Bead Kit, to warm to room temperature for 30 min prior to use (*see* **Note 5**).
2. Mix the beads thoroughly by vortexing for 15 s.
3. Aliquot 100 μl of beads for each capture into a single 1.5 ml tube (i.e., for one capture use 100 μl beads and for four captures use 400 μl beads). Enough beads for six captures can be prepared in a single tube.
4. Place the tube in a DynaMag-2 device. When the liquid becomes clear (should take less than 5 min), remove and discard the liquid being careful to leave all of the beads in the tube. Any remaining traces of liquid will be removed with subsequent wash steps.
5. While the tube is in the DynaMag-2 device, add twice the initial volume of beads of 1× bead wash buffer (i.e., for one capture use 200 μl of buffer and for four captures use 800 μl buffer).
6. Remove the tube from the DynaMag-2 device and vortex for 10 s.
7. Place the tube back in the DynaMag-2 device to bind the beads. Once clear, remove and discard the liquid.
8. Repeat above three steps for a total of two washes.
9. After removing the buffer following the second wash, resuspend by vortexing the beads in 1× the original volume using the 1× bead wash buffer (i.e., for one capture use 100 μl buffer and for four captures use 400 μl buffer).
10. Aliquot 100 μl of resuspended beads into new 0.2 ml tubes (i.e., one tube for each capture).
11. Place the tube in the DynaMag-2 device to bind the beads. Once clear, remove and discard the liquid.
12. The capture beads are now ready to bind the captured cDNA. Proceed immediately to the next step.

    Note: Do not allow the SeqCap capture beads to dry out. Small amounts of residual bead wash buffer will not interfere with binding of cDNA to the capture beads.

3.5.3 Bind cDNA to the Capture Beads

1. Transfer the hybridization samples to the capture beads prepared in the previous step. Ensure that at least 12.5 μl of each hybridization sample can be transferred to the capture beads. Mix thoroughly by pipetting up and down ten times.

2. Bind the captured sample to the beads by placing the tubes containing the beads and cDNA in a thermocycler set to +47 °C for 45 min (heated lid set to +57 °C). Mix the samples by vortexing for 3 s at 15-min intervals to ensure that the beads remain in suspension. It is helpful to have a vortex mixer located close to the thermocycler for this step.

3.5.4 Wash the Capture Beads plus Bound cDNA

1. After the 45-min incubation, add 100 μl of 1× wash buffer I heated to +47 °C to the 15 μl of capture beads plus bound cDNA.
2. Mix by vortexing for 10 s.
3. Transfer the entire content of each 0.2 ml tube to a 1.5 ml tube.
4. Place the tubes in the DynaMag-2 device to bind the beads.
5. Remove and discard the liquid once clear.
6. Remove the tubes from the DynaMag-2 device.
7. Add 200 μl of 1× stringent wash buffer heated to +47 °C.
8. Pipette up and down ten times to mix. Work quickly so that the temperature does not drop much below +47 °C.
9. Incubate at +47 °C for 5 min.
10. Repeat above six steps (starting with "Place the tubes in the DynaMag-2 device to bind the beads") for a total of two washes using 1× stringent wash buffer heated to +47 °C.
11. Place the tubes in the DynaMag-2 device to bind the beads.
12. Remove and discard the liquid once clear.
13. Add 200 μl of room-temperature 1× wash buffer I and mix by vortexing for 2 min (*see* **Note 13**). If liquid has collected in the tube's cap, tap the tube gently to collect the liquid into the tube's bottom before continuing to the next step.
14. Place the tubes in the DynaMag-2 device to bind the beads.
15. Remove and discard the liquid once clear.
16. Add 200 μl of room-temperature 1× wash buffer II.
17. Mix by vortexing for 1 min.
18. Place the tubes in the DynaMag-2 device to bind the beads.
19. Remove and discard the liquid once clear.
20. Add 200 μl of room-temperature 1× wash buffer III.
21. Mix by vortexing for 30 s.
22. Place the tubes in the DynaMag-2 device to bind the beads.
23. Remove and discard the liquid once clear.
24. Remove the tubes from the DynaMag-2 device.
25. Add 50 μl PCR-grade water to each tube of bead-bound captured sample.
26. Store the beads plus captured samples at −15 to −25 °C or proceed to the next step.

Note: There is no need to elute cDNA off the beads. The beads plus captured cDNA will be used as template in the post-capture LM-PCR.

3.6 Amplify the Captured Multiplex cDNA Sample Using LM-PCR

3.6.1 Resuspend the Post-LM-PCR Oligos

1. Briefly spin the lyophilized "post-LM-PCR Oligos 1 and 2," contained in the SeqCap EZ Accessory Kit v2, to allow the contents to pellet at the bottom of the tube. Please note that both oligos are contained within a single tube.
2. Add 480 µl PCR-grade water to the tube of centrifuged oligos.
3. Briefly vortex the resuspended oligos.
4. Spin down the tube to collect the contents.
5. The resuspended oligo tube should be stored at −15 to −25 °C.

3.6.2 Prepare the Post-capture LM-PCR Master Mix

Note: The post-capture LM-PCR master mix and the individual PCR tubes must be prepared on ice. Instructions for preparing individual PCR reactions are shown here. When assembling a master mix for processing multiple samples, prepare an excess volume of ~5 % to allow for complete pipetting (liquid-handling systems may require an excess of ~20 %). Note that each captured multiplexed cDNA sample requires two PCR reactions.

1. Assemble the post-capture LM-PCR master mix on ice (Table 19; *see* **Note 14**).
2. Pipette 30 µl of post-capture LM-PCR master mix into two reaction tubes or wells.
3. Vortex the bead-bound captured cDNA to ensure a homogenous mixture of beads.
4. Aliquot 20 µl of bead-bound captured cDNA as template into each of the two PCR tubes/wells.
5. Mix well by pipetting up and down.
6. Add 20 µl of PCR-grade water to the negative control.
7. Mix well by pipetting up and down five times.
8. Store the remaining bead-bound captured cDNA at −15 to −25 °C.

Table 19
Post-capture LM-PCR master mix

Post-capture LM-PCR master mix	Per individual PCR reaction (two reactions per cDNA sample) (µl)
KAPA HiFi HotStart Readymix (2×)	25
Post-LM-PCR oligos 1 and 2, 5 µM[a]	5
Total	30

[a]The post-LM-PCR oligos are contained within the SeqCap EZ Accessory Kit v2

3.6.3 Perform the Post-capture PCR Amplification

1. Place PCR tubes/plate in the thermocycler.
2. Amplify the captured cDNA using the following post-capture LM-PCR program (*see* **Note 2**):
   (a) Step 1: 45 s @ +98 °C.
   (b) Step 2: 15 s @ +98 °C.
   (c) Step 3: 30 s @ +60 °C.
   (d) Step 4: 30 s @ +72 °C.
   (e) Step 5: Go to **step 2**, repeat 13 times (for a total of 14 cycles).
   (f) Step 6: 5 min @ +72 °C.
   (g) Step 7: Hold @ +4 °C.
3. Store reactions at +2 to +8 °C until ready for purification, up to 72 h.

3.6.4 Purify the Amplified, Captured Multiplex cDNA Sample Using Agencourt AMPure XP Beads

1. Allow the DNA Purification Beads, contained in the SeqCap Pure Capture Bead Kit, to warm to room temperature for at least 30 min before use.
2. Pool the like amplified, captured multiplex cDNA sample libraries into a 1.5 ml microcentrifuge tube (approximately 100 μl). Process the negative control in exactly the same way as the amplified sample library.
3. Vortex the Agencourt AMPure XP Beads for 10 s before use to ensure a homogenous mixture of beads.
4. Add 180 μl Agencourt AMPure XP Beads to the 100 μl pooled amplified, captured multiplex cDNA sample library.
5. Vortex briefly.
6. Incubate at room temperature for 15 min to allow the cDNA to bind to the beads.
7. Place the tube containing the bead bound cDNA in a magnetic particle collector.
8. Allow the solution to clear.
9. Once clear, remove and discard the supernatant being careful not to disturb the beads.
10. Add 200 μl freshly prepared 80 % ethanol to the tube containing the beads plus cDNA. The tube should be left in the magnetic particle collector during this step.
11. Incubate at room temperature for 30 s.
12. Remove and discard the 80 % ethanol, and repeat the above three steps for a total of two washes with 80 % ethanol.
13. Following the second wash, remove and discard all of the 80 % ethanol.

14. Allow the beads to dry at room temperature, sufficiently for all the ethanol to evaporate.

    Note: Over-drying of the beads can result in yield loss.

15. Remove the tube from the magnetic particle collector.
16. Resuspend the cDNA using 52 μl of PCR-grade water.
17. Pipet up and down ten times to mix to ensure that all of the beads are resuspended.
18. Incubate at room temperature for 2 min.
19. Place the tube back in the magnetic particle collector and allow the solution to clear.
20. Remove 50 μl of the supernatant that now contains the amplified, captured multiplex cDNA sample library pool and transfer to a new 1.5 ml tube.

3.6.5 Determine the Concentration, Size Distribution, and Quality of the Amplified, Captured Multiplex cDNA Sample

1. Quantify the DNA concentration and measure the A260/A280 ratio of the amplified, captured multiplex cDNA sample and negative control using a NanoDrop Spectrophotometer.
   (a) The A260/A280 ration should be 1.7–2.0.
   (b) The LM-PCR yield should be ≥500 ng.
   (c) The negative control should not show significant amplification, which could be indicative of contamination.
2. Run 1 μl of the amplified, captured multiplex cDNA and negative control using an Agilent DNA 1000 chip (Fig. 4). Run the chip according to the manufacturer's instructions. Amplified, captured multiplex cDNA should exhibit an average fragment length between 150 and 500 bp.

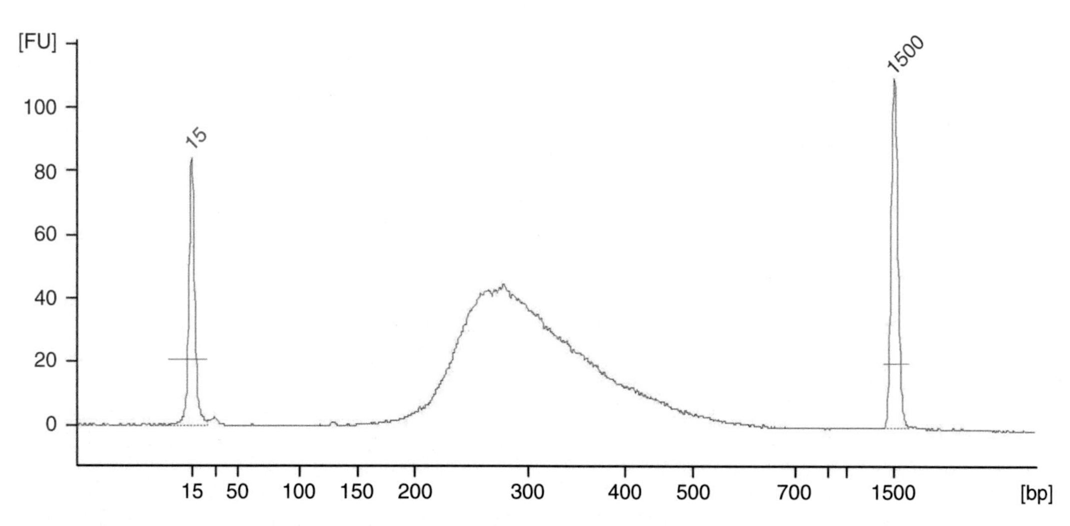

Fig. 4 Example of successfully amplified, captured multiplex sample library analyzed using an Agilent DNA 1000 chip

3. If the amplified, captured multiplex cDNA meets the requirements, proceed to sequencing (*see* **Note 15**). If the amplified, captured multiplex cDNA does not meet the A260/A280 ratio requirement, purify again using Agencourt AMPure XP Beads (or a Qiagen QIAquick PCR Purification column).

4 Notes

1. Custom designs with new lncRNA or subsets of lncRNA annotation can be created by contacting Roche NimbleGen to request a custom design. Small RNA transcripts (<50 nt) are not recommended as potential enrichment targets.
2. Read through the entire protocol and enter all necessary thermocycler programs prior to starting an experiment.
3. The SeqCap Pure Capture Bead Kit must not be frozen in order to avoid damage to the purification beads.
4. High-quality RNA is required for optimal cDNA synthesis yield. Care must be taken to avoid accidental sample exposure to RNases. Precautions include cleaning work areas and equipment with RNase removal agents, wearing gloves, changing gloves after touching potentially contaminated surfaces or equipment, and keeping reagents closed when not in use. RNA contamination with DNA or RNA extracted from impure tissue samples can cause inaccurate results.
5. AMPure XP beads and SeqCap Capture Beads cannot be used interchangeably. Ensure that SeqCap Capture Beads are used only to recover a library that has been hybridized to a probe pool. AMPure XP beads are used to purify the sample library such as during library preparation or after amplification steps.
6. Ensure that 80 % ethanol is *freshly prepared daily* for all bead wash steps.
7. Due to the multipurpose nature of the SeqCap Adapter Kits, adapters must be diluted from the stock adapter concentration for generating cDNA libraries as described in this protocol.
8. Due to the multipurpose nature of the SeqCap Adapter Kits, the pre-capture LM-PCR oligos contained in the SeqCap Adapter Kit are not used for the SeqCap RNA workflow. The 10× KAPA Library Amplification Primer Mix is used in place of the pre-capture LM-PCR oligos in the SeqCap Adapter Kit.
9. The pre-capture LM-PCR program calls for 11 PCR cycles; however the number of cycles may be adjusted higher to increase yield if 1 μg of amplified material is not consistently generated, or lower to decrease PCR duplicate rate if a large amount of material in great excess of 1 μg is consistently generated.

10. If desired, Qiagen QIAquick PCR purification columns may be used to clean up the pre-capture and post-capture LM-PCR reactions in place of Agencourt AMPure XP beads.
11. The SeqCap lncRNA Enrichment Kit has been tested with up to 12-plex pre-capture multiplexing.
12. The hybridization reaction can safely incubate for longer than 16–20 h and has been tested for up to 3 days; however the sample should be monitored for evaporation in particular when exposed to longer incubation periods.
13. Ensure that during washing, you progress from wash buffer I to wash buffer II to wash buffer III and not in the opposite order.
14. Perform two individual post-capture LM-PCR reactions for each captured multiplex cDNA sample, i.e., one cDNA sample is divided and amplified in two, 30 μl post-capture LM-PCR reactions.
15. A subset of ERCC controls are included in the lncRNA capture design. These controls can be used to estimate target enrichment performance prior to sequencing through qPCR, or after sequencing by analyzing the sequencing data. Prior to sequencing, qPCR with primers designed to the ERCC controls can be used to estimate target enrichment by comparing qPCR results from pre-capture vs. post-capture samples and/or analyzing results from post-capture samples by analyzing ERCC controls included in the capture design vs. ERCC controls excluded from the capture design. When analyzing sequencing results, estimate target enrichment by comparing ERCC controls included in the capture design vs. ERCC controls excluded from the capture design. More information about estimating target enrichment based on sequencing data is available by referring to the SeqCap RNA user's guide available from Roche NimbleGen.

    When proceeding to sequencing, it is recommended that users adhere to their standard sequencing practices including library QC steps and sample loading concentration. Sample libraries can be post-capture multiplexed (i.e., pooled immediately before sequencing); however care should be taken to ensure that any pooled libraries can be distinguished by their index sequences and that libraries are combined in equimolar concentrations, as desired.

Acknowledgement

NIMBLEGEN and SEQCAP are trademarks of Roche.
All other product names and trademarks are the property of their respective owners.
For life science research only. Not for use in diagnostic procedures.

References

1. Feng J, Bi C, Clark BS et al (2006) The Evf-2 noncoding RNA is transcribed from the Dlx-5/6 ultraconserved region and functions as a Dlx-2 transcriptional coactivator. Genes Dev 20:1470–1484
2. Rinn JL, Kertesz M, Wang JK et al (2007) Functional demarcation of active and silent chromatin domains in human HOX loci by noncoding RNAs. Cell 129:1311–1323
3. Barry G, Briggs JA, Vanichkina DP et al (2014) The long non-coding RNA Gomafu is acutely regulated in response to neuronal activation and involved in schizophrenia-associated alternative splicing. Mol Psychiatry 19:486–494
4. Sauvageau M, Goff LA, Lodato S et al (2013) Multiple knockout mouse models reveal lincRNAs are required for life and brain development. Elife 2:e01749
5. Guttman M, Donaghey J, Carey BW et al (2011) lincRNAs act in the circuitry controlling pluripotency and differentiation. Nature 477:295–300
6. Tian D, Sun S, Lee JT (2010) The long noncoding RNA, Jpx, is a molecular switch for X chromosome inactivation. Cell 143: 390–403
7. Matouk IJ, DeGroot N, Mezan S et al (2007) The H19 non-coding RNA is essential for human tumor growth. PLoS One 2:e845
8. Tufarelli C, Stanley JA, Garrick D et al (2003) Transcription of antisense RNA leading to gene silencing and methylation as a novel cause of human genetic disease. Nat Genet 34:157–165
9. Faghihi MA, Modarresi F, Khalil AM et al (2008) Expression of a noncoding RNA is elevated in Alzheimer's disease and drives rapid feed-forward regulation of beta-secretase. Nat Med 14:723–730
10. Guttman M, Garber M, Levin JZ et al (2010) Ab initio reconstruction of cell type-specific transcriptomes in mouse reveals the conserved multi-exonic structure of lincRNAs. Nat Biotechnol 28:503–510
11. Djebali S, Davis CA, Merkel A et al (2012) Landscape of transcription in human cells. Nature 489:101–108
12. Brunner AL, Beck AH, Edris B et al (2012) Transcriptional profiling of long non-coding RNAs and novel transcribed regions across a diverse panel of archived human cancers. Genome Biol 13:R75
13. Cabili MN, Trapnell C, Goff L et al (2011) Integrative annotation of human large intergenic noncoding RNAs reveals global properties and specific subclasses. Genes Dev 25:1915–1927
14. Mercer TR, Clark MB, Crawford J et al (2014) Targeted sequencing for gene discovery and quantification using RNA CaptureSeq. Nat Protoc 9:989–1009
15. Cabanski CR, Magrini V, Griffith M et al (2014) cDNA hybrid capture improves transcriptome analysis on low-input and archived samples. J Mol Diagn 16:440–451
16. Harrow J, Frankish A, Gonzalez JM et al (2012) GENCODE: the reference human genome annotation for The ENCODE Project. Genome Res 22:1760–1774

Chapter 9

ChIP-Seq: Library Preparation and Sequencing

Karyn L. Sheaffer and Jonathan Schug

Abstract

Chromatin immunoprecipitation with massively parallel DNA sequencing (ChIP-Seq) has been used extensively to determine the genome-wide location of DNA-binding factors, such as transcription factors, posttranscriptionally modified histones, and members of the transcription complex, to assess regulatory input, epigenetic modifications, and transcriptional activity, respectively. Here we describe methods to isolate chromatin from tissues, immunoprecipitate DNA bound to a protein of interest, and perform next-generation sequencing to identify a genome-wide DNA-binding pattern.

Key words Chromatin immunoprecipitation, Next-generation sequencing, DNA binding

1 Introduction

Gene expression is controlled by not only DNA sequence but also the interaction of DNA-binding proteins with higher order chromatin structure. Transcription factors play an integral part of the initiation of gene expression by interacting with transcriptional machinery to regulate gene promoters [1, 2]. Histone proteins are closely associated with DNA and posttranslational modifications are highly correlated with chromatin accessibility [3]. Additionally, many more small molecules have critical interactions with DNA [4]. Chromatin immunoprecipitation with massively parallel DNA sequencing (ChIP-Seq) provides a powerful method to map protein–DNA interactions genome-wide [5–7]. DNA-binding patterns merged with expression profiling is a powerful combination to investigate the impact of DNA binding of specific factors on gene expression [8, 9].

In this chapter we describe the steps to achieve successful ChIP-Seq results. First, a sequencing plan should be considered taking in account the nature of the protein of interest, cell type, controls, and sequencing depth required. Chromatin immunoprecipitation (ChIP) is carried out by isolating the cell type of interest and cross-linking DNA–protein interactions to preserve chromatin.

Yi Feng and Lin Zhang (eds.), *Long Non-Coding RNAs: Methods and Protocols*, Methods in Molecular Biology, vol. 1402, DOI 10.1007/978-1-4939-3378-5_9, © Springer Science+Business Media New York 2016

DNA–protein complexes are isolated by immunoprecipitation using an antibody that recognizes the protein of interest. Oligonucleotide adaptors are added onto the DNA isolated from these complexes to enable high-throughput sequencing. Here we provide a basic protocol and more detailed guidelines can be found through the ENCODE project [10].

2 Materials

2.1 DNA–Protein Cross-Linking

1. Formaldehyde.
2. Protease inhibitor cocktail.
3. ChIP whole-cell lysis buffer (10 mM Tris–HCl (pH 8), 10 mM NaCl, 3 mM $MgCl_2$, 1 % NP-40, 1 % SDS, 0.5 % DOC).

2.2 Chromatin Shearing and Input Analysis

1. Bioruptor (Diagenode).
2. PCR Purification Kit (Qiagen).
3. 2100 BioAnalyzer (Agilent Technologies).

2.3 Immunoprecipitation

1. Protease inhibitor cocktail.
2. ChIP dilution buffer: 16.7 mM Tris–HCl (pH 8.1), 167 mM NaCl, 0.01 % SDS, 1.1 % Triton-X 100, 1× protease inhibitor.
3. Dynabeads Protein-A or –G (Life Technologies).
4. TSE I: 20 mM Tris–HCl pH 8.1, 150 mM NaCl, 2 mM EDTA, 0.1 % SDS, 1 % TritonX-100.
5. TSE II: 20 mM Tris–HCl pH 8.1, 500 mM NaCl, 2 mM EDTA, 0.1 % SDS, 1 % TritonX-100.
6. ChIP buffer III: 10 mM Tris–HCl pH 8.1, 0.25 M LiCl, 1 mM EDTA, 1 % NP-40, 1 % deoxycholate.
7. TE: 10 mM Tris–HCl pH 8.1, 1 mM EDTA.
8. Elution buffer: 1 % SDS, 0.1 M $NaHCO_3$.

2.4 Sequencing Library Preparation

1. Multiplex Oligos (Illumina).
2. NEBNext ChIP-Seq DNA Sample Prep (NEB).
3. AMPure XP beads (NEB). Use at room temperature.

2.5 Sequencing Library Validation

1. 2100 BioAnalyzer (Agilent Technologies).
2. Illumina Library Quantification qPCR mix (Kapa Biosystems).

2.6 Next-Generation Sequencing

Illumina NextSeq 500, HiSeq 2000, or HiSeq 2500. An Illumina MiSeq could be used, but only for one or two samples at a time.

3 Methods

3.1 Check Feasibility

The ideal number of moles and molarity of the final library can be estimated by the following formula which provides a framework for assessing the feasibility of a particular experiment. This formula does not take into account splitting chromatin into aliquots for input libraries, multiple IPs, etc. The final molarity of most ChIP-Seq libraries is between 10 and 80 nM when working with a cell count between 500,000 and a few million.

$$\text{Library[moles]} = (\text{Nc} \times \text{Ns} \times \text{Ec} + \text{Nc} \times \text{Gbp}/\text{Fbp} \times \text{Lip}) \times \text{Eo} \times 2\text{\^{}Np} \times \text{Eb} / 6.022 \times 10^{23}$$

$$\text{Library[molarity]} = \text{Library[moles]}/15\ \mu\text{l}$$

where

- **Nc** is the number of cells.
- **Ns** is the number of sites in the diploid genome.
- **Ec** is the capture efficiency, between 0 and 1, for the ChIP process.
- **Gbp** is the size of the diploid genome in bp.
- **Fbp** is the average fragment size.
- **Lip** is the leakage rate of the unbound sites through the IP process.
- **Eo** is the overall efficiency, between 0 and 1, of DNA fragments passing through the washing and adapter ligation process.
- **Np** is the number of PCR cycles in the final amplification—18 cycles in this protocol.
- **Eb** is the efficiency, between 0 and 1, of the bead cleanup.

The term in parenthesis is the number of moles from *bound* sites plus *unbound* sites that manage to pass through the IP process. The usual number of bound sites, *Ns*, in the genome can range from a few hundred (for a very specific transcription factor) to perhaps 50,000–100,000 for a widely binding transcription factor or a modified histone with broad distribution. The effective number of bound sites is lower if occupancy has a stochastic component; that is, the target protein is found at the site in a only fraction of the cells at any given time.

The number of unbound fragments is estimated by the total diploid genome divided by the average fragment size. Note that for mouse or human samples, *Gbp/Fbp* is about 6e9/200 = 3e7. A PCR enrichment of 10× means a leakage rate of about 0.1, so *Gbp/Fbp**Lip* = 3e6. This is still much higher than even the upper estimate of 100,000 bound sites from above. Excellent IPs may yield enrichment values as high as 100 or more in which case the DNA fragments after IP are about 50/50 target/nontarget. Thus an excellent IP may make it difficult to make a library if there are very few sites. Some background leakage allows a sufficient molarity for the subsequent library prep process to succeed, especially when considering losses.

The term *Eo* encapsulates the losses in the IP, washing, end-repair, ligation, cleanup, etc. prior to PCR amplification. Mathematically it can be merged with the efficiency of the post-PCR bead cleanup, *Eb*, but this arrangement makes it a bit easier to analyze. It is important not to overload the antibodies with too much material as the excess will be wasted once the antibodies are saturated.

The formula also illustrates that the molarity of many of the intermediate steps is so low that it is difficult to measure them unless a large number of cells are used.

Using the formula above and the guidelines in Subheading 3.3 below, assess the feasibility of your experiment.

3.2 Generate the Sequencing Plan

When sequencing ChIP-Seq libraries the primary goal is to produce, in a cost-effective manner, adequate sequence reads which satisfy the statistical power necessary to identify the enriched binding regions. The main choices available to achieve these two goals are (1) the sequencing length, and (2) the sequencing depth.

3.2.1 Sequencing Length

For most ChIP-Seq experiments, the actual bases in the sequence reads are not important; rather, the goal is to obtain reads long enough to map uniquely in the target genome. Most analysis techniques cannot take effective advantage of reads that map to more than one location in the genome, so these are usually discarded. The locations reachable by a specific length of sequencing can be identified using the mappability tracks at the UCSC genome browser or by direct analysis of the genome [11]. The final length chosen also depends on what your sequencing service is offering. Sequencing lengths of 36, 50, 75, and 100 base pairs are common on Illumina platforms. For humans and complex model organisms 36–50 bp will suffice in many cases. Longer reads can be used if it is important to map more deeply into repetitive regions or to achieve more coverage per base to recover genome variants in the source organism. Paired-end sequencing can also be used to achieve unique mapping in more repetitive areas, but it is rarely worth the extra cost. However, paired-end sequencing can be useful if it is important to determine the length or exact end points of the fragment being sequenced.

The sequencing length must be compatible with the average fragment size. For example, if the average fragment size is 120 bp, and the sequencing length is 100 bp, then nearly half of the reads will contain adapter sequence at the 3′ end that will need to be trimmed before alignment.

3.2.2 Sequencing Depth

In many ChIP-Seq experiments the number of reads needed to ensure genome-wide coverage ranges between 10 and 50 million reads. Illumina's high-throughput sequencers, e.g., HiSeq 2500 in rapid mode, HiSeq 2000/2500 high-output mode, and NextSeq 500, generate between 150 and 400 million reads per lane or run. Thus the capacity of a single lane can be enough to cover between 3 and 40 samples. To take advantage of this capacity, libraries are barcoded, carefully checked for quality and molarity, and then combined in equimolar pools before sequencing. It is important to check with the manufacturer's recommendation for compatible barcodes as a bad combination can adversely affect base calling of the barcode and therefore the total data yield. If you are using a sequencing service or core, they may provide some or all of the necessary quality checks for this step. Consult with them to determine an exact plan. Your bioinformatician may also have suggestions or requirements for barcoding. See below for our recommendations.

Note that the uniquely alignable portion of reads commonly ranges from 60 to 85 % of the total number of reads; thus, your calculation for the number of reads needed should take this inefficiency into account.

Another important factor in determining sequencing depth is read redundancy. The library preparation protocol contains a PCR amplification step where the library is amplified between 1,000 and 65,000 times or more depending on the number of cycles. Many ChIP-Seq analysis packages or pipelines will rightly try to detect and remove PCR duplicates. Detection is usually accomplished by identifying common endpoints after alignment. When sonication is used to fragment the DNA, duplicate endpoints are more likely to come from PCR duplication rather than from independent cells yielding the same fragment. However, in a strong narrow peak, duplicate reads can come from independent fragments—a fact of which some analysis programs can take advantage. However, it is usually not worth sequencing a library to a depth much beyond its unique fragment count. This number usually cannot be known before sequencing, but initial sequencing data can be used to assess library complexity, and this analysis should be taken into account when considering deeper sequencing.

3.2.3 Barcoding/Pooling Plan

We introduced barcoding and pooling as a way to reduce costs, but it also provides some protection from lane-to-lane or run-to-run variability that could affect the results of the experiment. We generally find that technical replicates, i.e., resequencing the same library on a different lane, run, or machine, will give very similar results when performed properly. However, it is possible to see a CG content or fragment length bias that could affect the results of the data analysis by biasing for or against reads from different parts of the genome. One way to mitigate this possibility is to pool samples and sequence them together. In this manner, all the samples in the pool experience the same biases and therefore the biases will not have as great an affect on downstream data analyses.

The greatest protection is achieved by combining all samples for the experiment together in one large pool, but this may not be practical due to limitations of sample availability, barcode availability, etc. However, the final pooling plan should attempt to pool ChIP libraries with their corresponding inputs, and to keep samples from multiple conditions in the same pool as well.

3.3 DNA–Protein Cross-Linking

There are two limiting factors in the amount of starting material needed for ChIP-Seq: tissue/cell type and the chosen DNA-binding factor.

1. Generally it is recommended that a minimum of 2 mg fresh or frozen tissue or 250,000 cells is sufficient for one ChIP-Seq experiment. Chromatin yields from different tissue types can vary and chromatin isolation techniques may need to be optimized. In cases where there is little starting material, successful libraries can be obtained from pooling several IPs.
2. The specific DNA-binding factor of interest also determines the amount of starting material for ChIP-Seq. For proteins that bind at specific nucleotide sequences and are found at low frequency throughout the genome, such as transcription factors, 10 μg of starting material is recommended for successful sequencing libraries. For proteins that are abundant throughout the DNA, such as histones and histone modifications, 1–2 μg of chromatin is recommended.
3. Spin down tissue/cells at maximum speed for 10 s at room temperature and then aspirate supernatant.
4. Resuspend tissue/cell pellet in 500 μl 1× PBS and chop with fine scissors if necessary (*see* **Note 1**).
5. To cross-link protein to DNA, add 500 μl 2.22 % formaldehyde diluted in 1× PBS (*see* **Note 2**).
6. Incubate for 10 min at room temperature with constant mixing.
7. Stop cross-linking reaction by adding 59 μl of 2.5 M glycine. This provides a final concentration of 0.14 M.
8. Incubate for 5 min at room temperature with constant mixing.
9. Spin down cell material at maximum speed for 10 s at room temperature and aspirate the supernatant.
10. To wash the pellet, add 1 ml 1× PBS and flick to resuspend.
11. Spin down cell material at maximum speed for 10 s at room temperature and aspirate the supernatant.
12. Add 100–200 μl of cold ChIP whole-cell lysis buffer containing 1× protease inhibitor (*see* **Note 3**).
13. Snap freeze pellet in liquid nitrogen. The sample can be stored at −80 °C for several months.

3.4 Chromatin Shearing and Input Analysis

1. Use the Bioruptor sonication system to lyse cells and shear chromatin. Follow the recommended protocols found at www.diagenode.com (*see* **Note 4**).
2. To remove cell debris after full sonication procedure, spin samples for 15 min at maximum speed at 4 °C, and transfer supernatant to new tube.
3. Take 10 μl for input preparation. Snap freeze the remaining chromatin in liquid nitrogen and store at −80 °C.
4. To release input chromatin, add 90 μl 1× PBS and 3.5 μl 5 M NaCl. Mix well.
5. Incubate at 65 °C for 12–24 h.
6. To digest proteins away from DNA in the input chromatin, add 4 μl 1 M Tris–HCl (pH 7.5), 2 μl 500 mM EDTA, and 1 μl 10 mg/ml Proteinase K.
7. Incubate for 1 h at 45 °C.
8. To isolate remaining DNA, use PCR purification kit and elute in 50 μl elution buffer.
9. Analyze DNA fragmentation by measuring input DNA concentration and size range using the 2100 BioAnalyzer (Fig. 1). Samples need to contain DNA fragments within the 100–200 bp range before proceeding to the immunoprecipitation step.

3.5 Immunoprecipitation

1. Thaw sheared chromatin quickly at 37 °C and immediately place on ice.
2. Add appropriate amount of chromatin (1–10 μg) and 20 μl 50× protease inhibitor to 1 ml ChIP dilution buffer.
3. To immunoprecipitate DNA bound to the DNA-binding factor of interest, add 2 μg of a ChIP-grade antibody or control antibody, if appropriate (*see* **Note 5**).
4. Incubate for 12–14 h at 4 °C with constant mixing.
5. During this same time, prepare blocked Dynabeads Protein-A or -G (*see* **Note 6**).
   (a) Add 40 μl bead slurry to 1 ml ChIP dilution buffer and mix by inverting several times.
   (b) Spin at 2000 rpm (400 *g*) for 30 s at room temperature.
   (c) Allow beads to settle for 1 min and then remove supernatant with pipet tip. Be careful not to disturb agarose pellet.
   (d) Repeat two more times.
   (e) To block beads, resuspend beads in 68 μl cold ChIP dilution buffer, 10 μl 10 mg/ml BSA, and 2 μl 50× protease inhibitor. Do not use other blocking agents, such as salmon sperm or tRNA, because they will be incorporated into the library and be sequenced.
   (f) Incubate for 12–14 h at 4 °C with constant mixing.

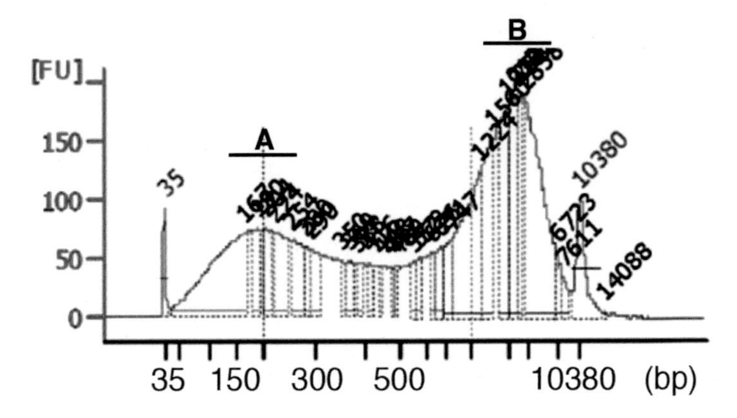

Fig. 1 Bioanalyzer trace of sheared input DNA. DNA size and quantity were determined using Agilent's High Sensitivity DNA Kit. Electropherogram shows control peaks at a lower marker (35 bp) and an upper marker (10380 bp), fluorescent units (FU), and base pairs (bp). Optimal sonication of chromatin produces DNA sizes between 150 and 2000 bp. This is a typical result with the majority of DNA with average sizes of 160 bp (*A*) or 1500 bp (*B*). (*A*) Small-size fragments with an average of 160 bp represent open chromatin regions that are more easily sheared with sonication. This DNA is enriched for open chromatin genomic regions, which are more likely to bind transcription factors. (*B*) Large-size fragments with an average of 1500 bp represent closed chromatin that is resistant to shearing. Loss of DNA within this size range may be due to over-sonication causing loss of open chromatin and which can lead to poor ChIP results. However, if you know that your ChIP target favors closed chromatin, additional sonication may be required to bring these long fragments into optimal sequence fragment size, at the expense of the open chromatin

6. Add 100 μl of blocked Dynabeads to each chromatin/antibody sample.
7. Incubate for 1 h at 4 °C with constant mixing.
8. Spin at 2000 rpm for 30 s at room temperature.
9. Allow beads to settle for 1 min and then remove supernatant with pipet tip.
10. Perform wash (**steps 10a–d**) with each of the following wash buffers in order:
    TSE I, TSE II, ChIP buffer III, TE.
    (a) Add 1 ml of the appropriate wash buffer to each sample.
    (b) Incubate for 5 min at room temperature with constant mixing.
    (c) Spin at 2000 rpm for 30 s at room temperature.
    (d) Allow beads to settle for 1 min and then remove supernatant with pipet tip.

11. To elute chromatin from beads, add 100 μl elution buffer to final pellet.
    (a) Incubate for 15 min at room temperature with constant mixing.
    (b) Spin at 2000 rpm for 30 s at room temperature.
    (c) Allow beads to settle for 1 min and then transfer the supernatant to a new tube.
12. Add an additional 100 μl elution buffer to pellet and repeat elution (**steps 11a–c**). Combine eluates.
13. Add 8 μl 5 M NaCl to eluate (200 μl) and incubate at 65 °C for 12–24 h.
14. To digest proteins in the input chromatin, add 8 μl 1 M Tris–HCl (pH 7.5), 4 μl 500 mM EDTA, and 1 μl 10 mg/ml Proteinase K.
15. Incubate for 1 h at 45 °C.
16. To isolate remaining DNA, use PCR cleanup kit and elute in 50 μl elution buffer.
17. It is important to validate that the immunoprecipitation worked by testing for enrichment of genomic regions that are known to bind your protein of interest. Enrichment is determined by performing qPCR on a nonspecific genomic region, such as 28S, compared to a known genomic region that binds the protein of interest. The ChIP sample should contain a higher quantity of the known genomic region compared to the input sample (Fig. 2). Enrichment is calculated from the Ct values: $2\^[(28S_{ChIP} - YourGene_{ChIP}) - (28S_{Input} - YourGene_{Input})]$ (*see* **Note 7**).

3.6 Sequencing Library Preparation

The provided protocol provides steps multiple DNA purification and size selection options. An input sample should be run as a positive control, as this should achieve a library every time. We strongly recommend sequencing the input library to provide a control for data analysis.

Before starting this stage of the protocol, recheck your experimental design for the multiplexing plan for samples. As discussed in the next-generation sequencing step, most Illumina sequencers generate more data per lane than are needed for a single sample. To reduce the cost of the experiment, multiple samples can be combined into a pool, and sequenced together in a single lane. The data for each sample is then computationally identified using a unique barcode, which has been assigned to each sample in the pool. Barcodes must be chosen to form a compatible set, though this is most important for small pools of 2–6 samples. Make careful note of which indexes are used for which samples and be prepared to communicate this to your bioinformatician.

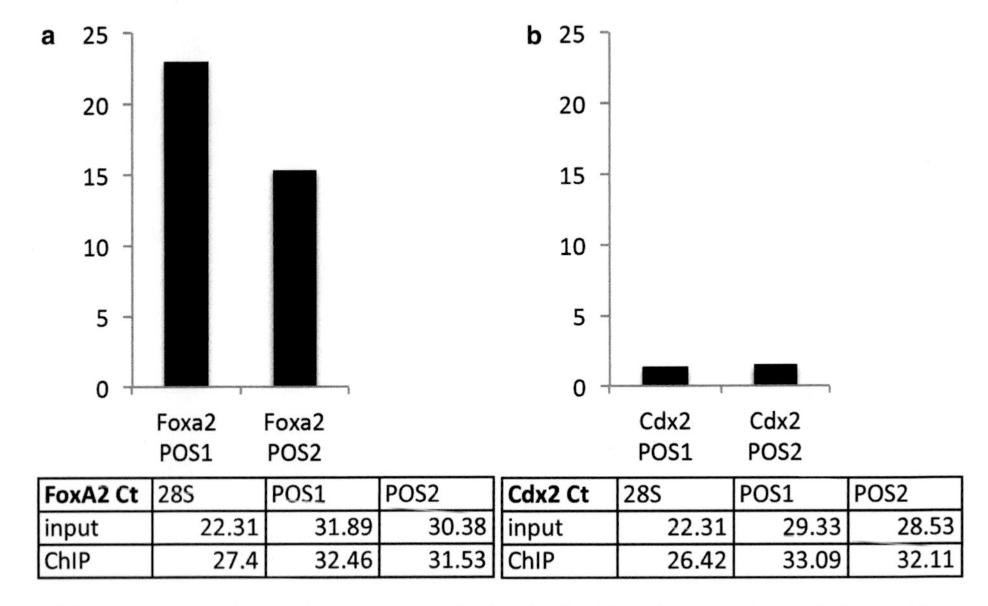

FoxA2 Ct	28S	POS1	POS2
input	22.31	31.89	30.38
ChIP	27.4	32.46	31.53

Cdx2 Ct	28S	POS1	POS2
input	22.31	29.33	28.53
ChIP	26.42	33.09	32.11

Fig. 2 qPCR ChIP validation: (**a**) ChIP was performed using the FoxA2 antibody on chromatin isolated from colon epithelia. Genomic regions POS1 and POS2 show robust enrichment of 22.9- and 15.3-fold, respectively, in ChIP samples relative to input. Ct values are shown in *chart* under graph. (**b**) ChIP was performed using the Cdx2 antibody. No enrichment was found at POS1 (1.3-fold) or POS2 (1.4-fold) in ChIP samples compared to input

Sequencing libraries can be made using one of many commercially available kits. In the following protocol, we use the NEBNext ChIP-Seq DNA Sample Prep Kit.

1. To perform end repair of DNA fragments, add 5 μl of end repair reaction buffer and 1 μl end repair enzyme mix to 44 μl of IP DNA and input DNA (10 ng in 44 μl).
2. Incubate for 30 min at 20 °C.
3. Purify DNA using AMPure XP beads and elute in 44 μl:
   (a) Warm beads to room temperature and vortex until they are well dispersed.
   (b) Add 1.8× volume of beads to DNA sample and mix by pipetting ten times.
   (c) Incubate the tubes at room temperature for 12 min.
   (d) Place tubes on magnetic stand and let sit for 2 min until the liquid appears clear.
   (e) Remove supernatant from each tube and discard.
   (f) Add 200 μl 80 % EtOH, incubate for 1 min, then remove supernatant, and discard.
   (g) Repeat EtOH wash once.
   (h) Let tubes stand at room temperature for 10 min to dry. Make sure that all EtOH is removed.

(i) Remove tubes from magnetic stand and resuspend beads in appropriate volume of elution buffer (EB). Mix well by pipetting ten times.

(j) Incubate at room temperature for 5 min.

(k) Place tubes on magnetic stand for 2 min until the liquid appears clear.

(l) Transfer supernatant to new tube. Be careful to avoid beads.

4. To prepare the DNA fragments for ligation to adapters by adding an "A" base to the 3′ end, add 5 μl of dA-tailing reaction buffer and 1 μl Klenow exonuclease to 44 μl DNA from previous step.
5. Incubate for 30 min at 37 °C.
6. Purify DNA using AMPure XP beads and elute in 19 μl as described in **step 3** of Subheading 3.5.
7. To ligate adapters to DNA fragments, dilute multiplexing adapter 1:20 in EB.
8. Prepare the following reaction mix: 6 μl 5× quick DNA ligase buffer, 1 μl adapter oligo mix, and 4 μl quick DNA ligase.
9. Add reaction mix to 19 μl of DNA sample from **step 6** of Subheading 3.5.
10. Incubate for 15 min at 20 °C.
11. To size select 200 bp libraries using AMPure XP beads, add 70 μl EB to the previous sample and add 70 μl (0.7×) resuspended beads and mix well by pipetting ten times (*see* **Note 8**).
12. Incubate for 5 min at room temperature.
13. Place tube on magnetic stand for 5 min until solution is clear and transfer supernatant to a new tube.
14. Add 15 μl (0.15×) resuspended AMPure XP beads and mix well by pipetting ten times.
15. Incubate for 5 min at room temperature.
16. Place tube on magnetic stand for 3 min until solution is clear and discard supernatant.
17. Add 200 μl 80 % EtOH, incubate for 1 min at room temperature, then remove supernatant, and discard.
18. Repeat EtOH wash once.
19. Let tubes stand at room temperature for 10 min to dry. Make sure that all EtOH is removed.
20. Remove tubes from magnetic stand and resuspend beads in 50 μl of EB. Mix well by pipetting ten times.
21. Place tubes on magnetic stand for 2 min until the liquid appears clear.
22. Transfer supernatant to new tube. Be careful to avoid beads.

23. To enrich the adapter-modified DNA fragments by PCR, add 2 µl of multiplex PCR primer 1, 2 µl multiplex PCR primer 2 with index, 50 µl Phusion HF PCR Master Mix, and 50 µl DNA from previous step.
24. Amplify using the following PCR protocol:
    (a) 30 s at 98 °C.
    (b) 10 s at 98 °C.
    (c) 30 s at 65 °C.
    (d) 30 s at 72 °C.
    (e) Repeat **steps b–d** for 18 cycles.
    (f) 5 min at 72 °C.
    (g) Hold at 4 °C.
25. Purify DNA using AMPure XP beads and elute in 15 µl as described in **step 3** of Subheading 3.5.

3.7 Sequencing Library Validation

The purpose of library validation is to ensure that the library is correctly prepared and of sufficient *cluster-able* molarity. When sonication is used to fragment DNA, the resulting fragments have random lengths, so the distribution of sizes in the library should be smooth. As discussed above, the inserts should be long enough to support the sequencing dimensions planned. The molarity should be high enough to be sequenced according to the Illumina's protocol. The nominal value is 2 nM, but in practice more ChIP-Seq libraries will be at 20–100 nM depending on the cell number, pattern of target binding, amount of material used, and number of PCR cycles. The Agilent 2100 BioAnalyzer (or equivalent from other manufacturers) measures the size distribution, but does not indicate if the library can actually be clustered. A second PCR-based test (Kapa Biosystems Library Quantification Kit) is suggested as it can precisely assess the molarity of fragments with the proper adapters in the library. Only these fragments will generate sequence data.

1. After amplification, libraries are validated for size and purity using the Agilent 2100 BioAnalyzer (Fig. 3). Depending on the quality of starting material and amount of chromatin that was immunoprecipitated, libraries may give various results illustrated in Fig. 3.
2. Precise quantification of the library is performed using the Kapa Library Quantification Kit for Illumina before proceeding to the sequencing reaction (Fig. 4).
    (a) Prepare library dilutions in EB (1/1000–1/100,000) depending on expected concentration determined by Bioanalyzer. Dilutions should allow the final concentration to fall within 0.0002–20 pM. One additional twofold dilution is recommended, with a further fourfold providing additional robustness. Each dilution should be measured in triplicate.

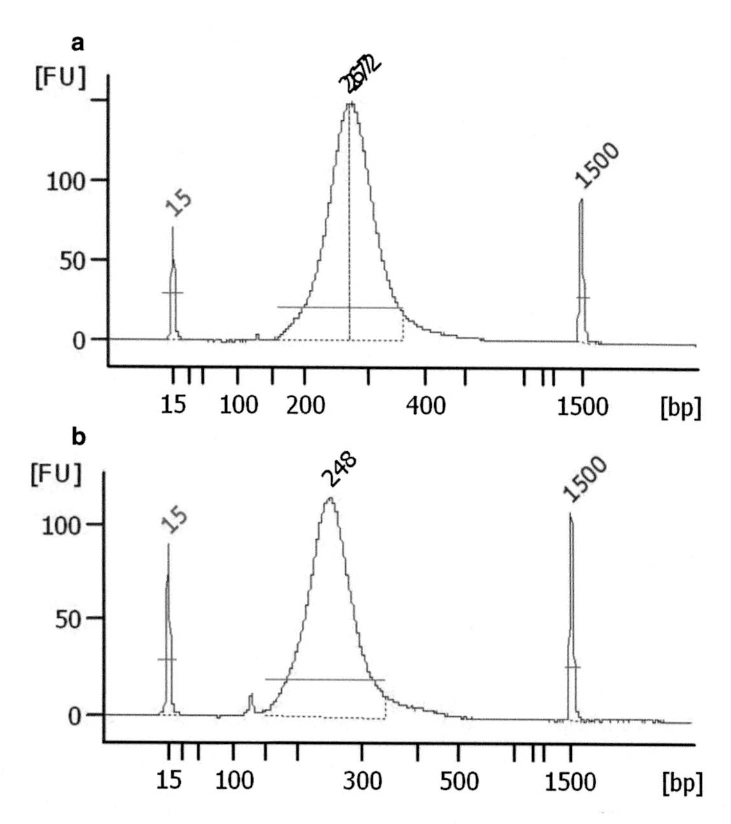

Fig. 3 Bioanalyzer results using Agilent DNA-1000 Kit. Electropherogram shows control peaks at a lower marker (15 bp) and an upper marker (1500 bp), fluorescent units (FU), and base pairs (bp). (**a**, **b**) Examples of multiplexed ChIP-Seq libraries that were successfully sequenced. (**a**) Library with high molarity and one sharp peak at the correct size (200–250 bp). (**b**) Library with one peak at the correct size (200–250 bp) and a smaller peak at 120 bp. A peak at 120 bp is from adaptors that do not have an insert. These will generate sequence reads and will reduce the number of usable reads. If this contamination is more than 10 % of the total DNA then library is considered of insufficient quality to continue to sequencing. This example has minimal contamination and was successfully sequenced

(b) Follow appropriate qPCR protocol provided by Kapa Biosystems.

(c) Use DNA standards provided in Kapa Library Quantification kit to create a standard curve, which converts average Ct to concentration (in pM). Adjust concentrations by size of library. Multiply appropriate dilution factor to calculate final library concentration.

3.8 Next-Generation Sequencing

Once the libraries have been precisely quantified using the Agilent Bioanalyzer and Kapa kit, they can be pooled in equimolar ratios. It is possible that sequencing will reveal a quantification failure, poor libraries, or other issues that would suggest repooling, so it is

Row	Parameter	Library 1		Library 2	
1	Average fragment length (Bioanalyzer)	340 bp		335 bp	
2	Estimated concentration (Bioanalyzer)	17 ng/μl = 80.9 nM		18 ng/μl = 87.0 nM	
3	Dilutions for qPCR	1/10K	1/20K	1/10K	1/20K
4	Triplicate Cq scores	9.24 9.25 9.21	10.21 10.12 ~~10.53~~	8.79 8.84 8.88	9.79 9.81 9.94
5	Average Cq score	9.23	10.17	8.84	9.85
6	ΔCq	0.93		1.01	
7	Average concentration for sample dilution calculated using standard curve (pM)	5.23	2.81	6.82	3.48
8	Average size-adjusted concentration for library dilution (pM)	6.96	3.74	9.20	4.69
9	Average final calculated concentration of undiluted library dilution (nM)	69.6	74.8	92.0	93.9
10	Deviation between final concentrations calculated from different dilutions	7.5%		2.1%	
11	Working concentration	72.2 nM = 15.2 ng/μl		92.9 nM = 19.2 ng/ μl	

Fig. 4 Working example of qPCR-based quantification of sequencing libraries (Kapa Biosystems Technical Sheets)

advisable to make a minimal volume of the pooled libraries. On the other hand, sequencing runs can fail, so enough pool should be made for a few runs if possible. We generally recommend making pools at 5 or 2 nM, which can be matched easily with Illumina's sequencing protocols. The protocols vary by machine and chemistry being used, so we refer the reader to the relevant Illumina manuals.

4 Notes

1. Tissue pieces need to be chopped into small pieces so that formaldehyde can permeate all cells within the incubation time. Cell pellets that are easily dissociated do not need extra manipulation.
2. Cross-linking conditions can affect the quality of results. The cross-linking must be strong enough to preserve the protein–DNA interactions of interest, but not so strong as to prevent the fragmentation of the chromatin by sonication. ChIP targets that are not bound directly to the DNA, but are bound indirectly via other proteins, may require additional stabilization such as XChIP [12].

3. Protease inhibitor should be added to the ChIP whole-cell lysis buffer immediately before tissue resuspension. Addition of phosphatase inhibitors may also be recommended depending on the modification status of the protein of interest. Resuspension volume depends on the amount of starting material with large visible cell pellets resuspended in the maximum volume.

4. Other sonication systems can be substituted. However, conditions listed in this protocol may not be transferable and optimization of the conditions must be performed.

   DNA shearing can also be performed using the nuclease, MNase, to randomly cut DNA [13]. Due to the nature of chromatin accessibility, MNase treatment can be performed so that only open chromatin regions are digested leaving DNA associated with nucleosomes intact. Only ChIP-Seq experiments using antibodies directed against histone proteins or histone modifications can utilize this alternative protocol. Applying digestion can leave transcription factor targets bound to a DNA fragment that is too short to map reliably or to measure with PCR. Even with histone modifications some care should be taken to ensure that the primer fits within a nucleosome wrap.

5. The quality of a ChIP experiment is highly dependent on the specificity of the antibody. Guidelines for the criteria and validation of ChIP-grade antibodies have been published by the ENCODE consortium [10].

   The data analysis phase of a ChIP-Seq experiment typically requires a control sample to identify the regions of enriched binding. The protocol described here uses input chromatin as the control. Input chromatin is a good measure of the chromatin accessibility under the cross-linking and sonication conditions used to prepare the ChIP-Seq sample. Some people may prefer to use an IgG antibody or a genetically modified organism with the target protein knocked out. We strongly recommend preparing an input library for each (set of) ChIP-Seq library prepared. ChIP can be performed using multiple antibodies from the same chromatin prep and the single input library can serve as a control for all of them.

6. Protein A and G have different binding strengths to immunoglobulin variations between species. A table illustrating antibody binding compatibility to protein A or G can be found at http://www.lifetechnologies.com/us/en/home/life-science/protein-expression-and-analysis/protein-sample-preparation-and-protein-purification/proteinspproteiniso-misc/protein-isolation/immunoprecipitation-using-dynabeads-protein-a-or-protein-g.html.

7. It is possible to make ChIP-Seq libraries without having a positive control to assess the ChIP; however, in our experience most such attempts have failed. The PCR test for enrichment

is important but does have some limitations and caveats: First, the negative control, 28S in this protocol, should be chosen with some care. While 28S is rarely bound by tissue-specific transcription factors, it can be bound by general transcription factors, modified histones, and members of certain transcription complexes. In this case, alternate negative controls should be identified. Multiple negative (and positive) controls can be used for a more robust assessment of enrichment.

Second, we generally recommend that the enrichment be 10× or more, though this may not be achievable or may depend on the distribution of the target protein. The goal is to get an enrichment that is readily identifiable in the sequencing data at a reasonable sequencing depth. If the enrichment were only 2× then the enrichment would not be significantly above the normal background fluctuations until the sequencing depth was quite high. This increases the cost of the experiment and, furthermore, may not be possible if the complexity of the library is not high enough to support deeper sequencing.

Third, the enrichment is being measured on DNA that has not been subjected to the size selection that will occur later in library preparation. It is possible to obtain strong enrichment in this PCR test that is derived largely from longer DNA fragments that do not appear in the final library.

Fourth, if the target is not well studied, then it is not uncommon for the few known sites to not be the strongest sites in the genome. This is especially true when the positive controls were identified from different tissues, developmental stages, etc. The PCR enrichment for these sites may never be very strong but nevertheless the IP has worked well.

8. Size selection is essential for removing excess adaptors and providing optimal sequencing efficiency. It is recommended that ChIP-Seq libraries fall within the size range of 200–250 bp. Size selection can also be achieved using a 2 % agarose gel.

References

1. Hahn S, Young ET (2011) Transcriptional regulation in Saccharomyces cerevisiae: transcription factor regulation and function, mechanisms of initiation, and roles of activators and coactivators. Genetics 189(3):705–736. doi:10.1534/genetics.111.127019
2. Farnham PJ (2009) Insights from genomic profiling of transcription factors. Nat Rev Genet 10(9):605–616. doi:10.1038/nrg2636
3. Rothbart SB, Strahl BD (2014) Interpreting the language of histone and DNA modifications. Biochim Biophys Acta 1839(8):627–643. doi:10.1016/j.bbagrm.2014.03.001
4. Rodriguez R, Miller KM (2014) Unravelling the genomic targets of small molecules using high-throughput sequencing. Nat Rev Genet 15(12):783–796. doi:10.1038/nrg3796
5. Mundade R, Ozer HG, Wei H, Prabhu L, Lu T (2014) Role of ChIP-seq in the discovery of transcription factor binding sites, differential gene regulation mechanism, epigenetic marks and beyond. Cell Cycle 13(18):2847–2852. doi:10.4161/15384101.2014.949201
6. Furey TS (2012) ChIP-seq and beyond: new and improved methodologies to detect and characterize protein-DNA interactions. Nat Rev Genet 13(12):840–852. doi:10.1038/nrg3306

7. Zhang Z, Pugh BF (2011) High-resolution genome-wide mapping of the primary structure of chromatin. Cell 144(2):175–186. doi:10.1016/j.cell.2011.01.003
8. Kieffer-Kwon KR, Tang Z, Mathe E, Qian J, Sung MH, Li G, Resch W, Baek S, Pruett N, Grontved L, Vian L, Nelson S, Zare H, Hakim O, Reyon D, Yamane A, Nakahashi H, Kovalchuk AL, Zou J, Joung JK, Sartorelli V, Wei CL, Ruan X, Hager GL, Ruan Y, Casellas R (2013) Interactome maps of mouse gene regulatory domains reveal basic principles of transcriptional regulation. Cell 155(7):1507–1520. doi:10.1016/j.cell.2013.11.039
9. Van Nostrand EL, Kim SK (2013) Integrative analysis of C. elegans modENCODE ChIP-seq data sets to infer gene regulatory interactions. Genome Res 23(6):941–953. doi:10.1101/gr.152876.112
10. Landt SG, Marinov GK, Kundaje A, Kheradpour P, Pauli F, Batzoglou S, Bernstein BE, Bickel P, Brown JB, Cayting P, Chen Y, DeSalvo G, Epstein C, Fisher-Aylor KI, Euskirchen G, Gerstein M, Gertz J, Hartemink AJ, Hoffman MM, Iyer VR, Jung YL, Karmakar S, Kellis M, Kharchenko PV, Li Q, Liu T, Liu XS, Ma L, Milosavljevic A, Myers RM, Park PJ, Pazin MJ, Perry MD, Raha D, Reddy TE, Rozowsky J, Shoresh N, Sidow A, Slattery M, Stamatoyannopoulos JA, Tolstorukov MY, White KP, Xi S, Farnham PJ, Lieb JD, Wold BJ, Snyder M (2012) ChIP-seq guidelines and practices of the ENCODE and modENCODE consortia. Genome Res 22(9):1813–1831. doi:10.1101/gr.136184.111
11. Derrien T, Estelle J, Marco Sola S, Knowles DG, Raineri E, Guigo R, Ribeca P (2012) Fast computation and applications of genome mappability. PLoS One 7(1):e30377. doi:10.1371/journal.pone.0030377
12. Tian B, Yang J, Brasier AR (2012) Two-step cross-linking for analysis of protein-chromatin interactions. Methods Mol Biol 809:105–120. doi:10.1007/978-1-61779-376-9_7
13. Wal M, Pugh BF (2012) Genome-wide mapping of nucleosome positions in yeast using high-resolution MNase ChIP-Seq. Methods Enzymol 513:233–250. doi:10.1016/B978-0-12-391938-0.00010-0

Chapter 10

Stellaris® RNA Fluorescence In Situ Hybridization for the Simultaneous Detection of Immature and Mature Long Noncoding RNAs in Adherent Cells

Arturo V. Orjalo Jr. and Hans E. Johansson

Abstract

RNA fluorescence in situ hybridization (FISH), long an indispensable tool for the detection and localization of RNA, is becoming an increasingly important complement to other gene expression analysis methods. Especially important for long noncoding RNAs (lncRNAs), RNA FISH adds the ability to distinguish between primary and mature lncRNA transcripts and thus to segregate the site of synthesis from the site of action.

We detail a streamlined RNA FISH protocol for the simultaneous imaging of multiple primary and mature mRNA and lncRNA gene products and RNA variants in fixed mammalian cells. The technique makes use of fluorescently pre-labeled, short DNA oligonucleotides (circa 20 nucleotides in length), pooled into sets of up to 48 individual probes. The overall binding of multiple oligonucleotides to the same RNA target results in fluorescent signals that reveal clusters of RNAs or single RNA molecules as punctate spots without the need for enzymatic signal amplification. Visualization of these punctate signals, through the use of wide-field fluorescence microscopy, enables the counting of single transcripts down to one copy per cell. Additionally, by using probe sets with spectrally distinct fluorophores, multiplex analysis of gene-specific RNAs, or RNA variants, can be achieved. The presented examples illustrate how this method can add temporospatial information between the transcription event and both the location and the endurance of the mature lncRNA. We also briefly discuss post-processing of images and spot counting to demonstrate the capabilities of this method for the statistical analysis of RNA molecules per cell. This information can be utilized to determine both overall gene expression levels and cell-to-cell gene expression variation.

Key words Exon, Intron, Fluorescence, In situ hybridization, FISH, qPCR, Single-molecule detection, lncRNA, mRNA, Gene expression, Transcription burst, Nucleus

1 Introduction

The transcriptome in any given cell at any given time is the result of combinatorial transcription of select genes followed by posttranscriptional processing and modification [1]. The discovery of new coding and noncoding RNA variants continues to add to our understanding of the sophistication of the transcriptome. Long noncoding RNAs (lncRNAs; >200 nucleotides [nts]) are primarily

Yi Feng and Lin Zhang (eds.), *Long Non-Coding RNAs: Methods and Protocols*, Methods in Molecular Biology, vol. 1402, DOI 10.1007/978-1-4939-3378-5_10,

restricted to cellular nuclei where they play extensive roles in the central aspects of gene regulation, with wide-ranging effects on numerous cellular functions such as cell cycle progression, cellular differentiation, and metabolism [2–5]. High-throughput approaches to study gene expression, such as RNAseq and qPCR, have been instrumental for the discovery and independent measurements of expression levels of new RNAs [6]. However, such techniques require the extraction of RNA, with concomitant destruction of cellular integrity and loss of cell-specific information. To determine how the transcriptome correlates with the phenotype of individual cells and cell populations, gene expression analysis instead must be performed on intact cells within a population. Ideally, such analyses would facilitate the detection of multiple coding and noncoding RNAs alike, unraveling the intrinsic variation in gene expression and the gene expression networks to which they belong.

In situ hybridization (ISH) is one gene expression analysis method that has been employed to detect and determine the cellular distribution of both DNA and RNA in cells and tissue. ISH long remained a cumbersome process due to the need for long cDNA- or gDNA-derived probes, and signal amplification to detect cellular transcripts, whether it be through colorimetric, radioactive, or fluorometric imaging [7].

Major advances for RNA FISH made in the Singer and Tyagi groups, both using short synthetic and site-specifically labeled oligodeoxynucleotide probes (oligos), enabled the reliable detection of single molecules of RNA in fixed, cultured cells (*summarized in ref.* 8). Major advantages both with Singer's use of three to five multiply labeled 50-mers and with Tyagi's 48 singly labeled 20-mers that bind to the same mRNA target are that they can be designed to avoid repetitive RNA sequences and to hybridize with similar Tm, as well as that they are more efficient at penetrating the cell matrix to reach their target RNAs. Together this leads to a non-amplified, yet strong and specific fluorescent signal with a high signal-to-background ratio. Other advancements such as the generation of an online probe designer and automation in the manufacturing of these probe sets (oligo synthesis, pre-labeling with fluorophores, and purification) have greatly reduced the time and complexity of probe preparation [8, 9].

Thanks to the inherent rapid binding kinetics of these short fluorescently labeled oligos, hybridization protocols have also been much simplified and adapted for single-nucleotide variant detection (SNV FISH) [10], for more rapid detection through TurboFISH [11], and even for combination with immunofluorescence [8, 12]. We expand on aspects of one such application here, intron chromosomal expression (ice) FISH [13]. This method relies on that RNA processing mostly occurs co-transcriptionally in specialized transcription and processing factories [6, 7, 14–18], such that active transcription can be revealed by using probe sets targeting intron sequences only. We anticipate the continual refinement of ISH methods and novel advances in techniques designed to distinguish highly similar RNAs.

Here, we provide a guide for successful RNA FISH assay design for the co-detection of both nascent transcripts and mature RNAs. The human genes for the primarily cytoplasmic MYC mRNA and H19 lncRNA, as well as the nuclear PVT1 and XIST lncRNAs, are used as examples. We present a streamlined RNA FISH protocol where two or more probe sets are employed to generate images that reveal transcription bursts and final localization of mature RNAs, and even single-molecule resolution for mature MYC, H19, and PVT1 RNAs in adherent cells.

These methods are suitable to determine the cell-to-cell distribution of both mRNAs and lncRNAs in fixed cell cultures and tissue sections and, as such, may be applied to interrogate the subcellular (and potentially subnuclear) location of the target RNAs. By comparing signals from exon and intron probe sets it can be determined if the site of action of lncRNAs is associated with or away from its site of transcription (and processing). Overlapping exon and intron signals additionally can provide cross-validation of each probe set. The cell-specific readout that transcription burst analysis provides is also suitable for measurements of dose- and time-dependent gene activation, allele-specific epigenetic control [19, 20], as well as gene-specific karyotyping of (engineered) cell lines. Lastly, we provide a brief overview of RNA spot detection and quantification through the means of image acquisition and post-analysis processes.

2 Equipment and Materials

2.1 Equipment

1. Wide-field fluorescence microscope (Nikon Eclipse Ti and NIS-Elements Ar Imaging Software, or equivalent), with a high numerical aperture (>1.3), 60–100× objective, and XYZ motorized stage for automated *z*-stacking capabilities.
2. Cooled CCD camera (at least −20 °C), ideally optimized for low-light-level imaging rather than for speed (13 μm pixel size or less is preferred) (*see* **Note 1**).
3. Strong light source, such as a mercury or metal-halide lamp over single-line light sources such as LEDs and lasers.
4. Filter sets appropriate for FAM, Quasar® 570, CAL Fluor® Red 610, Quasar 670 fluorophores (e.g., *those from Chroma for* FAM, catalog # 89000; Cy3™, catalog # SP102v1; Cy3.5™, catalog # SP103v2; Cy5.5™, catalog # 49022).
5. Cell culture hood and variable temperature CO_2 incubator.
6. Laboratory oven set at 37 °C.

2.2 Reagents and Cell Culture

All buffers and reagents are made with nuclease-free water.

1. A549 (human lung adenocarcinoma cell line, ATCC, catalog # CRM-CCL-185).
2. SK-BR-3 (human breast adenocarcinoma cell line, ATCC, catalog # HTB-30).
3. F-12K medium (Kaighn's Modification of Ham's F-12 Medium) supplemented with 10 % fetal bovine serum and penicillin-streptomycin A (A549 media).
4. McCoy's 5A medium supplemented with 10 % fetal bovine serum and penicillin-streptomycin (SK-BR-3 media).
5. 12-Well tissue culture plates.
6. Micro Cover Glasses, Round, No. 1 (VWR, catalog # 48380-046).

2.3 Formaldehyde Fixation and Cell Permeabilization

1. Phosphate-buffered saline (1× PBS).
2. Fixation buffer (4 % formaldehyde in 1× PBS) (*see* **Note 2**).
3. Nuclease-free water (not DEPC treated).
4. Ethanol, molecular biology grade, diluted to 70 % in nuclease-free water.

2.4 Alternative Methanol:Acetic Acid Fixation/Permeabilization

1. Methanol:glacial acetic acid (MeOH:AcOH) fixative (3:1, v/v).

2.5 Hybridization

1. Hybridization buffer (Biosearch Technologies, Inc. catalog # SMF-HBD2-10). Deionized formamide (Life Technologies, catalog #AM9342). Formamide must be added to the hybridization buffer at a 10 % final concentration. The hybridization buffer can be aliquoted and stored at −20 °C (*see* **Note 2**).
2. Stellaris RNA FISH Probes (Biosearch Technologies, catalog numbers; H19 exons Quasar 570: VSMF-2162-5; MYC exons Quasar 570: VSMF-2230-5; PVT1 exons Quasar 670: VSMF-2307-5; XIST exons Quasar 570: VSMF-2430-5; all intron probes were custom designed, catalog numbers MYC FAM: SMF1025-5; H19, and PVT1 CAL Fluor Red 610: SMF-1082-5; and XIST Quasar 670: SMF-1065-5).
3. Tris–EDTA buffer solution (10 mM Tris–HCl, 1 mM disodium EDTA, pH 8.0).
4. Humidified chamber: 150 mm tissue culture plate, Parafilm®.

2.6 Washing

1. Wash buffer A (Biosearch Technologies, catalog number SMF-WAD1-60): Add formamide to a 10 % final concentration.

2. Wash buffer B (Biosearch Technologies, catalog number SMF-WBD1-20): Wash buffer B must be diluted to the appropriate concentration with water.
3. DAPI nuclear counterstain (5 ng/ml DAPI in wash buffer A).

2.7 Mounting

1. Vectashield® Mounting Medium (Vector Laboratories, catalog #H-1000).
   Alternative GLOX (GLucose OXidase) anti-fade [21].
   (a) GLOX buffer (0.4 % glucose in 2× SSC and 10 mM Tris–HCl, pH 8.0): 10 % glucose stock solution (powdered glucose [Sigma-Aldrich, catalog # 158968] dissolved in nuclease-free water), 1 M Tris–HCl, pH 8.0.
   (b) GLOX anti-fade: 100 μL GLOX buffer + 1 μL catalase from bovine liver (Sigma-Aldrich, catalog # C3155) + 1 μL glucose oxidase from *Aspergillus niger* (Sigma-Aldrich, catalog # G0543) diluted to 3.7 mg/mL in 50 mM sodium acetate, pH ~5.0.
2. Microscope slides.
3. Clear nail polish.

3 Methods

3.1 Design of Probes

The human RNA targets of focus are from the myelocytomatosis oncogene MYC, the neighboring plasmacytoma variant 1 oncogene PVT1, the oncofetal lncRNA gene H19, and XIST encoding the X inactive specific transcript (Fig. 1). These targets illustrate how alternative transcription initiation, splicing, and polyadenylation [22–24] contribute to the diversity of RNA variants. Note that the 3′ end of some lncRNAs is generated by RNase P cleavage, and that they lack a polyA tails [25].

An advantage of RNA FISH over immunofluorescence is that probe set specificity can be well controlled up front by using available genome and transcriptome information. Probe sets can be chosen to detect single, many, or all RNA variants from a certain gene, from several related genes, or even from several viral serotypes. Probe sets for variant detection are thus referred to as inclusive (many or all) or exclusive (single).

MYC, PVT1, and H19 belong to a triumvirate of aggressive oncogenes [26–30]. As a general effector of transcription and potentiator of cell proliferation and differentiation, MYC protein is stabilized by the PVT1 lncRNA, and as such, the encoding genes on 8q24 are frequently found co-amplified in tumors. In turn, MYC activates the transcription of both PVT1 and H19. Note also that both pre- and mature lncRNAs from PVT1 and H19 also harbor miRNAs. The XIST lncRNA is abundantly expressed in female cells and intimately tied to dosage compensation through inactivation of one of the X-chromosomes [31, 32].

1. Target sequence selection: In this initial pre-design process, RNA variants from each encoding gene are evaluated for differential transcription start site use, splicing, and polyadenylation. By using the sequence common to all variants as the target, the same number of individual oligos in the probe set can be predicted to bind each RNA variant. Such "inclusive" probe sets therefore reveal fluorescent RNA spots with maximal uniform spot intensity, independently of subcellular localization and half-lives, and in turn enable accurate spot counting. "Exclusive" probe sets can be generated by targeting discrete variant-specific exon sequences. When combined, differently labeled inclusive and exclusive probe sets can yield otherwise difficult-to-extract data on the actual number of different splice variants in each cell and their cell-to-cell distribution.

   The genes of interest (Fig. 1) are initially inspected at NCBI's website (www.ncbi.nlm.nih.gov/gene) (*see* **Note 3**). In the case of MYC (gene ID: 4609; 8q24.1), two RefSeq mRNAs are presented with three exons, whereas several shorter (incomplete) mRNAs are found for the same gene at Ensembl (ENSG00000136997). Two transcription start sites can be discerned, but only one true polyadenylation signal and site. Hence, the common CDS (opened by a noncanonical CTG) of the full-length mRNA is chosen for design (nts 526-1890 of

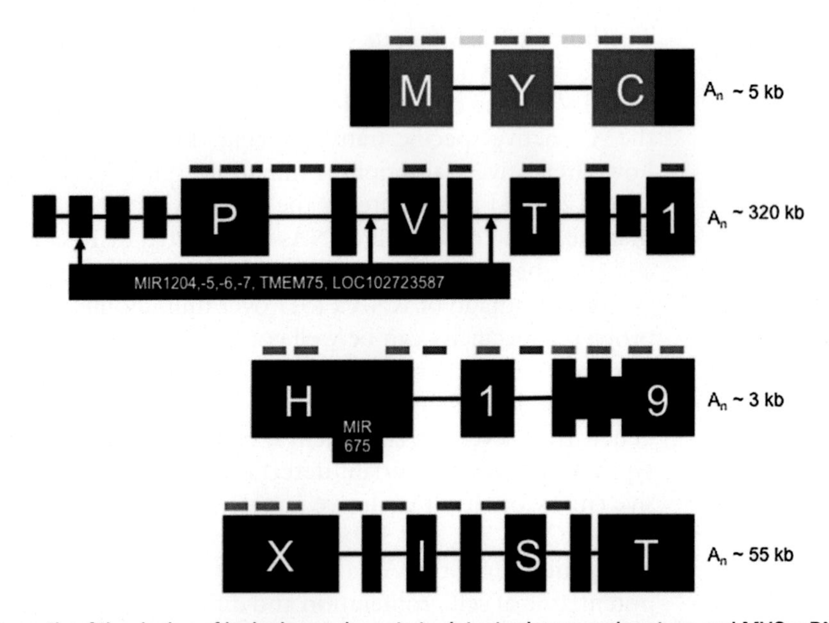

Fig. 1 A schematic of the design of inclusive probe sets to detect primary and mature and MYC mRNAs, as well as primary and mature H19, PVT1, and XIST lncRNAs. Noncoding sequences are represented in *black* and coding sequences in *blue*. Exons common to all variants are shown as *thick lines*, alternatively spliced exons as *medium thick lines*, and introns as *thin lines*. All four exon probe sets are inclusive and target the common regions of the mature RNA variants. The probe sets are color coded according to the dye label: MYC, H19, and XIST exons with Quasar 570: *orange*; PVT1 exons with CAL Fluor Red 610: *purple*; MYC introns with FAM: *green*; and H19, PVT1, and XIST introns with Quasar 670: *dark blue*

NM_002467.4). (XM_005250922.1 excludes a single CAG codon at the 5′ end of exon 2.) To further ensure specificity of this probe set to MYC mRNAs, the common RefSeq sequence is used to BLAST the human genome/transcriptome at NCBI (www.ncbi.nlm.nih.gov/blast/). This exercise reveals segments of NM_000389.4 with homology to the mRNAs from the MYCN and MYCL1 genes. In addition to information about potential cross-hybridization, the presence of potentially transcribed pseudogenes may also be uncovered by the BLAST process. The final target sequence for the generation of an inclusive exon probe set that will detect both NM_000389.4 and XM_005250922.1 mRNAs was maintained. Both introns (NG_007161.1, nts 5555-7178, 7951-9326) were chosen for the intron probe set.

The PVT1 locus spans 350 kb (gene ID: 5280; 8q24.1), and harbors in addition to the PVT1 lncRNA RefSeq another lncRNA and four miRNAs within the introns of the PVT1 gene. At Ensembl several PVT1 variants are presented at varying support levels. In the absence of a published comprehensive overview of PVT1 lncRNA variants and their expression, the RefSeq lncRNA (NR_003367.2, nts 1-1221, 1426-1699), excluding one frequently skipped exon, was chosen as a target for the exon probe set. At the same time, ≥15 unique exons leav19.7 pte room for the design of exclusive probe sets. The first 8 kb of the intron common to the major variants was chosen for the intron probe set (NC_000008.11 nts 127,890,999-127,898,944). (8 kb is the maximal target sequence length accepted by the probe designer.)

For H19 (gene ID: 283120; 11p15.5), one RefSeq lncRNA and two miRNAs are presented over five major exons. Alternatively spliced transcripts at varying support levels are shown at Ensembl. The gene and exon structure is well conserved in mouse, supporting the choice of the full-length lncRNA (NR_002196.1 nts 1-2322), excluding the miRNA sequences, as a target. For design of the intron probe set, 10 nts from each flanking exon were added to the target introns (NG_016165.1, nts 6307-6426, 6538-6656, 6746-6849, 6949-7053).

For XIST (gene ID: 7503; Xq13.2), one six-exon RefSeq lncRNA is presented. Alternative transcripts at varying support levels are shown at Ensembl. The full-length RNA is the major variant [33], and so the first 8 kb of mature lncRNA NR_001564.2 and up to 2 kb of introns 1–5 (NG_016172.1 nts 16373-18374, 20315-22316, 24517-26356, 26385-28286) were chosen as targets.

As a final step, differentially spliced exons in target sequences for inclusive exon and intron probe sets are each replaced by a single "n" before entry into the designer. This last step prevents the design of oligos across the splice site, which would otherwise only bind certain or nonexistent variants (*see* **Note 4**).

2. Probe set design: The Stellaris probe set name, gene name, and selected target sequence (with or without FASTA header) are entered into the designated fields in the Stellaris Probe Designer (http://www.biosearchtech.com/stellarisdesigner/). Choosing the organism (in this case human) allows the designer to utilize genome-specific information to mask against repetitive sequences, such as Alu elements. To start, a masking level of 5 (the highest) is chosen. The maximum number of oligos chosen for an exon probe sets is 48, and for intron sets 32. The output for all chosen targets exceeded the recommended 25 minimum number of oligos, except for the H19 introns (12 oligos) (*see* **Note 4**).

3.2 Cell Culture

The procedure for the culture [33] and fixation detailed here is for the male human lung adenocarcinoma cells (A549) and the mammary gland adenocarcinoma SK-BR-3 cell lines in 12-well cell culture plates, but can be adapted to other cell lines with minor modifications. A549 is haplotriploid [34] with the relevant karyotype: 8,8,8; 11,11,11; A10 [t(11;8) (q13;q8.24)]. SK-BR-3 is hypertriploid [35] with three X chromosomes.

Volumes should be adjusted accordingly when adapting this protocol for use in cell culture dishes or multi-well plates of a different size.

1. Seed cells on sterile, 18 mm round #1 cover glass in a 12-well plate. Plating density should range from 30,000 to 50,000 for A549 cells and 70,000 to 90,000 for SK-BR-3 cells per well.
2. Incubate cells at 37 °C and 5 % CO_2.
3. Allow cells to grow to approximately 80 % confluency prior to fixation, usually 2–3 days.

3.3 Formaldehyde Fixation and Cell Permeabilization

1. Aspirate the growth medium from each well, and wash with 1 mL of 1× PBS.
2. Aspirate the 1× PBS, and add 1 mL of fixation buffer. Allow the cells to incubate in the fixation buffer for 10 min at room temperature. Aspirate the fixation buffer, and wash twice with 1 mL of 1× PBS, aspirating between washes.
3. To permeabilize the cells, immerse the cells in 1 mL of 70 % ethanol for at least 1 h at 4 °C.
4. The cells can be stored submerged in 70 % ethanol at 4 °C for up to a week prior to hybridization.

3.4 Methanol–Acetic Acid Fixation/Permeabilization

This alternative method of fixation and permeabilization provides faster results, but may alter the overall cellular morphology and may not be compatible with simultaneous immunofluorescence [8, 10–12, 36].

1. Aspirate the growth medium from each well, and wash with 1 mL of 1× PBS.
2. Aspirate the 1× PBS, and add 1 mL of MeOH:AcOH solution. Allow the cells to incubate in the fixative solution for 10 min at room temperature.
3. The cells can be stored submerged in MeOH:AcOH at 4 °C and should be used within 48 h.

3.5 Hybridization

Reconstitute the Stellaris RNA FISH Probes in Tris–EDTA buffer solution to create a probe stock of 12.5 μM. To ensure that the probes are completely resuspended, thoroughly pipette up and down, then vortex, and centrifuge briefly. Make sure that the hybridization buffer and wash buffers are properly diluted and supplemented with formamide. To prepare the hybridization solution, add 0.5 μL of probe stock solution to 50 μL of hybridization buffer, then vortex, and centrifuge. This creates a working probe solution of 125 nM.

1. Aspirate the 70 % ethanol from the cover glass containing adherent cells within the 12-well plate. Add 1 mL of wash buffer A, and allow the cells to incubate at room temperature for 2–5 min.
   (a) For cells fixed and permeabilized with MeOH:AcOH, aspirate the MeOH:AcOH from the cover glass containing adherent cells within the 12-well plate. Add 1 mL of wash buffer A, and allow the cells to incubate at room temperature for 2–5 min.
2. Create a humidified chamber using a 150 mm tissue culture plate. Evenly line the bottom of the tissue culture plate with a flat, water-saturated paper towel. Place a 10 × 10 cm piece of Parafilm on top of the water-saturated paper towel. This chamber will help prevent evaporation of the probe solution from under the cover glass.
3. Within the humidified chamber, dispense 50 μL of the hybridization solution (containing probe) onto the Parafilm. Use forceps to gently transfer the cover glass, cell side down, onto the 50 μL droplet of hybridization solution. It is important that both the paper towel and Parafilm are completely level so that the hybridization solution will disperse evenly under the cover glass. Avoid the formation of bubbles.
4. Cover the humidified chamber with the tissue culture lid, and seal it with Parafilm.
5. Place the humidified chamber in a dark 37 °C oven for at least 4 h. The incubation can be continued overnight up to 16 h, thus allowing for an entire day of imaging, if necessary, on the next day (*see* **Note 5**).

3.6 Washing

1. Add 1 mL of wash buffer A to a fresh 12-well plate. Remove the humidified chamber from 37 °C, and gently transfer the cover glass (cells side up) to the 12-well plate containing wash buffer. Allow the cells to incubate in the dark at 37 °C for 30 min.
2. Aspirate the wash buffer A, and then add 1 mL of DAPI nuclear counterstain. Allow the cells to incubate in the dark at 37 °C for 30 min.
3. Aspirate the DAPI counterstain solution, and then add 1 mL of wash buffer B. Allow the cells to incubate at room temperature for 2–5 min.

3.7 Mounting

1. Place a small drop (approximately 15 μL) of Vectashield Mounting Medium onto a microscope slide. Gently lift the cover glass, and briefly touch the bottom edge to a Kimwipe® (or equivalent tissue) to remove excess buffer. Then place the cover glass, with the cells facing down, onto the drop of mounting medium. If necessary, GLOX anti-fade may be used as an alternative (*see* **Note 6**).
2. Gently wick away excess mounting media from the perimeter of the cover glass. Seal the cover glass perimeter with a thin coat of clear nail polish and allow it to dry. As needed, gently wash away any dried salt off the cover glass with water.
3. For best results, image the samples on the same day.

3.8 Imaging

1. Use a wide-field fluorescence microscope with a 60 or 100× oil objective to obtain single-molecule resolution. Acquire z-sections with 0.3 μm spacing that span the entire thickness of the cell. This ensures that each individual RNA spot is captured (*see* **Note 7**).
2. The exposure times can range from 1 to 2 s (*see* **Notes 8** and **9**).

3.9 Image Processing

1. ImageJ (http://rsbweb.nih.gov/ij/) is a widely accessible and useful collection of software for post-processing image analysis. Each three-dimensional stack is exported out of the Nikon software and imported into ImageJ. The three-dimensional stack is then merged into a two-dimensional image using the Maximum Intensity Projection feature. Here, images can be overlaid with DAPI images and/or another RNA target image from the same field of view (Figs. 2, 3, and 4). There are a variety of free or proprietary software to help facilitate quantification of single-molecule RNA FISH. We use the software developed by the Arjun Raj lab at University of Pennsylvania rajlab.seas.upenn.edu/StarSearch/launch.html [10, 13] (*see* **Note 10**). Alternate software has been discussed elsewhere [8].

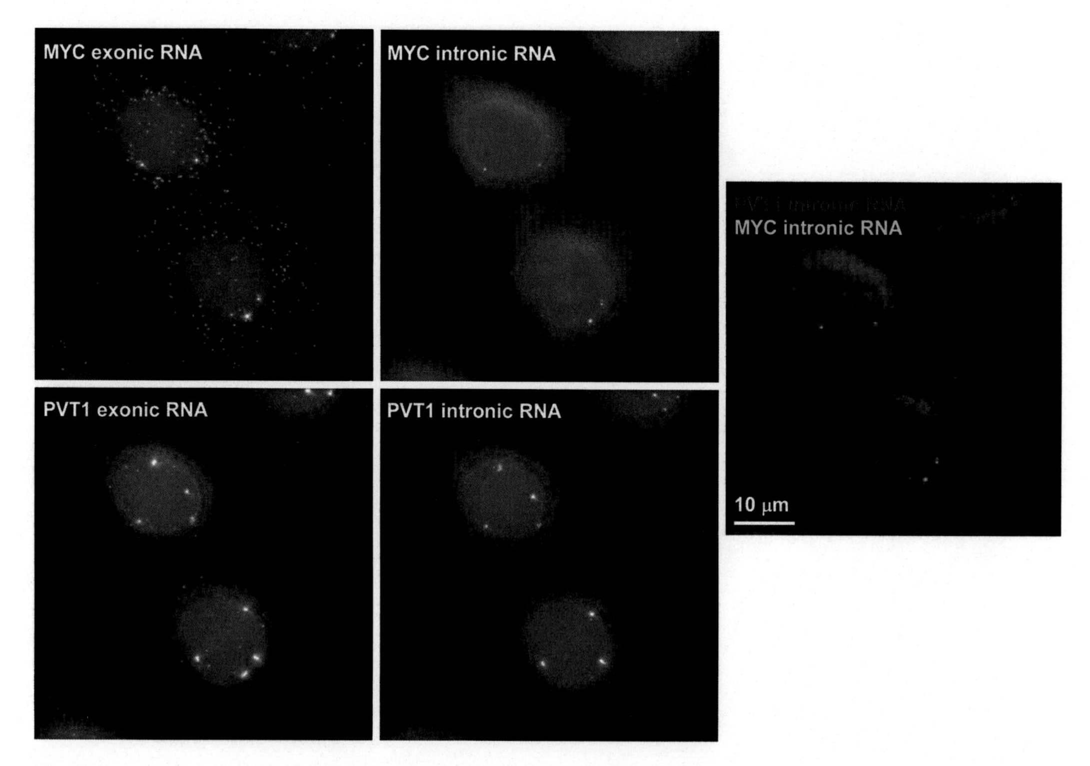

Fig. 2 Quadruplex Stellaris RNA FISH detection of MYC and PVT1 RNAs. Fixed A549 cells with DAPI-stained nuclei (*blue*) were probed simultaneously for mature MYC mRNAs, MYC introns, mature PVT1 lncRNAs, and PVT1 introns, and imaged with a 60×/1.4 NA oil objective. The MYC exon probe set revealed ~100 cytoplasmic mRNA spots. The exon and intron MYC probe sets both showed two bright and two weak (longer exposures) co-localized nuclear foci indicating variable activity at the MYC loci. The PVT1 exon set revealed predominantly nuclear single-molecule spots as well as four bright nuclear foci, and the PVT1 intron set revealed three to four bright nuclear foci. When superimposed (*right panel*), the MYC (*pseudo-colored green*) and PVT1 (*pseudo-colored red*) intron signals are adjacent and overlapping, consistent with the active expression of two neighboring genes. The difference in signal between MYC and PVT1 indicates independent control. The two types of signal by the PVT1 exon set also indicates that this lncRNA may have a clustered *cis*-function (or delayed miRNA processing), in addition to a nuclear single-molecule *trans*-function. Addition of probe sets targeting unique exons could add detail to the function of subsets of PVT1 lncRNAs

4 Notes

1. In general, spots observed in a wide-field fluorescence microscope are too dim to view through the eyepiece. This emphasizes the need for a cooled CCD camera, which greatly minimizes background noise. Furthermore, we discourage using a confocal microscope as the primary means of imaging; results tend to be inconsistent.
2. Formaldehyde and formamide are teratogens that are easily absorbed through the skin and should be used in a chemical fume hood. Be sure to warm the formamide to room temperature before opening the bottle.

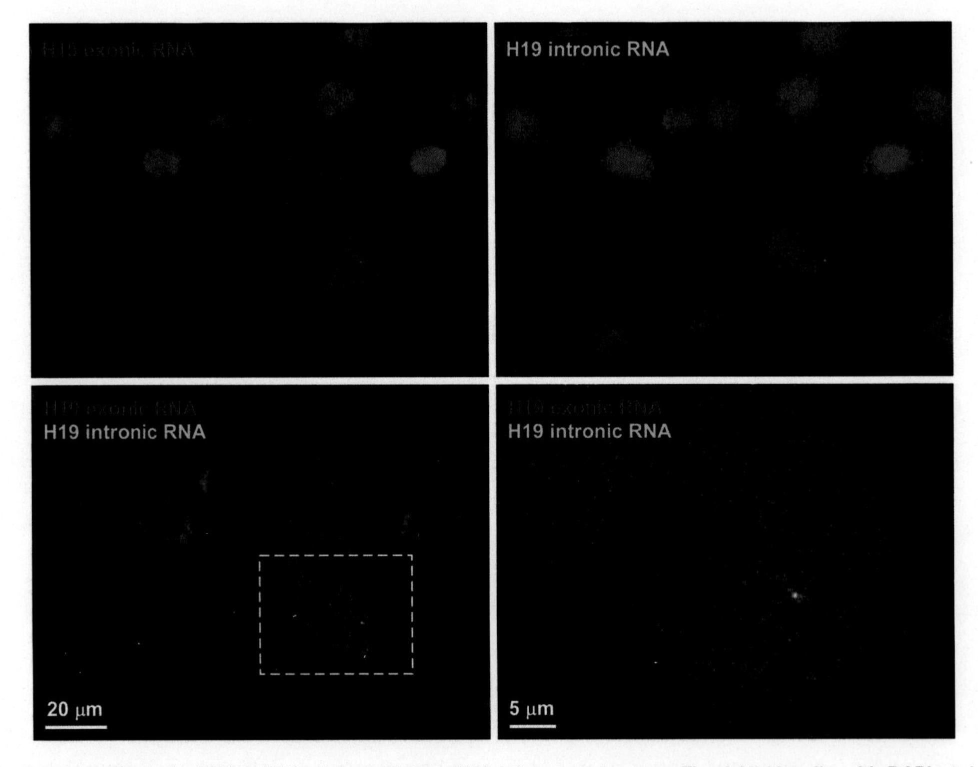

Fig. 3 Duplex Stellaris RNA FISH to detect H19 lncRNA introns and exons. Fixed A549 cells with DAPI-stained nuclei (*blue*) were probed simultaneously for mature H19 lncRNA (*upper left panel*; *pseudo-colored red*) and introns (*upper right panel*; *pseudo-colored green*), and were imaged with a 60×/1.4 NA oil objective. Single nuclear transcription bursts are revealed by both H19 probe sets (*lower panels*; *yellow*) as expected for the paternally imprinted gene [19]. The cytoplasmic punctate spots revealed by the exon probe set are consistent with the reported localization of the mature H19 lncRNA [37]

3. Similar sequence information can be obtained from the UCSC genome browser (www.genome.ucsc.edu) and Ensembl (www.ensembl.org). Benefits with the Ensembl annotation are modifiers for the quality confidence and completeness of annotated transcripts. Variant specific lncRNA information is also available elsewhere (www.lncipedia.org, lncrnadb.org).
4. To prevent the faulty design of probes, any nucleotide redundancy must be represented as "n" in the sequence. Other IUPAC letters are not allowed in the designer. It is also recommended that output oligo sequences from lower masking levels (3 and below) are BLASTed against the human transcriptome at NCBI. By furnishing the oligos with individual FASTA headers (>1, >2, etc.), the batch BLAST option can be utilized. Both the oligos that produce hits with <4 mismatches and the hit transcripts are tabulated. Transcript hits that are represented more than once are identified and the oligos responsible for the hits should be excluded from the final

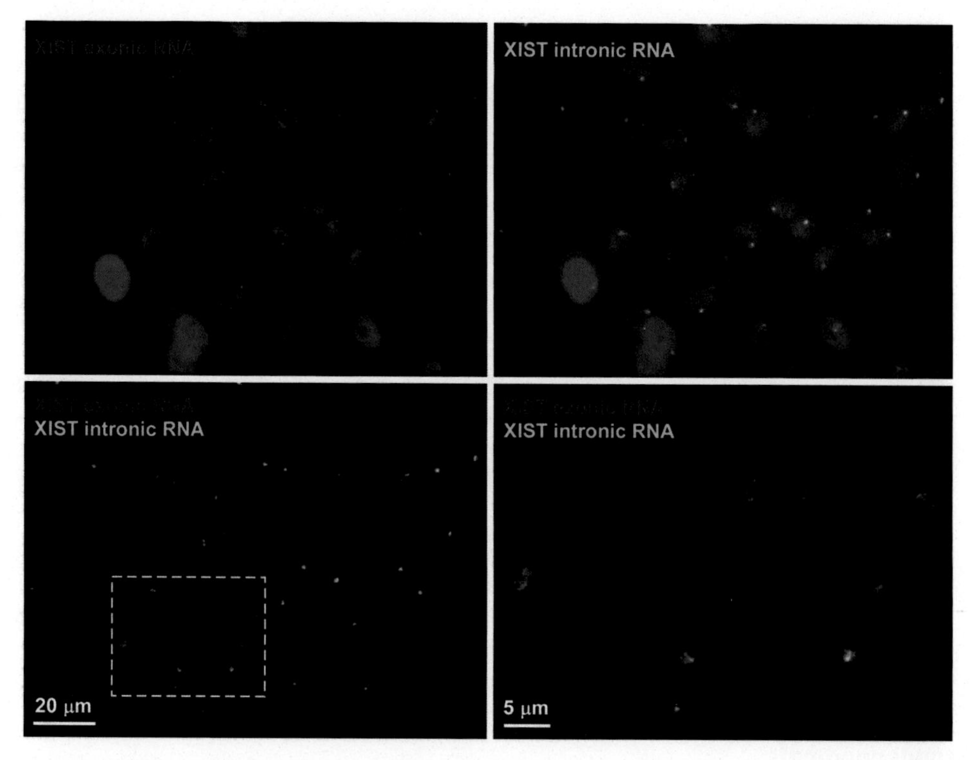

Fig. 4 Duplex Stellaris RNA FISH to detect XIST lncRNA introns and exons. Fixed SK-BR-3 cells with DAPI-stained nuclei (*blue*) were probed simultaneously for mature XIST lncRNA (*upper left panel*; *pseudo-colored red*) and introns (*upper right panel*; *pseudo-colored green*), and cells were imaged with a 60×/1.4 NA oil objective. The partially overlapping location (*lower two panels*) of the nuclear foci indicates that XIST gene transcription and X-chromosome inactivation in SK-BR-3 cells occur in *cis* on two of the three X-chromosomes

probe set to prevent off-target binding (Fig. 1). For the eight probe sets herein, all designed oligos were retained. Probe sets with <25 oligos generally are only recommended for detection of RNAs found in clusters, such as in transcription bursts or in Cajal bodies, and then in cells with minimal autofluorescence.

5. In general, hybridization times of 4–16 h are sufficient for quality results with Stellaris FISH Probe sets. However, for probe sets that initially demonstrate a low signal-to-background ratio, or that contain fewer that the recommended 25 oligos, both the optimal hybridization time and probe concentration must be determined experimentally.
6. The GLOX anti-fade is incompatible with the fluorescein (FAM) dye.
7. Our microscope is equipped with a motorized stage that enables the automatic capture of a series of z-sections at 0.3 μm. For each field of view, each different RNA target can be imaged sequentially. The ability to multiplex is dependent on the availability of filter sets with minimal spectral overlap.

8. In general, probe sets labeled with higher wavelength emitting fluorophores (e.g., Quasar 670) allow for the observation of reduced autofluorescence. However, these fluorophores generally require longer exposure times, ranging from 2 to 5 s, due to less efficient excitation light and emission detection.
9. Imaging via an unused filter at lower wavelengths (e.g., FAM/FITC) allows for the visualization of inherent cellular autofluorescence. This can greatly aid in the discrimination of true spots of single-RNA molecules versus false spots of autofluorescence.
10. The RNA quantification program processes a single TIFF *z*-stack and applies a rough linear Laplacian filter to assess the spatial derivation of the image with regard to intensity. A Gaussian smoothing filter helps subtract background and enhance spots of appropriate size and intensity. The convolved filter generates a graphical output that displays the relative signal-to-background ratio of every spot detected. The threshold automatically defined can also be manually adjusted, if necessary (i.e., low signal to background).

Acknowledgements

We gratefully acknowledge the continual support of Dr. Ron Cook and members of the Stellaris team at LGC Biosearch Technologies.

For research use only. Not for use in diagnostic procedures. Stellaris® is a trademark of LGC Biosearch Technologies. Products and technologies appearing in this application note may have trademark or patent restrictions associated with them. Please see http://www.biosearchtech.com/legal for a full legal disclosure.

References

1. ENCODE Project Consortium (2012) An integrated encyclopedia of DNA elements in the human genome. Nature 489:57–74. doi:10.1038/nature11247
2. Rinn J, Guttman M (2014) RNA function. RNA and dynamic nuclear organization. Science 345:1240–1241. doi:10.1126/science.1252966
3. Maamar H, Cabili MN, Rinn J et al (2013) linc-HOXA1 is a noncoding RNA that represses Hoxa1 transcription in cis. Genes Dev 27:1260–1271. doi:10.1101/gad.217018.113
4. Rinn JL, Chang HY (2012) Genome regulation by long noncoding RNAs. Annu Rev Biochem 81:145–166. doi:10.1146/annurev-biochem-051410-092902
5. Clark MB, Choudhary A, Smith MA et al (2013) The dark matter rises: the expanding world of regulatory RNAs. Essays Biochem 54:1–16. doi:10.1042/bse0540001
6. Tilgner H, Knowles DG, Johnson R et al (2012) Deep sequencing of subcellular RNA fractions shows splicing to be predominantly co-transcriptional in the human genome but inefficient for lncRNAs. Genome Res 22:1616–1625. doi:10.1101/gr.134445.111
7. Brown JM, Buckle VJ (2010) Detection of nascent RNA transcripts by fluorescence in situ hybridization. Methods Mol Biol 659:33–50. doi:10.1007/978-1-60761-789-1_3

8. Coassin SR, Orjalo AV Jr, Semaan SJ et al (2014) Simultaneous detection of nuclear and cytoplasmic RNA variants utilizing Stellaris® RNA fluorescence in situ hybridization in adherent cells. Methods Mol Biol 1211:189–199. doi:10.1007/978-1-4939-1459-3_15
9. Orjalo AV Jr, Johansson HE, Ruth JR (2011) Stellaris™ fluorescence *in situ* hybridization (FISH) probes: a powerful tool for mRNA detection. Nat Methods 8:I–III. doi:10.1038/nmeth.f.349
10. Levesque MJ, Ginart P, Wei Y et al (2013) Visualizing SNVs to quantify allele-specific expression in single cells. Nat Methods 10:865–867. doi:10.1038/nmeth.2589
11. Shaffer SM, Wu MT, Levesque MJ et al (2013) Turbo FISH: a method for rapid single molecule RNA FISH. PLoS One 8:e75120. doi:10.1371/journal.pone.0075120
12. Tripathi V, Fei J, Ha T et al (2015) RNA fluorescence in situ hybridization in cultured mammalian cells. Methods Mol Biol 1206:123–136. doi:10.1007/978-1-4939-1369-5_11
13. Levesque MJ, Raj A (2013) Single-chromosome transcriptional profiling reveals chromosomal gene expression regulation. Nat Methods 10:246–248. doi:10.1038/nmeth.2372
14. Gribnau J, de Boer E, Trimborn T et al (1998) Chromatin interaction mechanism of transcriptional control *in vivo*. EMBO J 17:6020–6027. doi:10.1093/emboj/17.20.6020
15. Fanucchi S, Shibayama Y, Burd S et al (2013) Chromosomal contact permits transcription between coregulated genes. Cell 155:606–620. doi:10.1016/j.cell.2013.09.051
16. Hacisuleyman E, Goff LA, Trapnell C et al (2014) Topological organization of multichromosomal regions by the long intergenic noncoding RNA Firre. Nat Struct Mol Biol 21:198–206. doi:10.1038/nsmb.2764
17. Papantonis A, Cook PR (2013) Transcription factories: genome organization and gene regulation. Chem Rev 113:8683–8705. doi:10.1021/cr300513p
18. Senecal A, Munsky B, Proux F et al (2014) Transcription factors modulate c-Fos transcription bursts. Cell Rep 8:1–9. doi:10.1016/j.celrep.2014.05.053
19. Rachmilewitz J, Goshen R, Ariel I et al (1991) Parental imprinting of the human H19 gene. FEBS Lett 309:25–28. doi:10.1016/0014-5793(92)80731-U
20. Ohno M, Aoki N, Sasaki H (2001) Allele-specific detection of nascent transcripts by fluorescence in situ hybridization reveals temporal and culture induced changes in Igf2 imprinting during pre-implantation mouse development. Genes Cells 6:249–259. doi:10.1046/j.1365-2443.2001.00417.x
21. Yiddish A, Forkey JN, McKinney SA et al (2003) Myosin V walks hand-over-hand: Single fluorophore imaging with 1.5-nm localization. Science 300:2061–2065. doi:10.1126/science.1084398
22. Derti A, Garrett-Engele P, Macisaac KD et al (2012) A quantitative atlas of polyadenylation in five mammals. Genome Res 22:1173–1183. doi:10.1101/gr.132563.111
23. Hoque M, Ji Z, Zheng D et al (2013) Analysis of alternative cleavage and polyadenylation by 3′ region extraction and deep sequencing. Nat Methods 10:133–139. doi:10.1038/nmeth.2288
24. Zhang S, Han J, Zhong D et al (2014) Genome-wide identification and predictive modeling of lincRNAs polyadenylation in cancer genome. Comput Biol Chem 52:1–8. doi:10.1016/j.compbiolchem.2014.07.001
25. Wiles JE, Freer SM, Spector DL (2008) 3′ End processing of a long nuclear-retained noncoding RNA yields a tRNA-like cytoplasmic RNA. Cell 135:919–932. doi:10.1016/j.cell.2008.10.012
26. Pachnis V, Belayew A, Tilghman SM (1984) Locus unlinked to alpha-fetoprotein under the control of the murine raf and Rif genes. Proc Natl Acad Sci USA 81:5523–5527, www.pnas.org/content/81/17/5523
27. Barsyte-Lovejoy D, Lau SK, Boutros PC et al (2006) The c-Myc oncogene directly induces the H19 noncoding RNA by allele-specific binding to potentiate tumorigenesis. Cancer Res 66:5330–5337. doi:10.1158/0008-5472.CAN-06-0037
28. Huppi K, Pitt JJ, Wahlberg BM et al (2012) The 8q24 gene desert: an oasis of non-coding transcriptional activity. Front Genet 3:69. doi:10.3389/fgene.2012.00069
29. Johnsson P, Morris KV (2014) Expanding the functional role of long noncoding RNAs. Cell Res 24:1284–1285. doi:10.1038/cr.2014.104
30. Tseng YY, Moriarity BS, Gong W et al (2014) PVT1 dependence in cancer with MYC copy-number increase. Nature 512:82–86. doi:10.1038/nature13311
31. Clemson CM, McNeil JA, Willard HF et al (1996) XIST RNA paints the inactive X chromosome at interphase: evidence for a novel RNA involved in nuclear/chromosome structure. J Cell Biol 132:259–275. doi:10.1083/jcb.132.3.259
32. Byron M, Hall LL, Lawrence JB (2013) A multifaceted FISH approach to study endogenous RNAs and DNAs in native nuclear and cell structures Curr Prot Hum Genet 4.15.1–4.15.21. doi: 10.1002/0471142905.hg0415s76
33. Davis JM (ed) (2002) Basic cell culture. Oxford University Press, New York

34. Peng KJ, Wang JH, Su WT et al (2010) Characterization of two human lung adenocarcinoma cell lines by reciprocal chromosome painting. Zool Res 31:113–121. doi:10.3724/SP.J.1141.2010.02113
35. Fogh J, Fogh JM, Orfeo T (1977) One hundred and twenty-seven cultured human tumor cell lines producing tumors in nude mice. J Natl Cancer Inst 59:221–226. doi:10.1093/jnci/59.1.221
36. Yan F, Wu X, Crawford M et al (2010) The search for an optimal DNA, RNA, and protein detection by in situ hybridization, immunohistochemistry, and solution-based methods. Methods 52:281–286. doi:10.1016/j.ymeth.2010.09.005
37. Brannan CI, Dees EC, Ingram RS, Tilghman SM (1990) The product of the H19 gene may function as an RNA. Mol Cell Biol 10:28–36. doi:10.1128/MCB.10.1.28

Chapter 11

Simultaneous RNA–DNA FISH

Lan-Tian Lai, Zhenyu Meng, Fangwei Shao, and Li-Feng Zhang

Abstract

A highly useful tool for studying lncRNAs is simultaneous RNA–DNA FISH, which reveals the localization and quantitative information of RNA and DNA in cellular contexts. However, a simple combination of RNA FISH and DNA FISH often generates disappointing results because the fragile RNA signals are often damaged by the harsh conditions used in DNA FISH for denaturing the DNA. Here, we describe a robust and simple RNA–DNA FISH protocol, in which amino-labeled nucleic acid probes are used for RNA FISH. The method is suitable to detect single-RNA molecules simultaneously with DNA.

Key words Fluorescence in situ hybridization (FISH), Single-molecule RNA FISH, Amino-labeled probes, Simultaneous RNA–DNA FISH, lncRNA

1 Introduction

Long noncoding RNAs (lncRNAs) are nonprotein-coding transcripts that are longer than 200 nucleotides. A long growing list of lncRNAs is identified in the current research [1]. It is speculated that many lncRNAs are components of nuclear architecture and players of epigenetic regulation. Therefore, it is important to study how lncRNAs interact with their DNA targets in the cellular context. Simultaneous RNA–DNA fluorescence in situ hybridization (RNA–DNA FISH) is a highly useful technique to address this question.

FISH not only reveals the subcellular localization of nucleic acids, but also provides quantitative information. For example, single-molecule RNA FISH (SMRF) can detect single-mRNA molecules in a single cell [2]. Furthermore, the DNA FISH signal intensity of telomeres can be used to measure individual telomere length in a given cell [3]. However, simultaneous RNA–DNA FISH was technically challenging. In a simple combination of RNA FISH and DNA FISH, the RNA signals are

Yi Feng and Lin Zhang (eds.), *Long Non-Coding RNAs: Methods and Protocols*, Methods in Molecular Biology, vol. 1402,
DOI 10.1007/978-1-4939-3378-5_11, © Springer Science+Business Media New York 2016

often damaged in DNA FISH by the harsh conditions (high temperature and low pH) used for denaturing the genomic DNA [4, 5].

We developed a robust method to protect the RNA signals in RNA–DNA FISH by introducing multiple protein components (immunostains) into the signal detection steps of RNA FISH [4]. This, followed by a formaldehyde fixation step, provides robust protection on the RNA signals during the subsequent DNA FISH. Introducing immunostains into RNA FISH is critical because formaldehyde acts as a fixative in the cellular context mainly by cross-linking the amino groups from the lysine residues of the protein matrix. Although introducing multiple layers of immunostains into RNA FISH provides robust protection on the RNA signal, it makes the protocol lengthy and tedious. It also unnecessarily amplifies the RNA signals and raises the background noise. Therefore, we further improved the method by adding amino groups directly onto the nucleic acid probes used in RNA FISH [6]. The improved method is simple and robust. Using amino-labeled RNA FISH probes, we successfully detected single-mRNA molecules simultaneously with DNA. Here, we describe the protocol in details.

2 Materials

2.1 Equipment

1. Cytospin centrifuge.
2. Reverse-phase HPLC equipped with C18 (10 × 250 mm).
3. Automated DNA synthesizer (Mermade 4, BioAutomation).

2.2 Commercial Reagents

1. Molecular trap pack.
2. Phosphoramidites and reagents used in oligo nucleotide synthesis: dT-CE Phosphoramidite, dmf-dG-CE Phosphoramidite, dA-CE Phosphoramidite, Ac-dC-CE phosphoramidite, Activator, Cap mix A, Cap mix B, Oxidizing solution, 3 % DCA/DCM Deblocking Mix.
3. 200 nmole size of CPG resin in Mermade column.
4. Nick translation kit.
5. Cy3-dUTP.
6. Aminoallyl-dUTP (Jena Bioscience, Cat# NU-803S).
7. Cytology funnels.
8. Superfrost/Plus microscopic slides.
9. 18 × 18 mm #1.5 cover slips.
10. DAPI-containing mounting media with antifade.

11. Transparent nail polish.
12. Rubber cement.
13. Coplin jars.

2.3 Buffers and Solutions

All solutions for RNA FISH are prepared with RNase-free water.

1. Prepare phosphoramidite solutions including the four natural nucleotides, amino-dT ($^{NH}{}_2T$), terminal amino linker (NH_2-C6), and fluorescein phosphoramidites (FAM). To each phosphoramidite, add 1.492 mL of anhydrous acetonitrile to directly prepare 0.067 M of solution.
2. Deprotection solution I: 40 % w/w Methylamine/water solution. 5.7 mL of methylamine is dissolved in 6 mL of water.
3. Deprotection solution II: 28 % Ammonium hydroxide solution.
4. HPLC buffer: 0.1 M TEAA buffer, pH 7.0. Dissolve 5.72 mL of acetic acid in 500 mL water. Add 10.12 g of triethylamine to the acetic acid solution. Adjust pH to 7.0 by adding either triethylamine or acetic acid. Make up the solution to 1 L and filter the solution by a 0.45 μm filter unit.
5. Cy5 coupling solution: Dissolve 1 mg of Cy5 NHS ester in 20 μL anhydrous DMSO. Dilute Cy5/DMSO solution by 2 mL of acetonitrile. 400 μL of *N,N*-Diisopropylethylamine is added right before the solution is applied to DNA resin.
6. Hybridization buffer: 50 % formamide, 2× SSC, 2 mg/mL BSA, 10 % dextran sulfate-500 K.
7. CSK buffer: 100 mM NaCl, 300 mM Sucrose, 3 mM $MgCl_2$, 10 mM PIPES (pH 6.8). The solution is autoclaved and stored at 4 °C.
8. 4 % (w/v) Paraformaldehyde (PFA) in 1× PBS: Dissolve paraformaldehyde in 4 mM NaOH, add in 10× PBS, adjust pH to 7.4, and bring up the solution to the final volume with water. The solution should be stored in dark at 4 °C for no more than 3 weeks.
9. 2 % PFA: Dilute from 4 % PFA using PBS.
10. 70, 80, 90, and 100 % ethanol (v/v in water).
11. 20× SSC.
12. 10 % Formamide, 2× SSC with 4 mM Vanadyl ribonucleoside.
13. PBS with 0.5 % Triton X-100.
14. PBS with 0.2 % Tween-20.
15. 70 % Formamide, 2× SSC.
16. 50 % Formamide, 2× SSC.

3 Methods

3.1 Probe Preparation

3.1.1 Synthesis of a Single Oligo Nucleotide Probe Dually Labeled with Fluorescein and Amino Group

A single fluorescently labeled oligo probe can be used in RNA FISH to detect RNA transcripts with repetitive sequences. In this section, to illustrate the probe synthesis process, we describe the procedure of synthesizing a fluorescein and amino dually labeled oligo probe with the sequence 5′-FITC-(^{NH_2}TAACCC)$_6$-3′. This probe was used to detect the telomeric repeat-containing RNA (*Terra*) [6].

1. Install ^{NH_2}T phosphoramidite solution on DNA synthesizer as the first unnatural nucleobase (*see* **Note 1**). Input the sequence of *Terra* probe to the synthesizer software with ^{NH_2}T as 1. Start a 200 nmole scale of DNA synthesis via standard solid-phase phosphoramidite chemistry on DNA synthesizer (*see* **Notes 2** and **3**). Leave the probe on CPG resin for FAM labeling in the next step.
2. Install FAM phosphoramidite solution on DNA synthesizer as the second unnatural nucleobase. Input DNA sequence as 5′-2 T-3′ to the synthesizer software. Start the synthesis and follow the standard synthetic procedure on DNA synthesizer (*see* **Note 4**). T is only used to represent the oligonucleotides already on GPG resin. In this step, only FAM is added to DNA probes as one nucleobase.
3. To cleave and deprotect DNA probe from CPG resin, load 250 μL of deprotection solution I onto the resin twice. Leave the solution on the resin for 10 min each time. Combine the two solutions and stand the solution for 2 h at room temperature. Simultaneously, protection groups on all the nucleobases are removed.
4. Dry solution from **step 3** by vacuum concentrator to yield a slightly yellow pellet.
5. Dissolve the probe crudes from **step 4** in 400 μL water to prepare samples for HPLC purification.
6. Inject 200 μL solution from **step 5** to reverse-phase HPLC. Elute the desired probe by acetonitrile over TEAA buffer (5–35 % over 30 min). The peak with absorption at both 260 and 490 nm is collected.
7. Repeat **step 6** to purify the rest half of probe crudes.
8. Combine the two collections and remove the solvent by lyophilization. The probe is dissolved in hybridization buffer and stored at −20 °C. The working concentration of the probe is 1 μM.
9. For quality control, the dually labeled probe is characterized by ESI-MS and is quantified based upon UV–Vis absorption of either DNA (260 nm) or fluorescein dye (490 nm).

Table 1
Sequences of amino-labeled probe set for single-molecule RNA FISH of EGFP mRNA

Probe	Sequence[a]	Probe	Sequence[a]
1	G*T*CCAGC*T*CGACCAGGA*T*GG	11	CGA*T*GCCC*T*TCAGC*T*CGATG
2	C*T*GAACT*T*GTGGCCG*T*TTAC	12	ATGT*T*GCCG*T*CCTCC*T*TGAA
3	AGC*T*TGCCG*T*AGGTGGCA*T*C	13	GTAGT*T*GTAC*T*CCAGC*T*TGT
4	GTGG*T*GCAGA*T*GAACT*T*CAG	14	CA*T*GATA*T*AGACG*T*TGTGGC
5	CAC*T*GCACGCCG*T*AGG*T*CAG	15	A*T*GCCG*T*TCTTC*T*GCTTG*T*C
6	GACT*T*GAAGAAG*T*CGTGC*T*G	16	GCGGATC*T*TGAAG*T*TCACC*T*
7	TGGACG*T*AGCCT*T*CGGGCA*T*	17	CGC*T*GCCG*T*CCTCGA*T*GTTG
8	CT*T*GAAGAAGATGG*T*GCGC*T*	18	TGTGA*T*CGCGC*T*TCTCG*T*TG
9	CGGGTC*T*TGTAG*T*TGCCG*T*C	19	G*T*CACGAAC*T*CCAGCAGGAC
10	GT*T*CACCAGGGTG*T*CGCCC*T*	20	G*T*CCA*T*GCCGAGAG*T*GA*T*CC

[a]Cy5 fluorophore is attached to 5′-end of each probe. Thymines highlighted in *italics* are ^{NH_2}T

3.1.2 Synthesis of Cy5 and Amino Dually Labeled Probe Set for SMRF

A set of 20–40 fluorescently labeled oligo probes is required in SMRF to detect single-RNA molecules. In this section, we describe the procedure of synthesizing a set of 20 Cy5 and amino dually labeled oligo probes, which was used to detect the single molecules of EGFP mRNA [6].

1. Synthesize 20 DNA probes with sequences listed in Table 1 on 200 nmole scale via standard solid-phase synthesis. Use ^{NH_2}T phosphoramidite solution to incorporate ^{NH_2}T at indicated positions (Table 1) by following the same procedure in Subheading 3.1.1, **step 1**.
2. Adding an amino C_6-alkyl chain: Install NH_2-C_6 phosphoramidite solution on DNA synthesizer as the second unnatural nucleobase. Add the NH_2-C_6 chain to 5′-terminal of each probe, via phosphoramidite chemistry on DNA synthesizer by following the same step as Subheading 3.1.1, **step 2**. Take off MMT protection group from terminal amine (*see* **Note 5**).
3. Transfer CPG resin of each probe from **step 2** into a 2 mL Eppendorf. Swell the resin in ACN for 20 min at room temperature. Centrifuge to set down the resin at the bottom of the Eppendorf and remove ACN completely.
4. Add 60 μL of Cy5 coupling solution to the resin from **step 3**. Vortex the mixture vigorously at room temperature for 2 h to conjugate Cy5 dye to alkyl amino terminal of DNA probes (*see* **Note 6**).

5. Centrifuge to set down the resin at the bottom of the Eppendorf and remove Cy5 coupling solution.
6. Repeat **steps 4** and **5** (*see* **Note 7**).
7. Cleave Cy5-labeled probes by incubating the resin in 200 μL of deprotection solution II. The solution is shaken gently for 1 h to cleave probes from resin. After the resins are removed by filtration, the blue-colored solution is kept in the dark at room temperature for 24 h to complete the deprotection of nucleobases.
8. Gently blow the surface of solution with nitrogen to remove most of the ammonium hydroxide.
9. Dilute solution from **step 8** with 200 μL water to prepare sample for HPLC purification.
10. Inject the solution from **step 9** to reverse-phase HPLC. Desired probes are eluted with ACN over TEAA buffer (5–50 % over 30 min). Collect the peak with UV–Vis absorptions at both 260 nm and 630 nm. Remove the solvent by lyophilization. The probes are dissolved in hybridization buffer at desired concentrations (*see* **Note 8**) and stored at −20 °C.
11. For quality control, the probes are characterized by ESI-MS and are quantified by UV–Vis absorption of both DNA (260 nm) and Cy5 dye (630 nm).

3.1.3 Nick Translation

In general, the probe used in RNA FISH or DNA FISH can be prepared by nick translation. The template DNA used in the nick translation reaction can be an appropriate BAC clone. For example, if a BAC clone contains no other transcribed regions than the target RNA, then the BAC can serve as the template DNA in nick translation to generate RNA FISH probes for the RNA target.

1. Add the following components (Nick Translation Kit) into a 0.2 mL tube and adjust the reaction volume to 50 μL with water:
   - 2.5 μg template DNA.
   - 7.5 μL dATP (0.4 mM).
   - 7.5 μL dCTP (0.4 mM).
   - 7.5 μL dGTP (0.4 mM).
   - 1.5 μL Cy3-dUTP (1 mM).
   - 1.5 μL Aminoallyl-dUTP (1 mM) (*see* **Note 9**).
   - 5 μL 10× Nick translation buffer.
   - 5 μL Enzyme mix.
2. Incubate the reaction at 15 °C for 2 h and 65 °C for 10 min.
3. Add 25 μg of Cot-1 DNA into the reaction mix.
4. Ethanol precipitation of the DNA in the reaction mix.
5. Resuspend the DNA pellet in 50 μL of hybridization buffer.
6. Store the probe in the dark at −20 °C.

3.2 Slide Preparation

3.2.1 Cytospin

The cytospin method helps to deposit cells cultured in suspension onto microscope slides. It also helps to separate cell clusters into single cells, for example the colonies of embryonic stem cells.

1. Trypsinize the cells and resuspend them in a concentration of 8×10^5 cells per mL in PBS (*see* **Note 10**).
2. Cytospin 100 μL cells onto microscope slides using cytology funnels at 254 *g* for 10 min.
3. Air-dry the slides for 2–3 min.
4. Rinse the slides in ice-cold PBS for 5 min in a Coplin jar.
5. Treat the slides with CSK buffer for 3 min (optional, *see* **Note 11**).
6. Fix the sample in 2 % PFA for 10 min at room temperature (*see* **Note 12**).
7. Store the slides in 70 % ethanol at 4 °C until use.

3.2.2 Cell Culture on Cover Slips

Cells can be directly cultured on cover slips before FISH experiments. In this way, better protection on cellular components and structures can be achieved because the cells are directly fixed on the cover slips without trypsinization and other physical manipulations.

1. Sterilize cover slips and lay the cover slips onto the bottom of a tissue culture container. Directly culture cells on the cover slips until the cell culture reaches a desired cell density.
2. Briefly rinse the cover slips in PBS twice.
3. Fix the cells in 2 % PFA for 10 min at room temperature (*see* **Note 12**).
4. Store the cover slips in 70 % ethanol at 4 °C until use.

3.3 FISH

3.3.1 Single-Molecule RNA FISH (SMRF)

The protocol of SMRF is adapted from the method described by Raj and colleagues [2]. A detailed protocol was published [7].

In this section, we describe the details of how to handle the cells directly cultured on cover slips. Readers are referred to Subheading 3.3.2 for details of how to handle the cells deposited onto microscope slides by cytospin.

1. Warm the probe to room temperature (*see* **Note 8**).
2. Aspirate 70 % ethanol off the cover slip.
3. Wash the cover slip three times with 2 mL of wash buffer (10 % formamide, 2× SSC with 4 mM Vanadyl ribonucleoside) each.
4. Pipette 10 μL of the probe onto a new microscope slide and place the cover slip with the cell-side down onto the probe.
5. Incubate the slide in a dark and humidified chamber at 30 °C overnight.
6. Slowly remove the cover slip and wash the cover slip with 2 mL of wash buffer (10 % formamide, 2× SSC with 4 mM Vanadyl ribonucleoside) at 30 °C for 30 min.

7. Aspirate the wash buffer.
8. Fix the cover slip in 4 % PFA at room temperature for 10 min.
9. Briefly rinse the cover slip twice with PBS. The sample is ready for the subsequent DNA FISH.

3.3.2 RNA FISH

In this section, we describe the details of how to handle the cytospin slides. Readers are referred to Subheading 3.3.1 for details of how to handle the cells directly cultured on cover slips.

1. Dehydrate the slide through 80, 90, and 100 % ethanol sequentially for 2 min each.
2. Air-dry the slide at room temperature for 2–3 min.
3. Denature the probe at 75 °C for 10 min, and pre-anneal the probe at 42 °C for 10–60 min (*see* **Note 13**). It is important to coordinate **steps 2** and **3**, so that when the slide is dried, the probe is also ready to be applied onto the slide.
4. Deposit 6 μL denatured probe onto the slide. The probe should be directly applied onto the "area" of cells. After the probe is added, a cover slip is laid on top of the area.
5. Probe hybridization is carried out at 42 °C in a dark and humid environment overnight.
6. Remove the cover slip gently.
7. Wash the slide three times (5 min for each wash) in 50 % formamide, 2× SSC at 45 °C with slight agitation.
8. Wash the slide three times (5 min for each wash) in 2× SSC at 45 °C with slight agitation.
9. Rinse the slide briefly in PBS.
10. Fix the slides in 4 % PFA at room temperature for 10 min.
11. Rinse the slides again briefly in PBS. The sample is ready for the subsequent DNA FISH.

3.3.3 DNA FISH

1. Permeabilize the sample with PBS 0.5 % Triton X-100 for 15 min (optional).
2. For DNA FISH using the chromosome paint, treat the slides with freshly prepared 0.1 M HCl 0.25 % Tween-20 for 1 min and wash with 2× SSC for 5 min (optional, *see* **Note 14**).
3. Denature the sample in 70 % formamide, 2× SSC at 75 °C for 10 min.
4. Dehydrate the sample through 80, 90, and 100 % ice-cold ethanol sequentially for 2 min each.
5. Air-dry the slides at room temperature for 2–3 min.
6. Denature the probe at 75 °C for 10 min, and pre-anneal the probe at 42 °C for 10–60 min (*see* **Note 13**). It is important to

coordinate **steps 5** and **6**, so that when the slide is dried, the probe is also ready to be applied onto the slide.

7. Add 6 μL probe, lay the cover slip on top of the probe, and seal the four edges of the cover slip with rubber cement.
8. Probe hybridization is carried out overnight at 42 °C in a dark and humid environment.
9. Remove the rubber cement and then the cover slip carefully.
10. Wash the sample three times (5 min for each wash) in 50 % formamide, 2× SSC at 45 °C with slight agitation.
11. Wash the sample three times (5 min for each wash) in 2× SSC at 45 °C with slight agitation.
12. Rinse briefly in PBS with 0.2 % Tween-20.
13. Briefly dry the sample for less than 1 min.
14. Apply mounting media with DAPI. Lay the cover slip.
15. Seal the four edges of the cover slip with minimal amount of nail polish. The sample is ready to be examined under microscopes.

4 Notes

1. Solutions of phosphoramidites, activator, and acetonitrile are dried over molecular trap packs 2 days before synthesis.
2. 200 nmole size of CPG resin in Mermade column are used for solid-phase synthesis unless indicated otherwise.
3. ^{NH2}T and natural nucleoside phosphoramidites are incorporated with standard coupling time, 60 s.
4. 5′-Fluorescein phosphoramidite is incorporated with a 3-min coupling time.
5. Turn on "DMT-off" function during the initial setup of DNA sequence on the synthesizer. 4-Monomethoxytrityl group on terminal amine is removed by deblocking mix when the probes are still attached to CPG resin.
6. Vigorously vortex the resins are highly recommended to increase the coupling yields. Cover the Eppendorf with aluminum foil to protect the resin from light.
7. After Cy5 coupling, the resin turns to deep blue color.
8. In SMRF, it is important to test a wide range of probe concentrations (5–50 nM) to determine the optimized probe concentration for each experiment [7].
9. Not all the amino-labeled dNTPs are compatible substrates for DNA polymerase I in the nick translation reaction. If a different amino-labeled dNTP is chosen for the nick translation reaction,

it is important to confirm whether the amino-labeled dNTP is a compatible enzyme substrate in the nick translation reaction.

10. The concentration needs to be optimized for different cell types. 8×10^5 cells per mL is suitable for mouse ES cells and fibroblasts.
11. CSK treatment permeabilizes cells and improves probe penetration in FISH. This step is necessary for detecting nuclear RNA signals. The treatment time needs to be optimized for individual experiments.
12. In RNA–DNA FISH experiments, the sample is fixed twice in PFA solutions, one before RNA FISH (pre-fixation), and the other one after RNA FISH (post-fixation). To make sure that the amino probes used in RNA FISH can be fixed efficiently with the matrix of cellular proteins in the post-fixation step, we pre-fix the cells in a milder condition (2 % PFA).
13. If double-stranded DNA is used as probe for FISH, it is necessary to denature the probe. Probe denaturation also helps to reduce the secondary structures of single-stranded probes. In some experiments, unlabeled Cot-1 DNA (a fraction of repetitive DNA fragments isolated from genomic DNA) is mixed with the probe. After probe denaturation, the probe is pre-annealed with Cot-1 DNA at 42 °C to block unspecific probe hybridization during FISH. The time for the pre-annealing step needs to be optimized for individual experiments.
14. For some difficult DNA targets, briefly subjecting the cells to 0.1 M HCl 0.25 % Tween-20 enhances the accessibility of the probe to the DNA target.

Acknowledgement

L-FZ is supported by the Singapore National Research Foundation under its Cooperative Basic Research Grant administered by the Singapore Ministry of Health's National Medical Research Council. FS is supported by Nanyang Assistant Professor Fellowship (M4080531) and Ministry of Education Tier 2 Research Grant (M4020163). The authors would also like to acknowledge the Nanyang Technological University College of Science Collaborative Research Award.

References

1. Wilusz JE, Sunwoo H, Spector DL (2009) Long noncoding RNAs: functional surprises from the RNA world. Gene Dev 23:1494–1504
2. Raj A, van den Bogaard P, Rifkin SA et al (2008) Imaging individual mRNA molecules using multiple singly labeled probes. Nat Methods 5:877–879
3. Lansdorp PM, Verwoerd NP, van de Rijke FM et al (1996) Heterogeneity in telomere length of human chromosomes. Hum Mol Genet 5(5):685–691

4. Lai L-T, Lee PJ, Zhang L-F (2013) Immunofluorescence protects RNA signals in simultaneous RNA–DNA FISH. Exp Cell Res 319:46–55
5. Zhang L-F, Huynh KD, Lee JT (2007) Perinucleolar targeting of the inactive X during S phase: evidence for a role in the maintenance of silencing. Cell 129:693–706
6. Basu R, Lai L-T, Meng Z et al (2014) Using amino-labeled nucleotide probes for simultaneous single molecule RNA–DNA FISH. PLoS ONE 9(9):e107425
7. Raj A, Tyagi S (2010) Detection of individual endogenous RNA transcripts in situ using multiple singly labeled probes. Methods Enzymol 472:365–386

Chapter 12

Visualizing Long Noncoding RNAs on Chromatin

Michael Hinten, Emily Maclary, Srimonta Gayen, Clair Harris, and Sundeep Kalantry

Abstract

Fluorescence in situ hybridization (FISH) enables the detection of specific nucleic acid sequences within single cells. For example, RNA FISH provides information on both the expression level and localization of RNA transcripts and, when combined with detection of associated proteins and chromatin modifications, can lend essential insights into long noncoding RNA (lncRNA) function. Epigenetic effects have been postulated for many lncRNAs, but shown for only a few. Advances in in situ techniques and microscopy, however, now allow for visualization of lncRNAs that are expressed at very low levels or are not very stable. FISH-based detections of RNA and DNA coupled with immunological staining of proteins/histone modifications offer the possibility to connect lncRNAs to epigenetic effects. Here, we describe an integrated set of protocols to detect, individually or in combination, specific RNAs, DNAs, proteins, and histone modifications in single cells at a high level of sensitivity using conventional fluorescence microscopy.

Key words Immunofluorescence, Fluorescence in situ hybridization, RNA FISH, DNA FISH, Long noncoding RNAs, Epigenetic, Chromatin, Histone modifications

1 Introduction

LncRNAs are increasingly invoked as factors that establish epigenetic profiles [1, 2]. Some lncRNAs are hypothesized to regulate gene expression at specific loci [3]. Other lncRNAs are postulated to regulate gene expression chromosome-wide, as in the case of the inactive X-chromosome [1, 4]. Refinements of fluorescence in situ hybridization (FISH) techniques have resulted in the ability to visualize lncRNAs at their sites of synthesis on the chromosome or site of function in the genome [5]. Combined with immunofluorescence (IF) detection of epigenetic factors and histone modifications, FISH techniques can permit the formulation of specific hypotheses of lncRNA function. For example, colocalization of epigenetic modifiers or histone modifications with lncRNA transcripts can suggest the recruitment of epigenetic factors to discrete loci in the genome.

Yi Feng and Lin Zhang (eds.), *Long Non-Coding RNAs: Methods and Protocols*, Methods in Molecular Biology, vol. 1402, DOI 10.1007/978-1-4939-3378-5_12, © Springer Science+Business Media New York 2016

Here, we describe detection via FISH of lncRNAs both with probes that do not distinguish the DNA strand of synthesis and with strand-specific probes. We additionally describe methods for IF-based detection of proteins and histone modifications and DNA FISH procedure to mark specific DNA loci of interest in cultured and embryonic cells. These protocols can be used individually or in combination. We provide examples of cells profiled by IF and RNA FISH and RNA FISH followed by DNA FISH interrogating the X-linked Xist and Tsix lncRNAs that are expressed from and function *in cis* on the X-chromosome along with histone H3 lysine 27 trimethylation (H3-K27me3).

2 Materials

2.1 Sample Preparation: Cultured Cells

1. Sterile, gelatinized glass cover slips (*see* **Note 1**).
2. 0.2 % Gelatin: 0.2 g Gelatin per 100 ml of ultrapure DNase/RNase-free water. Autoclave to dissolve and sterilize.
3. 6-Well tissue culture dish.
4. Cytoskeletal buffer (CSK): 100 mM NaCl, 300 mM sucrose, 3 mM magnesium chloride, 10 mM PIPES ($C_8H_{18}N_2O_6S_2$, pH 6.8; be sure to avoid using the sodium-salt version of PIPES). For PIPES, make 0.25–0.5 M stock; bring into solution by adding NaOH while measuring pH.
5. CSK buffer with 0.4 % Triton X-100: Make stock of 10–20 % Triton X-100 solution in ultrapure DNase/RNase-free water before adding to CSK.
6. 4 % Paraformaldehyde (PFA; Electron Microscopy Science 16 % Paraformaldehyde in aqueous solution (Fisher, #50-980-487), diluted in ultrapure DNase/RNase-free water and with a final concentration of 1× phosphate-buffered saline [PBS]).
7. 70 % Ethanol made with filtered ddH_2O.
8. Plate sealer tape.

2.2 Sample Preparation: Mouse Embryos/Embryo Fragments

1. Sterile, gelatinized glass cover slips (*see* **Note 1**).
2. 0.2 % Gelatin: 0.2 g Gelatin per 100 ml of ultrapure DNase/RNase-free water. Autoclave to dissolve and sterilize.
3. Pipette for plating embryos or embryo fragments: A glass Pasteur pipette pulled to a width only slightly larger than the embryos/embryo fragments.
4. 1× PBS with 6 mg/ml of bovine serum albumin (BSA): To make 50 ml, combine 4 ml BSA (Invitrogen, 7.5 g /100 ml #15260037), 5 ml 10× PBS, and 41 ml ultrapure DNase/RNase-free water.

5. Fixation and permeabilization solution: 1 % PFA with 0.05 % Tergitol in a final concentration of 1× PBS.
6. 1 % PFA.
7. 70 % Ethanol made with filtered ddH_2O.
8. 6-Well dish, or similar container for storage (*see* **Note 2**).
9. Plate sealer tape.

2.3 Immunofluorescence (IF)

1. 1× PBS.
2. 6-Well dish, or similar chamber to be used for washing cover slips (*see* **Note 2**).
3. Blocking buffer: 1× PBS with 0.5 mg/ml BSA, 50 μg/ml tRNA, 80 units/ml RNase inhibitor (such as Invitrogen's RNaseOUT), and 0.2 % Tween-20 (make and use a 10 % Tween-20 stock). Pre-warm to 37 °C.
4. Primary antibody of choice.
5. Small glass plate (*see* **Note 3**).
6. Parafilm.
7. Forceps.
8. IF chamber: A small humid chamber for incubating slides, humidity provided by 1× PBS (*see* **Note 4**).
9. Incubator set to 37 °C.
10. 1× PBS with 0.2 % Tween-20.
11. Fluorescently conjugated secondary antibody: AlexaFluor (Invitrogen) secondary antibodies work well with this protocol; AF488, AF555, and AF647 have very similar absorption and emission spectra to the Fluorescein-12, Cy3, and Cy5 dyes used for FISH, respectively. These antibodies can be used in conjunction with FISH probes for multicolor imaging with the same set of fluorescence microscope filters.
12. 4′,6-Diamidino-2-phenylindole, dihydrochloride (DAPI) (Life Technologies, 10 mg) (for IF without subsequent RNA FISH).
13. Mounting medium (for IF without subsequent RNA FISH, *see* **Note 5**).
14. Microscope slides (for IF without subsequent RNA FISH).
15. 2 % PFA (for IF followed by RNA FISH only).

2.4 RNA and DNA FISH: Probe Labeling for Double-Stranded Probes (See Note 6)

1. BioPrime DNA Labeling System (Life Technologies, #18094011) (*see* **Note 7**).
2. Ultrapure DNase/RNase-free water.
3. TE buffer: 10 mM Tris–HCl, 1 mM EDTA, pH 8 (*see* **Note 8**).
4. DNA template for labeling: Large templates, such as fosmids or BACs, work best for labeling by random priming. BACs are

recommended for DNA FISH probes, as the larger template size will maximize the signal despite low copy number of DNA compared to transcribed RNA in the cell. Ensure that the template preparation is extremely pure; contaminating bacterial DNA will lead to high levels of background fluorescent signal.

5. Custom dNTP mixture:
   (a) For use with Cy3 or Cy5 fluorescently labeled dCTP (*see* below): 2 mM dATP, 2 mM dGTP, 2 mM dTTP, 1 mM dCTP.
   (b) For use with fluorescein-labeled dUTP (fluorescein-12-dUTP; *see* below): 2 mM dCTP, 2 mM dATP, 2 mM dGTP, 1 mM dTTP.
6. 1 mM Fluorescently labeled nucleotide: Fluorescein-12-dUTP (Roche)-labeled probes excite maximally at 495 nm and emit maximally at 521 nm. Cy3-labeled dCTP (GE Healthcare)-labeled probes excite maximally at 550 nm and emit maximally at 570 nm, while Cy5-labeled dCTP (GE Healthcare)-labeled probes absorb maximally in the far-red end of the spectrum, at 649 nm, and emit maximally at 670 nm. The spectra for these dyes do not overlap significantly, and labeled probes can be combined with other FISH probes and used with DAPI, which absorbs maximally at 358 nm and emits maximally at 461 nm when bound to double-stranded DNA, or fluorescently conjugated antibodies for multicolor imaging.
7. G-50 ProbeQuant Micro Columns (GE Healthcare).
8. Tabletop centrifuge.
9. 20 mg/ml Yeast tRNA: Reconstitute lyophilized yeast tRNA (Invitrogen, #15401-029) in ultrapure DNase/RNase-free water. Aliquot and store at −20 °C.
10. 3 M Sodium acetate, pH 5.2 (*see* **Note 8**).
11. Molecular biology-grade 100 % ethanol.
12. Heat block set to 37 °C.

2.5 RNA FISH: Probe Labeling for Strand-Specific Probes (See Note 6)

1. MAXIscript T3/T7 Kit (Ambion, AM1326).
2. Linear DNA template, PCR amplified with T3 or T7 promoter sequence upstream of the sequence to be labeled: To detect RNA transcribed from the desired strand, the probes must be synthesized complementary to the target transcript; the T3 or T7 RNA polymerase promoter sequences must therefore be incorporated at the appropriate ends of the template DNA. See MAXIscript kit instructions for details on template preparation.
3. Custom NTP mixture (*see* **Note 9**):
   (a) For use with Cy3 or Cy5 fluorescently labeled CTP: 2 mM ATP, 2 mM GTP, 2 mM TTP, 1 mM CTP.

(b) For use with fluorescein fluorescently labeled UTP (fluorescein-12-UTP: 2 mM CTP, 2 mM ATP, 2 mM GTP, 1 mM TTP.

4. 100 nmol Fluorescently labeled nucleotide: Fluorescein-12-UTP (Roche), Cy3-CTP (GE Healthcare), or Cy5-CTP (GE Healthcare) (*see* notes on emission spectra in Subheading 2.4).
5. 0.5 M EDTA (*see* **Note 8**).
6. 20 mg/ml Yeast tRNA: Reconstitute lyophilized yeast tRNA in ultrapure DNase/RNase-free water. Aliquot and store at −20 °C.
7. 5 M Ammonium acetate (*see* **Note 8**).
8. 100 % Molecular biology-grade ethanol.
9. Ultrapure DNase/RNase-free water.
10. 37 °C Heat block.
11. Mini Quick Spin RNA Columns (Roche, #11814427001).
12. Tabletop centrifuge.

2.6 RNA and DNA FISH: Probe Precipitation

1. Fluorescently labeled probes in ethanol (*see* Subheadings 3.4 and 3.5): A combination of double-stranded and strand-specific probes may be used.
2. 20 mg/ml Yeast tRNA: Reconstitute lyophilized yeast tRNA in ultrapure DNase/RNase-free water. Aliquot and store at −20 °C.
3. 1 mg/ml COT-1 DNA (Invitrogen) (Optional, *see* **Note 10**).
4. 10 mg/ml Salmon sperm DNA.
5. 5 M Ammonium acetate (*see* **Note 8**).
6. Ultrapure DNase/RNase-free water.
7. 100 % Molecular biology-grade ethanol.
8. Tabletop centrifuge.
9. 70 % Ethanol: Dilute 100 % ethanol using filtered ddH_2O.
10. Deionized formamide.
11. 20× Sodium saline citrate (SSC) (Invitrogen).
12. 2× Hybridization solution: Two parts ultrapure DNase/RNase-free water, one part 20× SSC, two parts 50 % dextran sulfate (*see* **Note 11**).
13. Vacuum centrifuge.
14. Heat block set to 90 °C.

2.7 RNA FISH: Sample Hybridization

1. 100 % Molecular biology-grade ethanol: Use filtered ddH_2O to make stocks of 70, 85, and 95 % ethanol.
2. Fluorescently labeled probes (*see* Subheadings 2.4–2.6).

3. 6-Well dish, or similar chamber to be used for dehydration and for washing cover slips (*see* **Note 2**).
4. Small glass plate (*see* **Note 3**).
5. Parafilm.
6. Forceps.
7. Hybridization chamber: A small humid chamber for incubating slides, humidity provided by 2× SSC/50 % deionized formamide (*see* **Note 4**).
8. Incubator set to 37 °C, for overnight incubation.
9. 2× SSC/50 % deionized formamide: 5 ml 20× SSC, 25 ml deionized formamide, 20 ml filtered ddH_2O.
10. 2× SSC: 5 ml 20× SSC, 45 ml filtered ddH_2O.
11. 1× SSC: 2.5 ml 20× SSC, 47.5 ml filtered ddH_2O.
12. DAPI.
13. Incubator set to 39 °C, for washes.
14. Mounting medium (*see* **Note 5**).
15. Microscope slides.

2.8 DNA FISH: Sample Hybridization

1. 1× PBS.
2. 1 % PFA with 0.5 % Tergitol and 0.5 % Triton X-100.
3. 100 % Molecular biology-grade ethanol: Use filtered ddH_2O to make stocks of 70, 85, and 95 % ethanol.
4. 6-Well dish, or similar chamber to be used for dehydration and for washing cover slips (*see* **Note 2**).
5. RNase A solution: 1.25 μg/μl RNase A, diluted in 2× SSC.
6. 2× SSC/70 % deionized formamide: 5 ml 20× SSC, 35 ml deionized formamide, 10 ml filtered ddH_2O.
7. Heat block set to 95 °C.
8. Fluorescently labeled double-stranded probe (*see* Subheadings 2.5 and 2.6).
9. Small glass plate (*see* **Note 3**).
10. Parafilm.
11. Forceps.
12. Hybridization chamber: A small humid chamber for incubating slides, humidity provided by 2× SSC/50 % deionized formamide (*see* **Note 4**).
13. Incubator set to 37 °C, for overnight incubation.
14. 2× SSC/50 % deionized formamide: 5 ml 20× SSC, 25 ml deionized formamide, 20 ml filtered ddH_2O.
15. 2× SSC: 5 ml 20× SSC, 45 ml filtered ddH_2O.

16. 1× SSC: 2.5 ml 20× SSC, 47.5 ml filtered ddH_2O.
17. DAPI.
18. Incubator set to 39 °C, for washes.
19. Mounting medium (*see* **Note 5**).
20. Microscope slides.

3 Methods

3.1 Sample Preparation: Cultured Cells

1. Grow cells to desired confluency on sterile, gelatinized glass cover slips (*see* **Note 1**) in the bottom of a 6-well tissue culture dish.
2. Prepare reagents for fixation and permeabilization: Make and sterile filter CSK and CSK with 0.4 % Triton X-100 buffers and chill to 4 °C. Prepare 4 % PFA. All subsequent steps should be performed in the tissue culture hood. The CSK and CSK+0.4 % Triton X-100 buffers should be kept chilled on ice during the procedure.
3. Aspirate media from all six wells of a 6-well dish
4. Pipette 2 ml of ice-cold CSK buffer in each well of the 6-well dish.
5. After 30 s, aspirate the CSK buffer: By the time CSK buffer has been pipetted into each of the six wells, the first well has incubated in the CSK buffer for ~30 s, so simply pipette CSK buffer into the six wells and then aspirate right afterward from the first well onward.
6. Pipette 2 ml of ice-cold CSK+0.4 % Triton X-100 buffer in each well of the 6-well dish.
7. After 30 s, aspirate the CSK+0.4 % Triton X-100. *See* details in **step 3**.
8. Pipette 2 ml of ice-cold CSK buffer in each well of the 6-well dish.
9. After 30 s, aspirate the CSK buffer. *See* details in **step 3**.
10. Pipette 2 ml of 4 % PFA in each well of the 6-well dish. Incubate the cells in PFA for 10 min.
11. Aspirate the PFA and pipette in 5 ml of cold 70 % ethanol in each well. Repeat the ethanol wash a total of three times, to remove all traces of PFA.
12. Seal dish with plate sealer tape and store cells at −20 °C (*see* **Note 12**).

3.2 Sample Preparation: Mouse Embryos

1. Plate embryos or dissected embryo fragments on sterile, gelatinized glass cover slips (22×22 cm square cover slips, *see* **Note 1**) in 1× PBS 6 mg/ml BSA using finely pulled Pasteur pipette.

2. Aspirate excess 1× PBS 6 mg/ml BSA and let dry for about 20–30 min.
3. Fix and permeabilize with 50 μl of 1 % PFA in 1× PBS with 0.05 % Tergitol for 5 min (to get rid of excess solution on cover slip after the 5-min permeabilization/fixation, simply tap solution off onto a paper towel).
4. Fix again with 50 μl of 1 % PFA (no Tergitol) in 1× PBS for 5 min. Drain the excess solution onto a paper towel.
5. Place cover slips with plated embryos in a 6-well dish. Rinse with three changes of 70 % ethanol. Seal dish with plate sealer tape and store in 70 % ethanol at −20 °C until use (*see* **Note 12**).

3.3 Immunofluorescence (IF) (Fig. 1)

1. Begin with fixed, permeabilized cells or embryo samples, plated on gelatinized glass cover slips and stored in 70 % ethanol (*see* Subheadings 3.1 and 3.2).
2. Make blocking buffer and warm to 37 °C.
3. Place sample cover slip in a 6-well dish that contains 2 ml of 1× PBS in each well (*see* **Note 2**).
4. Wash briefly with three changes of 1× PBS to remove ethanol.
5. Wash with 1× PBS three times, 3 min each on a rocker.
6. Wrap a glass plate tightly with parafilm for incubating cover slips for subsequent steps.
7. Block slides for 30 min at 37 °C in 50 μl pre-warmed blocking buffer in a humid chamber: Place a 50 μl drop of blocking buffer on the parafilm-wrapped glass plate and invert the cover slip, sample side down, into the blocking buffer. Place the parafilm-wrapped plate in the humid chamber, and incubate for 30 min at 37 °C. All incubations in blocking buffer, primary antibody, or secondary antibody should be set up in this manner.
8. Carefully lift cover slip from blocking buffer with forceps and place into a 50 μl droplet of diluted primary antibody on a parafilm-wrapped plate. Incubate with 50 μl primary antibody diluted in pre-warmed blocking buffer (dilution based on primary antibody you are using) in a humid chamber at 37 °C for 1 h.
9. Remove cover slip from primary antibody solution and place, sample-side up, in a 6-well dish. Wash three times with 1× PBS/0.2 % Tween-20 for 3 min on a rocker.
10. Incubate in 50 μl pre-warmed blocking buffer on a parafilm-wrapped plate in a humid chamber for 5 min at 37 °C.
11. Incubate with 50 μl secondary antibody diluted in pre-warmed blocking buffer in humid chamber at 37 °C for 30 min.

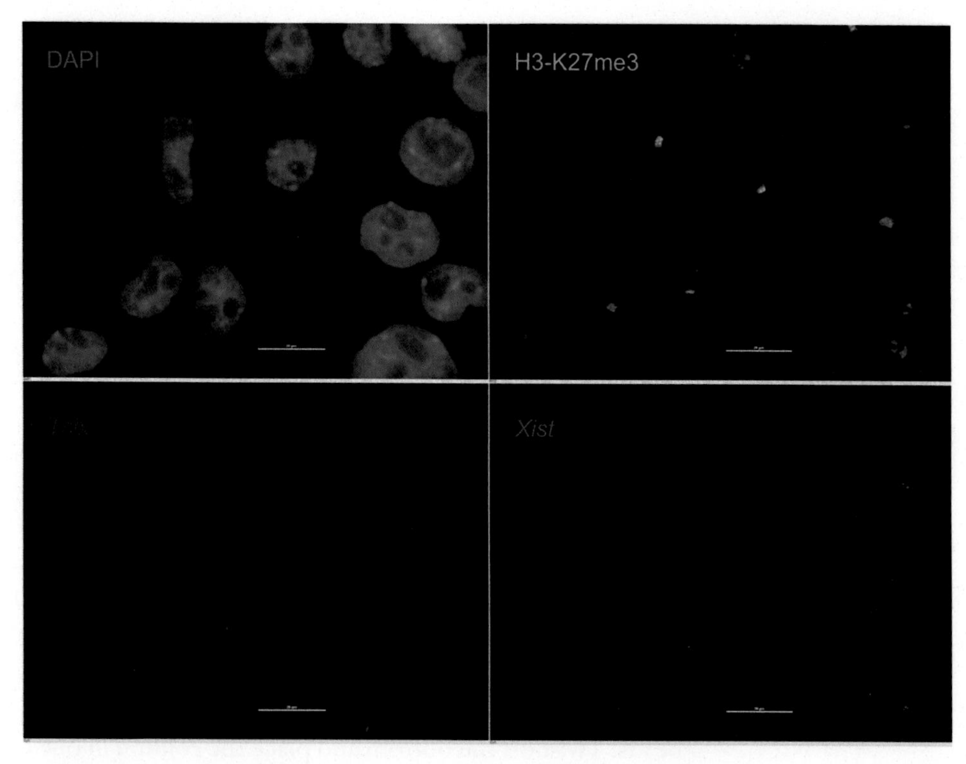

Fig. 1 Immunofluorescence followed by strand-specific RNA FISH in female mouse trophoblast stem cells. Histone H3 lysine 27 trimethylation (H3-K27me3; in *green*) is enriched on the inactive X-chromosome that is marked by Xist lncRNA accumulation (*purple*). The active X-chromosome is marked by nascent expression of the Xist antisense Tsix lncRNA (*red pinpoints*). Nuclei are stained *blue* with DAPI

Antibody dilution depends on secondary antibody used; AlexaFluor-conjugated secondary antibodies should be used at a 1:300 dilution.

12. Remove cover slip from secondary antibody and wash three times with 1× PBS/0.2 % Tween-20 for 3 min each on a rocker.
    (a) If processing samples only for IF, the first wash of **step 11** should contain a 1:100,000 to 1:200,000 dilution of DAPI (*see* **Note 13**). Then, rinse once briefly with PBS/0.2 % Tween-20 and wash two more times for 5–7 min each while rocking to remove excess DAPI. Remove cover slip from dish, tap off excess liquid, and then mount on a slide, sample-side down, in mounting medium. Image samples or store at −20 °C for later imaging.
    (b) If samples will be stained by RNA FISH probes following IF, pipette a 100 μl drop of 2 % PFA on a glass plate

wrapped in parafilm. Place the cover slip sample-side down in PFA and place in the humid chamber. Incubate the sample for 10 min at room temperature, and then proceed to RNA FISH hybridization (Subheading 3.7).

3.4 RNA FISH: Probe Labeling for Double-Stranded Probes

1. Dissolve 50–100 ng DNA template in TE buffer in a final volume of 19 μl.
2. On ice, add 20 μl 2.5× random primer solution (supplied with BioPrime labeling kit).
3. Denature DNA by heating for 5 min in a 90 °C heat block.
4. Immediately cool on ice for 5 min.
5. Add 5 μl of custom dNTP mixture.
6. Add 5 μl of 1 mM fluorescently labeled nucleotide.
7. Add 2 μl (80 units) Klenow fragment (supplied with the BioPrime Labeling Kit, *see* **Note 14**).
8. Incubate at 37 °C overnight. Cover the tube with aluminum foil to protect the probe from light during this step.
9. Add 5 μl of stop buffer.
10. Prepare G-50 column: Vortex briefly to mix the Sephadex beads. Twist off the bottom and place the column in one of the provided collection tubes. Spin at 750 × *g* for 1 min.
11. Move column to a new microcentrifuge tube and pipette your probe into the center of the column. Spin at 750 × *g* for 1 min. Measure eluted volume by pipette (*see* **Note 15**).
12. Add 10 μl of 20 mg/ml yeast tRNA.
13. Add 3 M sodium acetate to a final concentration of 0.3 M.
14. Add 2.5 volumes of 100 % ethanol.
15. Spin at 4 °C for 20 min at top speed, ~21,100 × *g*.
16. Carefully aspirate the supernatant using the vacuum apparatus (*see* **Note 16**).
17. Resuspend the pellet in 360 μl of RNase/DNase-free ultrapure ddH_2O. Make sure that the pellet is fully dissolved before moving to the next step.
18. Add 40 μl 3 M sodium acetate.
19. Add 1 ml 100 % ethanol.
20. Store labeled probe at −20 °C in the dark (*see* **Note 17**).

3.5 RNA FISH: Probe Labeling for Strand-Specific Probes

1. Start with 1 μg template DNA.
2. Add RNase/DNase-free ultrapure ddH_2O so that the total volume at the end of **step 6** will be 20 μl.
3. Add 2 μl 10× transcription buffer (supplied with MAXIscript kit).
4. Add 5 μl NTP mixture (supplied with MAXIscript kit, *see* **Note 9**).

5. Add 2 μl of fluorescently labeled NTP.
6. Add 2 μl T3 (60 units) or T7 (30 units) enzyme mix (supplied with MAXIscript kit, *see* **Note 9**).
7. Mix thoroughly.
8. Incubate for 2 h at 37 °C.
9. Add 1 μl of TURBO DNase (supplied with MAXIscript kit) and mix well.
10. Add 1 μl of 0.5 mM EDTA.
11. Prepare Quick Spin RNA Column: Vortex briefly to mix the Sephadex beads. Remove cap, twist off the bottom tip, and then place the column in a microcentrifuge tube. Spin at 1000 × *g* for 1 min.
12. Move column to a new microcentrifuge tube and pipette your probe into the center of the column (*see* **Note 15**). Spin at 1000 × *g* for 1 min. Measure your new volume by pipette.
13. Add 10 μl of 20 mg/ml yeast tRNA.
14. Add 5 M ammonium acetate to a final concentration of 0.5 M.
15. Add 2.5 volumes of 100 % ethanol.
16. Spin at 4 °C for 20 min at top speed, 21,100 × *g*.
17. Carefully aspirate the supernatant using the vacuum apparatus (*see* **Note 16**).
18. Resuspend the pellet in 360 μl RNase/DNase-free ultrapure ddH_2O. Make sure that the pellet is fully dissolved before moving to the next step.
19. Add 40 μl of 5 M ammonium acetate.
20. Add 1 ml of 100 % ethanol.
21. Store labeled probe at −20 °C in the dark (*see* **Note 17**).

3.6 RNA FISH: Probe Precipitation (Figs. 1 and 2)

All spin steps should be completed at top speed, ~21,100 × *g*.

1. Aliquot 100 μl of fluorescently labeled probe in ethanol. A combination of double-stranded and strand-specific probes may be precipitated together (*see* **Note 18**). Vortex all probe stocks before aliquoting.
2. Add 15 μl of 20 mg/ml tRNA (300 μg). Mix thoroughly by pipetting (*see* **Note 19**).
3. Add 15 μl of 1 mg/ml mouse COT-1 DNA (15 μg). Mix thoroughly by pipetting. (Optional: *See* **Notes 10** and **19**. If not using COT-1 DNA, add 15 μl of RNase/DNase-free ultrapure water.)
4. Add 15 μl of 10 mg/ml sheared boiled salmon sperm DNA (150 μg). Mix thoroughly by pipetting (*see* **Note 19**).
5. Add 20 μl 5 M ammonium acetate. Vortex briefly to mix (*see* **Note 20**).

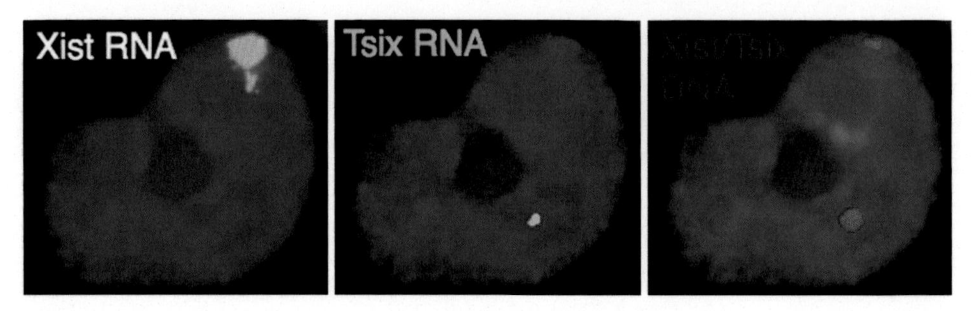

Fig. 2 RNA followed by DNA FISH in female mouse epiblast stem cells. Strand-specific RNA FISH detection of Xist lncRNA (*white*) and the Xist antisense Tsix lncRNA (*green*) followed by DNA FISH detection of the Xist/Tsix locus. Nuclei are stained *blue* with DAPI

6. Add 45 μl of DNase/RNase-free ultrapure water. Vortex briefly to mix (*see* **Note 20**).
7. Add 250 μl of 100 % ethanol. Vortex briefly to mix (*see* **Note 20**).
8. Spin for 20–30 min at 4 °C to precipitate probe.
9. Carefully aspirate the supernatant using a vacuum apparatus (*see* **Note 16**).
10. Add 500 μl of 70 % ethanol. Vortex thoroughly. Spin at room temp for 4 min.
11. Carefully aspirate the supernatant using a vacuum apparatus (*see* **Note 16**).
12. Add 500 μl of 100 % ethanol. Vortex thoroughly. Spin at room temp for 4 min.
13. Carefully aspirate the supernatant using a vacuum apparatus (*see* **Note 16**).
14. Dry in a speed vacuum without heat for 5 min, or alternatively air-dry. No ethanol should remain in the tube for the next step.
15. Add 50 μl of 100 % deionized formamide to the tube. Make sure that the pellet is submerged in the formamide.
16. Denature in a heat block at 80–90 °C for 10 min. At ~5 min, pipette the formamide up and down to make sure that the pellet goes into solution. Place back into the heat block.
17. Immediately cool on ice for 5 min.
18. Add 50 μl 2× hybridization solution. Mix thoroughly by pipetting up and down.
19. Preanneal at 37 °C for 1–1.5 h. (Optional: *See* **Note 10**. Preannealing is not required for strand-specific probes or for double-stranded probes precipitated without COT-1 DNA.)
20. Store probe at −20 °C in the dark.

3.7 RNA FISH: Sample Hybridization (Fig. 2)

1. Start with permeabilized and fixed samples plated on cover slips in 70 % ethanol (*see* Subheading 3.1 or 3.2), or samples previously processed for IF (*see* Subheading 3.3).
2. Dehydrate the cover slips by moving them through a room-temperature ethanol series (85, 95, and 100 % ethanol) for 2 min each.
3. Remove the cover slips from the well and air-dry at room temperature for 15 min (*see* **Note 21**).
4. Set up the FISH hybridization. Use 8–10 μl of probe per 22×22 mm cover slip. Pipette the probe onto a parafilm-wrapped glass plate (*see* **Note 3**), invert the cover slip onto the droplet of probe, and tap lightly with forceps to help the probe spread across the cover slip.
5. Hybridize overnight at 37 °C in a humid chamber (humidity provided by 2× SSC/50 % formamide) (*see* **Note 4**).
6. Make all wash solutions (2× SSC/50 % formamide, 2× SSC, and 1× SSC), and warm the solutions to 39 °C.
7. Carefully peel the cover slip off of the parafilm, re-invert, and place in a well containing pre-warmed 2× SSC/50 % formamide. Be sure that the cover slip goes into the well with the sample-side up.
8. Wash with pre-warmed 2× SSC/50 % formamide at 39 °C, three times for 7 min each.
9. Wash with pre-warmed 2× SSC at 39 °C, three times for 7 min each. Add DAPI into the last 2× SSC wash at a 1:100,000–1:200,000 dilution (*see* **Note 13**).
10. Rinse once quickly with pre-warmed 1× SSC.
11. Wash with pre-warmed 1× SSC at 39 °C, two times for 7 min each.
12. Use mounting medium to mount the cover slip onto a labeled microscope slide. Invert the slide onto a paper towel and press gently but firmly to remove excess mounting medium from under the cover slip.
13. Seal the cover slip and let dry thoroughly before viewing under a microscope.
14. Slides can be stored in the dark at −20 °C to preserve the fluorescent signal.

3.8 DNA FISH: Sample Hybridization

1. Begin with fixed, permeabilized cells or embryo samples, plated on gelatinized glass cover slips (22×22 cm square cover slips) and stored in 70 % ethanol (*see* Subheadings 3.1 and 3.2) or with samples previously stained for RNA FISH (*see* Subheading 3.7). For samples previously stained by RNA FISH, RNA FISH signals will be degraded during DNA FISH,

so samples should be imaged prior to beginning this protocol. Take note of the visual fields imaged, so that RNA FISH images can be aligned with DNA stains later.

2. If using samples previously stained for RNA FISH, use a razor blade to cut away the nail polish used to seal the cover slip to the slide. Then submerge the entire slide with the cover slip still attached in a solution of 2× SSC. While the sample is still submerged gently peel off the cover slip by letting the solution of 2× SSC infiltrate between the cover slip and the slide. For samples not previously processed for RNA FISH, proceed to **step 5**.
3. Wash the cover slip with 1× PBS three times quickly and then incubate in 1× PBS for 5 min at room temperature.
4. Refix the cell/embryos with 1 % PFA containing 0.5 % Tergitol and 0.5 % Triton X-100. Fix for 10 min at room temperature.
5. Dehydrate the cover slips by moving them through a room-temperature ethanol series (70, 85, and 100 % ethanol) for 2 min each.
6. Remove the cover slips from the well and air-dry at room temperature for 15 min (*see* **Note 21**).
7. RNase treatment: Treat with 1.25 μg/μl RNase A in 2× SSC. Invert slides onto a 100 μl drop of RNase A solution on parafilm stretched over a glass slide. Incubate at 37 °C for 30 min. RNase treatment allows the probe to gain access to the DNA.
8. Dehydrate the cover slips by moving them through a room-temperature ethanol series (85, 95, and 100 % ethanol) for 2 min each.
9. Remove the cover slips from the well and air-dry at room temperature for 15 min (*see* **Note 21**).
10. Denature the samples in a pre-warmed solution of 2× SSC/70 % formamide on a glass slide or glass plate stationed on top of a heat block set at 95 °C for 11 min.
11. Immediately dehydrate through a –20 °C ethanol series (70, 85, 95, and 100 % ethanol) for 2 min each.
12. Remove the cover slips from the well and air-dry at room temperature for 15 min (*see* **Note 21**).
13. Set up the FISH hybridization. Use 8–10 μl of probe per 22×22 mm cover slip. Pipette the probe onto parafilm-wrapped glass plate (*see* **Note 3**), invert the cover slip onto the droplet of probe, and tap lightly with forceps to help the probe spread across the cover slip. Be sure to avoid air bubbles that may result in uneven distribution of the probe.
14. Hybridize overnight at 37 °C in a humid chamber (humidity provided by 2× SSC/50 % formamide, *see* **Note 4**).

15. Make all wash solutions (2× SSC/50 % formamide, 2× SSC), and warm the solutions to 39 °C.
16. Carefully peel the cover slip off of the parafilm, re-invert, and place in a well containing pre-warmed 2× SSC/50 % formamide. Be sure that the cover slip goes into the well with the sample side up.
17. Wash with 2× SSC/50 % formamide at 39 °C twice, 7 min each.
18. Wash with pre-warmed 2× SSC plus a 1:100,000–1:200,000 dilution of DAPI for 7 min at 39 °C (*see* **Note 13**).
19. Wash with pre-warmed 2× SSC without DAPI.
20. Use mounting medium to mount the cover slip onto a microscope slide. Invert the slide onto a paper towel and press gently but firmly to remove excess mounting medium from under the cover slip.
21. Seal the cover slip with clear nail polish and let dry thoroughly before viewing under a microscope.
22. Image and store the slide at −20 °C. When combining RNA FISH with DNA FISH, RNA FISH-stained samples must be imaged first and the coordinates of the cells on cover slip recorded. Following DNA FISH, the same cells should be imaged, to allow for analysis of overlaid images.

4 Notes

1. To make gelatinized cover slips:
   (a) For tissue culture cells, sterilize cover slips by dipping in 100 % ethanol and running through a flame. Once the cover slip is sterilized in this manner, place it in a well of the 6-well dish. In the tissue culture hood, pipette 0.5–1 ml sterile 0.2 % gelatin onto the cover slip and spread around. Keep the gelatin on the surface of the cover slip and avoid letting the gelatin infiltrate between the cover slip and the bottom of the well, or the cover slip may stick to the tissue culture surface. Aspirate excess gelatin and let dry.
   (b) For embryos, place a 25–50 μl droplet of 0.2 % sterile gelatin solution on the cover slip, spread liberally around with pipette tip, and let dry.
2. Use 22 × 22 mm cover slips, which fit well within a single well of a 6-well dish. All the dehydration and washing steps can be performed in the wells of this dish.
3. Short glass plates designed for casting protein gels are a good size for the hybridization. For example, BioRad's Mini-

PROTEAN Short Plates fit well in the humid chamber described in **Note 4**.

4. For humid chambers, a microscope slide box (i.e., one that holds 100 slides) works well. Place paper towels soaked in 1× PBS in the bottom of the box to create a humid chamber for immunofluorescence. For RNA and DNA FISH hybridization procedures, create a humid chamber by placing paper towels soaked in 2× SSC/50 % formamide in the bottom of the box.
5. For mounting medium, use Vectashield (Vector Labs) or similar anti-fade mounting medium and seal cover slips with clear nail polish after mounting on slides.
6. For "double-stranded" probes (Subheadings 2.4 and 3.4), while the labeled probes are single stranded, the probes are created from a double-stranded fosmid or BAC template using random primers. As such, these probes will hybridize to both strands of DNA or RNA. Double-stranded probes can be used for DNA FISH and for RNA FISH for most genes. With strand-specific probes (Subheadings 2.5 and 3.5), only one strand is labeled, and the probe will detect RNA from only the complementary strand. These probes can be used to distinguish sense and antisense transcription from a single locus.
7. This kit was originally designed for the preparation of biotinylated probes. The protocol outlined above modifies the BioPrime labeling procedure to make fluorescently labeled probes. Do not use the supplied 10× dNTP mixture, which includes biotinylated nucleotides. Prepare your own dNTP mixture, and add in the fluorescently labeled nucleotides separately (Subheading 3.4, **steps 5** and **6**). Alternative random primer solutions have been tested, but do not work as well as the 2.5× solution provided with this kit.
8. While these solutions can be prepared from scratch, purchasing them is preferred to ensure their RNase/DNase-free purity.
9. A set of 10 mM NTPs and the T3 and T7 enzymes are included in the MAXIscript kit. Both the NTPs and enzymes can be replaced with other commercially available substitutes with no apparent loss of enzymatic activity.
10. COT-1 DNA can be added to double-stranded probes to help reduce background by hybridizing to repetitive sequences, if necessary. COT-1 DNA should not be used in strand-specific probes, as these probes are specific to the target sequence. While we do not routinely use mouse COT-1 in the precipitation of mouse probes, we do usually add species-specific COT-1 to double-stranded probes for other species (i.e., human, rat; but have not tested whether background staining would increase if COT-1 were omitted). If COT-1 is omitted, preannealing is not required.

11. 50 % Dextran sulfate is very viscous. Make the 2× hyb solution in a tube with graduations. If you add the dextran sulfate last, instead of pipetting it you can use a pipette tip to scoop it into the tube up to the appropriate measurement marking on the side of microtube. Vortex thoroughly to mix.
12. Fixed cells and embryos on cover slips in 70 % ethanol can be stored in −20 °C for at least 1 year. Seal with plate sealer tape to minimize evaporation of ethanol, and replace the ethanol in the wells occasionally (once every 2 months or so), as it will evaporate over time despite the plate sealer.
13. Dilute DAPI 1:500 in RNase/DNase-free ultrapure water, and store in the dark at −20 °C. Add 6–10 μl of this dilution into each well while washing samples.
14. The Klenow Fragment included with the BioPrime kit can be replaced with other commercially available Klenow enzymes with no apparent loss of enzymatic activity.
15. Be sure to pipet your probe into the center of the beads in the column during column purification. The probe must travel through the beads, not down the side.
16. The pellet is usually attached to the bottom of the tube, but be careful during aspiration of the supernatant. You can put a 10 μl pipette tip on the end of your aspiration hose to slow down the suction.
17. Ethanol stocks of probes should last up to 1 year when stored sealed and protected from light.
18. At this stage, multiple probes that were labeled with different fluorophores can be combined. To start, precipitate 100 μl ethanol stock of each probe together. If any of the probes are too bright or too faint, the initial starting volume of the ethanol stock of the probes can be adjusted.
19. The volumes of tRNA, Cot-1 DNA, and salmon sperm DNA (Subheading 3.6, **steps 2–4**) are determined by the final volume of probe; these will stay constant whether you are precipitating one or multiple probes together.
20. The volumes of ammonium acetate, H_2O, and 100 % ethanol (Subheading 3.6, **steps 5**–7) are determined by the starting volume of ethanol; these will need to be scaled up proportionately if you are precipitating multiple probes together. For example, if precipitating two probes together and using 100 μl of each ethanol stock, then the volumes of ammonium acetate, H_2O, and ethanol need to be doubled.
21. Do not simply aspirate the 100 % ethanol out and leave the cover slip to dry in the well; the glass cover slip may stick to the bottom of the well. Find a safe place to set the cover slip to dry. Laying them across the inside of a empty pipette tip box (for example, a p1000 tip box) works well.

Acknowledgements

This work was funded by an NIH National Research Service Award #5-T32-GM07544 from the National Institute of General Medicine Sciences to E.M.; an NIH Director's New Innovator Award (DP2-OD-008646-01) to S.K.; a March of Dimes Basil O'Connor Starter Scholar Research Award (5-FY12-119); and the University of Michigan Endowment for Basic Sciences.

References

1. Lee JT, Bartolomei MS (2013) X-inactivation, imprinting, and long noncoding RNAs in health and disease. Cell 152:1308–1323
2. Brockdorff N (2013) Noncoding RNA and Polycomb recruitment. RNA 19:429–442
3. Brown JD, Mitchell SE, O'Neill RJ (2012) Making a long story short: noncoding RNAs and chromosome change. Heredity 108:42–49
4. Maclary E, Hinten M, Harris C, Kalantry S (2013) Long nonoding RNAs in the X-inactivation center. Chromosome Res 21:601–614
5. Maclary E et al (2014) Differentiation-dependent requirement of Tsix long non-coding RNA in imprinted X-chromosome inactivation. Nat Commun 5:4209

Chapter 13

Non-isotopic Method for *In Situ* LncRNA Visualization and Quantitation

Botoul Maqsodi and Corina Nikoloff

Abstract

In mammals and other eukaryotes, most of the genome is transcribed in a developmentally regulated manner to produce large numbers of long noncoding RNAs (lncRNAs). Genome-wide studies have identified thousands of lncRNAs lacking protein-coding capacity. RNA *in situ* hybridization technique is especially beneficial for the visualization of RNA (mRNA and lncRNA) expression in a heterogeneous population of cells/tissues; however its utility has been hampered by complicated procedures typically developed and optimized for the detection of a specific gene and therefore not amenable to a wide variety of genes and tissues.

Recently, bDNA has revolutionized RNA *in situ* detection with fully optimized, robust assays for the detection of any mRNA and lncRNA targets in formalin-fixed paraffin-embedded (FFPE) and fresh frozen tissue sections using manual processing.

Key words lncRNA *in situ* hybridization, lncRNA visualization, Branched DNA assay, bDNA assay

1 Introduction

Genome-wide studies have been used successfully to identify that the majority of the human and mouse genomes are transcribed, yielding a vast and complex network of transcripts that includes thousands of long noncoding RNAs (ncRNAs), longer than 200 nt, and microRNA (miRNA), around 24 nt, with no protein-coding capacity [1–5]. Only a small number of lncRNAs have been studied in detail and are associated with a function [6, 7]. LncRNAs have been associated with numerous molecular functions such as modulating transcriptional patterns, regulating protein activities, serving structural or organizational roles, and altering RNA processing events [8]. They also have been identified as having a crucial role in cancer biology and as a driver of tumor suppression and oncogenic functions [9, 10].

The majority of lncRNAs are expressed at very low levels, some as low as one or less than one copy per cell [11], making it difficult

Yi Feng and Lin Zhang (eds.), *Long Non-Coding RNAs: Methods and Protocols*, Methods in Molecular Biology, vol. 1402,
DOI 10.1007/978-1-4939-3378-5_13,

to validate their expression levels and the function. A complete understanding of the function requires molecular and cell biology characterization and *in situ* validation. *In situ* quantitation of lncRNA has been limited by a number of problems. For example, *in situ* detection of lncRNA cannot be accomplished using antibody-based techniques such as IHC, whereas RNA expression assays have had limited utility due to long and complicated workflows (e.g., radioactive formats), low sensitivity (e.g., nonradioactive formats), and/or the inability to multiplex.

ViewRNA® technology is an established bDNA technique that allows routine visualization of any RNA or lncRNA in FFPE (single sections, tissue microarrays, fine-needle aspirates, animal, human, and plant) and fresh frozen tissues. ViewRNA technology is designed to amplify signal without amplifying background. The technology enables up to 8,000-fold signal amplification in manual assays allowing reliable visualization of low-, medium-, and high-expressing lncRNA targets.

2 Materials

Prepare all solutions using ultrapure water and analytical grade reagents. Diligently follow all waste disposal regulations when disposing waste materials.

2.1 NBF FFPE Block Components

1. Reagents: 50 mL 37 % formaldehyde, 450 mL distilled water, 3.25 g sodium phosphate, dibasic (Na_2HPO_4), 2 g sodium phosphate, monobasic (NaH_2PO_4).
2. Preparation: Combine all ingredients in a 500 mL glass bottle and mix well. Store at room temperature (RT).

2.2 PFA FFPE Block Components

1. Reagents: Deionized water, HCl (dilute), NaOH (1 N), paraformaldehyde powder, 1× PBS (0.137 M NaCl, 0.05 M NaH_2PO_4, pH 7.4).
2. Preparation: For 1 L of 4 % paraformaldehyde, add 800 mL of 1× PBS to a glass beaker on a stir plate in a ventilated hood. Heat while stirring to approximately 60 °C. Take care that the solution does not boil. Add 40 g of paraformaldehyde powder to the heated PBS solution. The powder will not immediately dissolve into solution. Slowly raise the pH by adding 1 N NaOH dropwise from a pipette until the solution clears. Once the paraformaldehyde is dissolved, the solution should be cooled and filtered. Adjust the volume of the solution to 1 L with 1× PBS. Recheck the pH, and adjust it with small amounts of dilute HCl to approximately 6.9. The solution can be aliquoted and frozen or stored at 2–8 °C for up to 1 month.

2.3 FFPE Slide Components

1. Non-Clipped X-tra® Slides (Leica Biosystems); Superfrost™ Plus Slides (Fisher). 50, 70, and 100 % ethanol. Dry incubator, oven, and ThremoBrite instrument (Abbott Molecular).

2.4 lncRNA In Situ Hybridization Components

1. ViewRNA ISH Tissue 2-Plex Assay, ViewRNA Probe(s) (TYPE 1 and/or TYPE 6), Tissue-Tek® Staining Dish (clear color), Tissue-Tek Clearing Agent Dish (green color), and Tissue-Tek Vertical 24 Slide Rack (Affymetrix).
2. 1,000 mL Glass beaker, forceps, hydrophobic barrier pen, rectangular cover glass 24×55 mm, aluminum foil, double-distilled water (ddH$_2$O), 100 % ethanol (200 proof), 37 % formaldehyde, 27–30 % ammonium hydroxide, DAPI (optional: for fluorescent detection), 10× PBS (pH 7.2–7.4), Gill's Hematoxylin I, and Xylene or Histo-Clear.
3. Mounting media: UltraMount Permanent Mounting Medium or ADVANTAGE Mounting Medium.
4. Equipment: Microplate shaker (optiona); ThermoBrite System with Humidity Strips (Abbott Molecular) or tissue culture incubator with >85 % humidity and 0 % CO_2, and three aluminum slide racks; ViewRNA Temperature Validation Kit (Affymetrix); waterproof remote probe thermometers validated for 90–100 °C; pipettes: P20, P200, P1000; fume hood; Isotemp™ hot plates; tabletop microtube centrifuge; water bath capable of maintaining 40 ± 1 °C; vortexer; dry incubator or oven capable of maintaining 60 ± 1 °C; bright-field and/or fluorescent microscope.

3 Methods

Carry out all procedures at room temperature unless otherwise specified.

3.1 FFPE Tissue Block Preparation

1. Immediately place freshly dissected tissues in ≥20 volumes of fresh 10 % neutral buffered formalin (NBF) or 4 % paraformaldehyde (PFA) for 16–24 h at RT.
2. Trim larger specimens to ≤3 mm thickness to ensure faster diffusion of the fixative into the tissue. Rinse, dehydrate, and embed in paraffin block.
3. Store FFPE tissue blocks at RT.

3.2 FFPE Tissue Slide Preparation

1. Section FFPE tissue to a thickness of 5 ± 1 μm. The maximum tissue area is 20 × 30 mm and should fit within the hydrophobic barrier (Fig. 1).
2. Mount sections onto one positively charged glass slide. Avoid other colored labels as they tend to give high background.

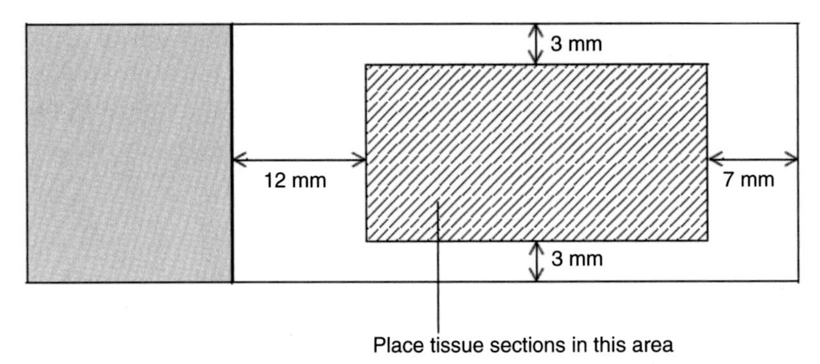

Fig. 1 FFPE tissue slide preparation. Tissue placement to avoid drying and nonspecific background

3. Air-dry freshly mounted sections at RT overnight or at 37 °C for 5 h. Bake slides at 60 °C for 1 h to immobilize tissue sections. Slides are ready to use in the ViewRNA protocol.
4. For short term store sections in a slide box at 4 °C for up to 2 weeks. For long term store sections in a slide box at −20 °C for up to 1 year.

3.3 Optimal Pretreatment Conditions for FFPE Tissue Slides for Manual Processing

1. Use the Pretreatment Lookup Table (Table 1) to find the optimal conditions and the range of tolerance for your sample. Always run a no-probe control slide to assess assay background.
2. If your tissue type is not listed in Pretreatment Lookup Table, and you have only limited slides available for pretreatment optimization, the Heat Pretreatment and Protease Incubation Times for Limited Sample (Table 2) provide the recommended pretreatment times.

3.4 Prepare Buffers and Reagents

1. Verify that the hybridization system is set to 40 °C and that it is appropriately humidified.
2. Prepare the following reagents: 3 L 1× PBS (300 mL 10× PBS + 2.7 L ddH_2O), 200 mL 4 % formaldehyde (178 mL 1× PBS + 22 mL 37 % formaldehyde), 4 L of wash buffer (add in the following order: 3 L ddH_2O + 36 mL Wash Comp 1 + 10 mL Wash Comp 2 and adjust total volume to 4 L with ddH_2O), 500 mL 1× pretreatment solution (in 1 L glass beaker add 5 mL 100× pretreatment solution + 495 mL ddH_2O), 200 mL storage buffer (60 mL Wash Comp 2 + 140 mL ddH_2O), 1 L of 0.01 % ammonium hydroxide (in a fume hood add 0.33 mL 30 % ammonium hydroxide + 999.67 mL ddH_2O), and 200 mL DAPI (final dilution 3.0 μg/mL in 1× PBS, store in the dark at 4 °C—if planning to use fluorescence detection).
3. Ensure availability of 600 mL 100 % ethanol, 1.4 L ddH_2O; 600 mL xylene or 400 mL Histo-Clear, 200 mL Gill's

Table 1
Pretreatment lookup table for manual processing

Tissue information		Optimal conditions (minutes)		Range of tolerance (minutes)
Species	Type	Heat treatment at 90–95 °C	Protease at 40 °C	(Heat treatment, protease)
Mouse	Brain	10	10	
	Heart	10	40	(20, 20)
	Kidney	20	20	(10, 20)
	Liver	20	20	(5, 40); (10, 20)
	Lung	10	20	
	Retina	10	10	
Rat	Kidney	10	20	(10, 10); (20, 20)
	Liver	10	20	(10, 40)
	Spleen	20	10	
	Thyroid	10	20	
Human	Brain	20	10	(10, 10); (10, 20)
	Breast	20	15	(20, 20); (25, 250); (30, 20); (25, 20)
	Colon	5	20	(5, 10)
	Kidney	20	10	(5, 20)
	Liver	20	20	(10, 20)
	Lung	10	20	
	Lymph node	10	20	
	Nasal polyp	5	5	
	Osteoarthritic tissue	20	20	
	Pancreas	10	10	(10, 20), (5, 10)
	Prostate	10	20	(5, 10), (20, 10), (10, 10)
	Salivary gland	10	10	(5, 10)
	Skin	50	10	
	Tonsil	10	20	
	Thyroid	10	20	
Salmon	Heart	10	10	
	Muscle	10	20	
Monkey	Mucosal rectum	10	20	

Hematoxylin I, and 200 mL of 3 μg/mL DAPI in 1× PBS (optional, for fluorescence detection store in the dark at 4 °C until use).

4. Thaw probe(s), briefly centrifuge to collect contents, and place on ice until use.
5. Prewarm 40 mL 1× PBS, Probe Diluent QT, PreAmplifier Mix QT, Amplifier Mix QT, and Label Probe Diluent QF buffers to 40 °C.
6. Bring Fast Red Tablets, Naphthol Buffer, Blue Buffer, and AP Enhancer Solution to RT.

Table 2
Heat pretreatment and protease incubation times for limited sample

Number of available slides	Heat treatment time (minutes)	Protease time (minutes)
3	5	10
	10	10
	10	20
5	5	10
	5	20
	10	10
	10	20
	20	10
7	5	10
	5	20
	10	10
	10	20
	20	10
	20	20
	0	0

3.5 Tissue Pretreatment for lncRNA Visualization

1. Start with prepared tissue (slides baked at 60 °C for 1 h from Subheading 3.2). Label the slides using a pencil.
2. Deparaffinize using xylene (work in a fume hood). Transfer the rack of baked slides to a green clearing dish containing 200 mL of xylene. Incubate the slides at RT in xylene for 5 min with frequent agitation. Repeat step twice with fresh xylene. Remove the slide rack from the xylene and wash the slides twice, each time with 200 mL of 100 % ethanol for 5 min with frequent agitation. Remove the slides from the rack and place them face up on a paper towel to air-dry for 5 min at RT.
3. Dab the hydrophobic barrier pen on a paper towel several times before use. Trace a hydrophobic barrier (Fig. 1). Allow the barrier to dry at RT for 20–30 min.
4. Heat pretreatment (use the optimal time for the tissue of interest—Table 1). Tightly cover the beaker containing the 500 mL of 1× pretreatment solution with aluminum foil, place it on a hot plate, and heat the solution to 90–95 °C. Load the slides into the vertical slide rack. Using a pair of forceps, submerge the slide rack into the heated 1× pretreatment solution. Cover the glass beaker with aluminum foil and incubate at 90–95 °C for the optimal time determined for your tissue. After the pretreatment, remove the slide rack with forceps, submerge it into a clear staining dish containing 200 mL of ddH_2O, and wash for 1 min with frequent agitation. Repeat the wash step one more time with another 200 mL of fresh ddH_2O. Transfer the slide rack to a clear staining dish containing 1× PBS.

IMPORTANT: Do not let the tissue sections dry out from this point forward. After heat pretreatment, sections can be stored covered in 1× PBS at RT overnight.

5. Dilute Protease QF 1:100 in prewarmed 1× PBS and briefly vortex to mix.
6. Remove each slide, flick, and tap on its edge and then wipe the backside on a laboratory wipe. Place the slides face up on a flat platform and immediately add 400 μL of the diluted Protease QF onto the tissue section. Transfer the slides to the hybridization system and incubate at 40 °C for the optimal time for the tissue.
7. Take out one slide at a time, and decant the protease. Immediately place each slide in the slide rack submerged in a clear staining dish filled with 200 mL of 1× PBS. Wash the slides twice, each time with 200 mL of fresh 1× PBS for 1 min with frequent agitation.
8. Transfer the slide rack to a clear staining dish containing 200 mL of 10 % NBF and fix for 5 min at RT under a fume hood. Wash the slides twice, each time with 200 mL of fresh 1× PBS for 1 min with frequent agitation.

3.6 Probe Hybridization to Target lncRNA Using ViewRNA Probes

1. Dilute ViewRNA Probe 1:40 in prewarmed Probe Diluent QT and briefly vortex to mix.
2. Remove each slide, flick, and tap on its edge and then wipe the backside on a laboratory wipe. Place the slides face up on a flat platform and immediately add 400 μL of diluted ViewRNA Probe. Transfer the slides to the hybridization system, and incubate at 40 °C for 2 h.
3. Take out each slide one at a time, and decant the probe solution. Immediately place each slide in the slide rack submerged in a clear staining dish filled with 200 mL of wash buffer. Wash the slides three times, each time with 200 mL of fresh 1× wash buffer for 2 min with constant and vigorous agitation.

3.7 Branched DNA Signal Amplification Using ViewRNA Reagents

1. Briefly swirl PreAmplifier Mix QT bottle to mix the solution. Remove each slide and flick to remove the wash buffer. Tap the slide on its edge and then wipe the backside on a laboratory wipe. Place slides face up on a flat platform and immediately add 400 μL of PreAmplifier Mix QT to each tissue section. Transfer slides to the hybridization system and incubate at 40 °C for 25 min.
2. Take out each slide one at a time, and decant the PreAmplifier Mix QT. Immediately place each slide in the slide rack submerged in a clear staining dish filled with 200 mL of wash buffer. Wash the slides three times, each time with 200 mL of fresh 1× wash buffer for 2 min with constant and vigorous agitation.

3. Briefly swirl Amplifier Mix QT bottle to mix the solution. Remove each slide and flick to remove the wash buffer. Tap the slide on its edge and then wipe the backside on a laboratory wipe. Place slides face up on a flat platform and immediately add 400 μL of Amplifier Mix QT to each tissue section. Transfer the slides to the hybridization system and incubate at 40 °C for 15 min.
4. Take out each slide one at a time, and decant the Amplifier Mix QT. Immediately place each slide in the slide rack submerged in a clear staining dish filled with 200 mL of wash buffer. Wash the slides three times, each time with 200 mL of fresh 1× wash buffer for 2 min with constant and vigorous agitation.
5. Dilute Label Probe 6-AP 1:1,000 in prewarmed Label Probe Diluent QF and briefly vortex to mix. Remove each slide, flick, and tap on its edge and then wipe the backside on a laboratory wipe. Place slides face up on a flat platform and immediately add 400 μL of diluted Label Probe 6-AP solution to each tissue section. Transfer slides to the hybridization system and incubate at 40 °C for 15 min.
6. Take out one slide at a time; decant the Label Probe 6-AP solution. Immediately place each slide in the slide rack submerged in a clear staining dish filled with 200 mL of wash buffer. Wash the slides three times, each time with 200 mL of fresh 1× wash buffer for 2 min with constant and vigorous agitation.
7. Prepare Fast Blue Substrate to use in the next step. In a 15 mL conical tube add 5 mL of Blue Buffer and 105 μL of Blue Reagent 1, and vortex. Add 105 μL of Blue Reagent 2 and vortex. Add 105 μL Blue Reagent 3 and briefly vortex. Protect from light by wrapping in aluminum foil until use.
8. Remove each slide, flick, and tap on its edge and then wipe the backside on a laboratory wipe. Place slides face up on a flat platform and immediately add 400 μL of Fast Blue Substrate. Transfer the slides to the hybridization system and incubate in the dark at RT for 30 min.
9. Take out one slide at a time; decant the Fast Blue Substrate. Immediately place each slide in the slide rack submerged in a clear staining dish filled with 200 mL of wash buffer. Wash the slides three times, each time with 200 mL of fresh 1× wash buffer for 2 min with constant and vigorous agitation.
10. Remove each slide, flick, and tap on its edge and then wipe the backside on a laboratory wipe. Place slides face up on a flat platform. Immediately add 400 μL of the AP Stop QT and incubate in the dark at RT for 30 min. Take out each slide one at a time, and decant the AP Stop QT solution.

Immediately place each slide in the slide rack submerged in a clear staining dish filled with 200 mL of PBS. Wash the slides three times, each time with 200 mL of fresh 1× PBS for 1 min by moving the slide rack up and down.

11. Dilute Label Probe 1-AP 1:1000 in prewarmed Label Probe Diluent QF and briefly vortex to mix. Remove each slide, flick, and tap on its edge and then wipe the backside on a laboratory wipe. Place slides face up on a flat platform and immediately add 400 μL of diluted Label Probe 1-AP solution to each tissue section. Transfer slides to the hybridization system and incubate at 40 °C for 15 min.
12. Take out one slide at a time; decant the Label Probe 1-AP solution. Immediately place each slide in the slide rack submerged in a clear staining dish filled with 200 mL of wash buffer. Wash the slides three times, each time with 200 mL of fresh 1× wash buffer for 2 min with constant and vigorous agitation.
13. Remove each slide, flick, and tap on its edge and then wipe the backside on a laboratory wipe. Place slides face up on a flat platform. Immediately add 400 μL of the AP-Enhancer Solution to each tissue section and incubate at RT for 5 min while preparing the Fast Red Substrate.
14. Prepare the Fast Red Substrate: In a 15 mL conical tube add 5 mL of naphthol buffer and one Fast Red Tablet. Vortex at high speed to completely dissolve the tablet. Protect from light until use by wrapping the tube in aluminum foil.
15. Decant the AP Enhancer Solution and flick the slide twice to completely remove any excess AP Enhancer Solution. Tap the slide on its edge then wipe the backside on a laboratory wipe. Immediately add 400 μL of Fast Red Substrate onto each tissue section. Transfer the slides to the hybridization system and incubate at 40 °C for 30 min.
16. Take out one slide at a time; decant the Fast Red Substrate. Immediately place each slide in the slide rack submerged in a clear staining dish filled with 200 mL of PBS. Move the slide rack up and down for 1 min.

3.8 Counterstain with Hematoxylin and/or DAPI

1. Transfer the slide rack to the clear staining dish containing the 200 mL of Gill's hematoxylin and stain for 30–60 s at RT.
2. Wash the slides three times, each time with 200 mL of fresh ddH_2O for 1 min by moving the slide rack up and down. Pour off the ddH_2O, refill with 200 mL of 0.01 % ammonium hydroxide, and incubate the slides for 10 s. (Unused 0.01 % ammonium hydroxide can be stored at RT for up to 1 month.) Wash the slides once more in 200 mL of fresh ddH_2O by moving the rack up and down for 1 min.

3. Optional—If you plan to view slides using a fluorescent microscope, move the slide rack into a clear staining dish containing 200 mL DAPI (3 μg/mL). Stain the slides for 1 min, and then rinse them in 200 mL of fresh ddH_2O by moving the slide rack up and down for 1 min. Remove each slide, flick, and tap on its edge and then wipe the backside on a laboratory wipe. Place them face up on a paper towel to air-dry in the dark. Ensure that slide sections are completely dry before mounting (~20 min).

3.9 Mount and Image Using DAKO Ultramount Mounting Medium

1. Place slide flat on countertop with specimen facing up. Dab the first 2–3 drops of Ultramount onto a paper towel to remove bubbles. Apply sufficient amount of Ultramount to completely cover the specimen with a thin layer (3–4 drops) of mounting medium.
2. Place slides horizontally in a 70 °C oven/incubator to dry the mounting medium (10–30 min). Drying time will depend on the amount of mounting medium applied. Image or store slides at RT.

3.10 Mount and Image Using ADVANTAGE Mounting Medium

1. Place a 24 × 55 mm cover glass horizontally onto a clean, flat surface. Dab the first 2–3 drops of mounting media onto a paper towel to remove bubbles. Add two drops of the ADVANTAGE medium directly onto the middle of the cover glass.
2. Use a pipette tip to draw out any air bubbles in the droplets. Invert the specimen slide and slowly place it onto the mounting medium at an angle. Make sure that the tissue comes into contact with the mounting medium first before completely letting go of the glass slide to overlap with the cover glass. After mounting, flip the slide over and place it on its edge on a laboratory wipe to soak up and remove excess mounting medium. Allow the slide to dry at RT in the dark for 15 min. Do not bake the slides to speed up the drying process.
3. To prevent bubble formation, seal all four edges of the cover glass with a flat black-colored nail polish (iridescent or colored nail polish can autofluoresce and interfere with fluorescent imaging). Image the results using a bright-field and/or fluorescence microscope. Store slides at RT.

4 Notes

1. Reagent preparation: Always scale reagents according to the number of assays to be run. Include one slide volume overage.
2. Avoid colored labels as they tend to give high background.

3. When dispensing reagents on slides, make sure that the tissue section is fully covered with solution.
4. Endogenous alkaline phosphatase: The bDNA assays use alkaline phosphatase to convert a chromogenic substrate into a colored signal. For this reason it is important to assess the level of endogenous alkaline phosphatase (AP) activity in your tissue of interest prior to performing the assay. Certain types of tissue (such as stomach, intestine, placenta, and mouse embryo) are known to possess high levels of endogenous AP activity that can interfere with the assay. To empirically determine the level of endogenous AP activity in your tissue type, perform the pretreatment protocol as instructed. After the protease treatment and fixation in 10 % NBF, wash the samples in 1× TBS (Sigma, T5912-1 L), and incubate the sections with either Fast Blue Substrate or Fast Red Substrate. If present, endogenous AP can be inactivated with 0.2 M HCl/300 mM NaCl at RT for 15 min just before the probe hybridization but after the sample has undergone protease treatment, 10 % NBF fixation, and two washes in 1× PBS.
5. ViewRNA is a hybridization-based assay. Therefore all temperatures must be closely monitored.
6. The most critical step in the assays is the tissue pretreatment. If tissue is not properly pretreated, RNA may not be exposed and available for detection (under treatment) or RNA may be floating out of the tissue (over treatment).
7. Always run a no-probe control to assess background and a housekeeping gene probe to assess RNA integrity in the sample.
8. When designing TMAs to be used in RNA ISH assay, it is important to understand that only one optimized condition can be used when running the assay. Therefore, if you want multiple tissue types within the same TMA block, you should choose tissues with similar optimal conditions.
9. Two-plex assays considerations: The advantage of using alkaline phosphatase-conjugated label probe for the enzymatic signal amplification is the availability of substrates with dual property, such as Fast Red and Fast Blue, which allows for both chromogenic and fluorescent detection of the targets. However, for a 2-plex assay in which both Label Probe 1 and Label Probe 6 are conjugated to the same alkaline phosphatase, the enzyme conjugates are unable to differentiate between Fast Red and Fast Blue if both substrates are added simultaneously. As a result, the enzymatic signal amplification has to be performed sequentially in order to direct substrate/color specificity to each target. Additionally, complete inactivation of the first alkaline phosphatase-conjugated label probe (LP6-AP) is

necessary, especially when employing fluorescence mode for the detection of the targets. Otherwise, the residual LP6-AP activity can also convert Fast Red substrate in subsequent step into a red signal even at locations where TYPE 1 target is not present, giving a false impression that the Fast Blue and Fast Red signals are co-localized. For this reason, it is absolutely necessary to quench any residual LP6-AP activity with the ViewRNA AP Stop QT prior to proceeding with the second label probe hybridization and development of the Fast Red color as this will ensure specific signals in fluorescent mode and brighter aqua blue dots in chromogenic mode.

10. Fast Red has a very broad emission spectrum and its bright signal can bleed into adjacent Cy5 channel if one uses the standard Cy3/Cy5 filter sets for imaging. For this reason, it is critical that the recommended filter set for Fast Blue detection is used to avoid spectral bleed through of the Fast Red signal into the Fast Blue channel and interfering with Fast Blue detection. For Fast Red Substrate, use Cy3/TRITC filter set: excitation: 530 ± 20 nm; emission: 590 ± 20 nm; dichroic: 562 nm. For Fast Blue Substrate, use custom filter set: excitation: 630 ± 20 nm; emission: 775 ± 25 nm; dichroic: 750 nm. For DAPI filter set: excitation: 387/11 nm; emission: 447/60 nm.

References

1. Zhang B, Arun G, Mao YS, Lazar Z, Hung G, Bhattacharjee G, Xiao X, Booth CJ, Wu J, Zhang C, Spector DL (2012) The lncRNA Malat1 is dispensable for mouse development but its transcription plays a cis-regulatory role in the adult. Cell Rep 2:111–123
2. Chodroff RA, Goodstadt L, Sirey TM, Oliver PL, Davies KE, Green ED, …, Ponting CP (2010) Long noncoding RNA genes: conservation of sequence and brain expression among diverse amniotes. Genome Biol 11(7): R72
3. Guttman M, Amit I, Garber M, French C, Lin MF, Feldser D, …, Lander ES (2009) Chromatin signature reveals over a thousand highly conserved large non-coding RNAs in mammals. Nature 458(7235): 223–227
4. Kapranov P, Willingham AT, Gingeras TR (2007) Genome-wide transcription and the implications for genomic organization. Nat Rev Genet 8(6):413–423
5. Ørom UA, Derrien T, Beringer M, Gumireddy K, Gardini A, Bussotti G, …, Shiekhattar R (2010) Long noncoding RNAs with enhancer-like function in human cells. Cell 143(1): 46–58
6. Mercer TR, Dinger ME, Sunkin SM, Mehler MF, Mattick JS (2008) Specific expression of long noncoding RNAs in the mouse brain. Proc Natl Acad Sci 105(2):716–721
7. Kogo R, Shimamura T, Mimori K, Kawahara K, Imoto S, Sudo T, …, Mori M (2011) Long noncoding RNA HOTAIR regulates polycomb-dependent chromatin modification and is associated with poor prognosis in colorectal cancers. Cancer Res 71(20):6320–6326
8. Wilusz JE, Sunwoo H, Spector DL (2009) Long noncoding RNAs: functional surprises from the RNA world. Genes Dev 23: 1494–1504
9. Dong R, Jia D, Xue P, Cui X, Li K, Zheng S, …, Dong K (2014) Genome-wide analysis of long noncoding RNA (lncRNA) expression in hepatoblastoma tissues. PloS one 9(1): e85599
10. Prensner JR, Chinnaiyan AM (2011) The emergence of lncRNAs in cancer biology. Cancer Discov 1(5):391–407
11. Mercer TR, Gerhardt DJ, Dinger ME, Crawford J, Trapnell C, Rinn JL et al (2012) Targeted RNA sequencing reveals the deep complexity of the human transcriptome. Nat Biotechnol 30(1):99–104

Chapter 14

Detection of Long Noncoding RNA Expression by Nonradioactive Northern Blots

Xiaowen Hu, Yi Feng, Zhongyi Hu, Youyou Zhang, Chao-Xing Yuan, Xiaowei Xu, and Lin Zhang

Abstract

With the advances in sequencing technology and transcriptome analysis, it is estimated that up to 75 % of the human genome is transcribed into RNAs. This finding prompted intensive investigations on the biological functions of noncoding RNAs and led to very exciting discoveries of microRNAs as important players in disease pathogenesis and therapeutic applications. Research on long noncoding RNAs (lncRNAs) is in its infancy, yet a broad spectrum of biological regulations has been attributed to lncRNAs. As a novel class of RNA transcripts, the expression level and splicing variants of lncRNAs are various. Northern blot analysis can help us learn about the identity, size, and abundance of lncRNAs. Here we describe how to use northern blot to determine lncRNA abundance and identify different splicing variants of a given lncRNA.

Key words Long noncoding RNA, RNA expression, Northern blots

1 Introduction

lncRNAs are operationally defined as RNA transcripts larger than 200 nt that do not appear to have coding potential [1–5]. Given that up to 75 % of the human genome is transcribed to RNA, while only a small portion of the transcripts encodes proteins [6], the number of lncRNA genes can be large. After the initial cloning of functional lncRNAs such as H19 [7, 8] and XIST [9] from cDNA libraries, two independent studies using high-density tiling array reported that the number of lncRNA genes is at least comparable to that of protein-coding genes [10, 11]. Recent advances in tiling array [10–13], chromatin signature [14, 15], computational analysis of cDNA libraries [16, 17], and next-generation sequencing (RNA-seq) [18–21] have revealed that thousands of lncRNA genes are abundantly expressed with exquisite cell type and tissue specificity in human. In fact, the GENCODE consortium within the framework of the ENCODE project recently reported 14,880

Yi Feng and Lin Zhang (eds.), *Long Non-Coding RNAs: Methods and Protocols*, Methods in Molecular Biology, vol. 1402,
DOI 10.1007/978-1-4939-3378-5_14,

manually annotated and evidence-based lncRNA transcripts originating from 9,277 gene loci in human [6, 21], including 9,518 intergenic lncRNAs (also called lincRNAs) and 5362 genic lncRNAs [14, 15, 20]. These studies indicate that (1) lncRNAs are independent transcriptional units; (2) lncRNAs are spliced with fewer exons than protein-coding transcripts and utilize the canonical splice sites; (3) lncRNAs are under weaker selective constraints during evolution and many are primate specific; (4) lncRNA transcripts are subjected to typical histone modifications as protein-coding mRNAs; and (5) the expression of lncRNAs is relatively low and strikingly cell type or tissue specific.

The discovery of lncRNA has provided an important new perspective on the centrality of RNA in gene expression regulation. lncRNAs can regulate the transcriptional activity of a chromosomal region or a particular gene by recruiting epigenetic modification complexes in either *cis*- or *trans*-regulatory manner. For example, Xist, a 17 kb X-chromosome-specific noncoding transcript, initiates X chromosome inactivation by targeting and tethering Polycomb-repressive complexes (PRC) to X chromosome in *cis* [22–24]. HOTAIR regulates the HoxD cluster genes in *trans* by serving as a scaffold which enables RNA-mediated assembly of PRC2 and LSD1 and coordinates the binding of PRC2 and LSD1 to chromatin [12, 25]. Based on the knowledge obtained from studies on a limited number of lncRNAs, at least two working models have been proposed. First, lncRNAs can function as scaffolds. lncRNAs contain discrete protein-interacting domains that can bring specific protein components into the proximity of each other, resulting in the formation of unique functional complexes [25–27]. These RNA-mediated complexes can also extend to RNA–DNA and RNA–RNA interactions. Second, lncRNAs can act as guides to recruit proteins [24, 28, 29], such as chromatin modification complexes, to chromosome [24, 29]. This may occur through RNA–DNA interactions [29] or through RNA interaction with a DNA-binding protein [24]. In addition, lncRNAs have been proposed to serve as decoys that bind to DNA-binding proteins [30], transcriptional factors [31], splicing factors [32–34], or miRNAs [35]. Some studies have also identified lncRNAs transcribed from the enhancer regions [36–38] or a neighbor loci [18, 39] of certain genes. Given that their expressions correlated with the activities of the corresponding enhancers, it was proposed that these RNAs (termed enhancer RNA/eRNA [36–38] or ncRNA-activating/ncRNA-a [18, 39]) may regulate gene transcription.

As a novel class of RNA transcripts, the expression level and splicing variants of lncRNAs are various. Northern blot analysis can help us learn about the identity, size, and abundance of lncRNAs. Here we describe how to use northern blot to determine lncRNA abundance and identify different splicing variants of a given lncRNA.

2 Materials

2.1 Materials for DIG-Labeled RNA Probe Synthesis

1. DNA template (*see* **Note 1**).
2. Restriction enzyme.
3. 10× DIG RNA labeling mix (Roche, 11277073910).
4. T7 RNA polymerase (10 U/μL) and 5× transcription buffer (Agilent, 600123).
5. Dnase I (NEB, M0303S, 2000 U/mL).
6. Agarose gel electrophoresis supplies for DNA fragment purification.
7. Quick Spin Columns for radiolabeled RNA purification Sephadex G-50 (Roche, 11274015001).
8. Gel Extraction Kit (Qiagen, 28704).

2.2 Materials for Separating RNA by Electrophoresis

1. Nucleic acid agarose.
2. 55 °C water bath.
3. 10× Denaturing gel buffer (Invitrogen, AM8676).
4. Heat block.
5. Gel electrophoresis apparatus.
6. 3-(N-morpholino) propanesulfonic acid (MOPS).
7. Sodium acetate.
8. 0.5 M EDTA.
9. 10× MOPS buffer: 200 mM MOPS, 50 mM sodium acetate, 20 mM EDTA, adjust pH to 7.0. To make 1× MOPS gel running buffer, mix one part of 10× MOPS buffer with nine parts of RNase-free water.
10. RNA loading buffer (Invitrogen, AM8552).
11. Ethidium bromide (only if RNA visualization is needed).
12. DIG-labeled RNA marker (Roche, 11373099910).

2.3 Materials for Transferring RNA to the Membrane

1. 20× SSC (Invitrogen, 15557-036).
2. Razor blade.
3. 3 M Filter paper.
4. Positively Charged Nylon Membrane (Roche).
5. Blunt end forceps.
6. Paper towel.
7. RNase-free flat-bottomed container as buffer reservoir.
8. Clean glass pasture pipet as roller.
9. Lightweight (150–200 g) object serving as weight during transfer.

10. Supports of the reservoir (i.e., a stack of books).
11. Stratalinker® UV Crosslinker.

2.4 Materials for Probe-RNA Hybridization

1. 20× SSC.
2. 10 % SDS.
3. DIG easy Hyb Granules (Roche, 11796895001).
4. 68 °C Shaking water bath.
5. Heat block.
6. Hybridization oven.
7. Hybridization bags.
8. Low-stringency buffer: 2× SSC with 0.1 % SDS.
9. High-stringency buffer: 0.1× SSC with 0.1 % SDS.

2.5 Materials for Detection of Probe-RNA Hybrids

1. Washing and blocking buffer set (Roche, 11585762001).
2. Anti-DIG-alkaline phosphatase antibody (Roche, 11093274910).
3. NBT/BCIP Stock Solution (Roche, 11681451001).
4. CDP-Star, Ready-to-Use (Roche, 12 041 677 001).
5. TE buffer: 10 mM Tris–HCL, 1 mM EDTA, adjust pH to approximately 8.

3 Methods (*See* Note 2)

lncRNA northern blot analysis aims to characterize lncRNA expression. The protocol includes five parts: (1) RNA probe synthesis and labeling; (2) RNA sample electrophoresis; (3) RNA transfer; (4) RNA-probe hybridization; and (5) RNA-probe hybrid detection.

3.1 DIG-Labeled RNA Probe Synthesis by In Vitro Transcription

The DIG-labeled RNA probe synthesis is very similar to the biotinylated RNA synthesis described in lncRNA pull-down assay. The differences between the two procedures are the following:

1. Since the probe needs to be complement to the target sequence, the probe RNA is transcribed from the 3′ end of the target sequence. We clone the gene of interest in reverse orientation to make the in vitro transcription template for northern probes.
2. Use DIG labeling mix in place for the biotin-label mix.

3.2 Separating RNA Samples by Electrophoresis

3.2.1 Gel Setup

1. Wipe the gel rack, tray, and combs with RNAZap, rinse with water, and let dry.
2. Weight 100 g agarose in a clean glass flask and mix with 90 mL RNase-free water. Melt the agarose completely by heating with a microwave. Put the flask with melted agarose in a 55 °C water bath.

3. In a fume hood (*see* **Note 3**), add 10 mL 10× denaturing gel buffer to the gel mix that is equilibrated to 55 °C. Mix the gel solution by gentle swirling to avoid generating bubbles. Slowly pour the gel mix into the gel tray, pop any bubbles, or push them to the edges of the gel with a clean pipet tip. The thickness of the gel should be about 6 mm. Slowly place the comb in the gel. Allow the gel to solidify before removing the comb.
4. Right before RNA electrophoresis, place the gel tray in the electrophoresis chamber with the wells near the negative lead and add 1× MOPS gel running buffer in the chamber until it is 0.5–1 cm over the top of the gel (*see* **Note 4**).

3.2.2 RNA Electrophoresis

1. Mix no more than 30 μg sample RNA with 3 volumes of RNA loading buffer (*see* **Notes 5** and **6**). To destruct any secondary structure of the RNAs, incubate the RNA with loading buffer at 65 °C for 15 min using a heat block. Spin briefly to collect samples to the bottom of the tube and put the tubes on ice (*see* **Note 7**).
2. Carefully draw the RNAs in the tip without trapping any bubbles at the end of pipet tip, place the pipet tip inside of the top of the well, slowly push samples into the well, and exit the tip without disturbing the loaded samples. If markers are needed, load one lane with DIG-labeled RNA marker.
3. Run the gel at 5 V/cm (*see* **Note 8**).
4. (Optional) Stain the gel with ethidium bromide and visualize the RNA under UV (*see* **Note 9**).

3.3 Transfer RNA from Agarose Gel to the Membrane

3.3.1 Material Preparation

1. Use a razor blade to trim the gel by cutting through the wells and discard the unused gel above the wells. For marking the orientation, make a notch at a corner.
2. Cut the membrane to the size slightly larger than the gel. Make a notch at a corner to align the membrane with gel in the same orientation. Handle the membrane with care—only touching the edges with gloved hands or blunt tip forceps.
3. Cut eight pieces of filter paper the same size of the membrane.
4. Cut a stack of paper towels that are 3 cm in height and 1–2 cm wider than the gel.
5. Pour 20× SSC into a flat-bottomed container that has bigger dimension of the agarose gel. This serves as the buffer reservoir and can also be used to wet the paper and membrane. Put the reservoir on a support (i.e., a stack of books) so that its bottom is higher than the paper towel stack.
6. Cut three pieces of filter paper that are large enough to cover the gel and long enough to reach over to the reservoir. These papers serve as the bridge to transfer buffer from the reservoir to the gel.

3.3.2 Transfer Setup

1. Stack paper towel on a clean bench and put three pieces of dry filter paper on top.
2. Wet two more pieces of filter paper and put on top of the dry filter paper. Gently roll out any bubbles between the filter paper layers.
3. Carefully put the membrane on top of the wet filter paper. Gently roll out any bubbles between the membrane and the filter papers.
4. Put the trimmed gel onto the center of the membrane with the bottom of the gel touching the membrane (i.e., the gel plane that faces down during electrophoresis will be in contact with the membrane), and align the notches of the gel and membrane. Roll out bubbles between the membrane and the gel.
5. Place three more pieces of pre-wet filter paper on top of the gel and roll out bubbles between filter paper layers.
6. Wet the three pieces of paper bridge and place them with one end on top of the stack and the other end in buffer reservoir. Make sure that there is no bubble between any layers of paper (*see* **Note 10**).
7. Place a 150–200 g object with the size similar to the gel on top of the stack.
8. Transfer the gel for 15–20 min per mm of gel thickness. It usually takes about 2 h (*see* **Note 11**).

3.3.3 RNA Cross-Link

Disassemble the transfer stack carefully and rinse the member with 1× MOPS gel running buffer to remove residual agarose. Blot off excessive liquid and immediately subject the membrane to cross-link treatment. Cross-link the RNA to the membrane with Stratalinker® UV Crosslinker using the autocross-link setting (*see* **Note 12**). Air-dry the membrane at room temperature. At this point, the membrane can be subjected to hybridization immediately or stored in a sealed bag between two pieces of filter paper at 4 °C for several months before hybridization.

3.4 Hybridization of DIG-Labeled Probes to the Membrane (See Note 13)

3.4.1 Prehybridization

1. Reconstitute the DIG easy Hyb Granules: Add 64 mL RNase-free water into one bottle of the DIG easy Hyb Granules, and stir for 5 min at 37 °C to completely dissolve the granules. DIG easy Hyb buffer will be used in prehybridization and hybridization. The reconstituted DIG easy Hyb buffer is stable at room temperature for up to 1 month.
2. For every 100 cm^2 membrane, 10–15 mL Hyb buffer should be used for prehybridization. Measure the appropriate amount of Hyb buffer for prehybridization, place it in a clean tube, and pre-warm it in a 68 °C water bath (*see* **Note 14**).

3. Put the membrane in a hybridization bag, add the pre-warmed Hyb buffer from the previous step, seal the bag properly, and incubate the membrane in Hyb buffer at 68 °C for at least 30 min with gentle agitation (*see* **Note 15**). Prehybridization can be up to several hours as far as the membrane remains wet.

3.4.2 Hybridization

1. For every 100 cm^2 membrane, 3.5 mL Hyb buffer is needed for hybridization. Measure the appropriate amount of Hyb buffer for hybridization, place it in a clean tube, and pre-warm it in a 68 °C water bath (*see* **Note 14**).
2. Determine the amount of RNA probe needed (*see* **Note 16**) and place it into a microcentrifuge tube with 50 μL RNase-free water. Denature the probe by heating the tube at 85 °C for 5 min and chill on ice immediately.
3. Mix the denatured probe with pre-warmed Hyb buffer by inversion.
4. Remove prehybridization buffer from the membrane and immediately replace with the pre-warmed hybridization buffer containing the probe.
5. Seal the bag properly and incubate the membrane in probe-containing Hyb buffer at 68 °C overnight with gentle agitation (*see* **Note 15**).
6. The next day, pre-warm the high-stringency buffer to 68 °C and pour low-stringency buffer in an RNase-free container at room temperature and make sure that it is enough to cover the membrane.
7. Cut open the hybridization bag, remove the Hyb buffer, and immediately submerge the membrane in the low-stringency buffer.
8. Wash the membrane twice in low-stringency buffer at room temperature for 5 min each time with shaking.
9. Wash the membrane twice in high-stringency buffer at 68 °C for 5 min each time with shaking (*see* **Note 17**).

3.5 Detection of DIG-Probe/Target RNA Hybrids

3.5.1 Localizing the Probe-Target Hybrid with Anti-DIG Antibody

1. Transfer the membrane from the last wash in high-stringency buffer to a plastic container with 100 mL washing buffer. Incubate for 2 min at room temperature and discard the washing buffer.
2. Add 100 mL blocking buffer onto the membrane and incubate for more than 30 min (up to 3 h) with shaking at room temperature.
3. Dilute anti-DIG-alkaline phosphatase antibody at the ratio of 1:5000 in blocking buffer and incubate the membrane in 20 mL diluted antibody for 30 min at room temperature with shaking.
4. Wash membrane twice with 100 mL of washing buffer for 15 min each time at room temperature.

*3.5.2 Visualizing Probe-Target Hybrids Using Chromogenic or Chemiluminescent Method (See **Note 18**)*

1. Equilibrate the membrane in 20 mL detection buffer for 3 min at room temperature. If using the chromogenic method, prepare the color substrate solution while equilibrating the membrane.
2. For chromogenic detection:
   (a) Put the membrane with the RNA side facing up in a container and incubate in 10 mL color substrate solution in the dark without shaking.
   (b) When the desired intensity for the band is observed, discard the color substrate solution and rinse the membrane in 50 mL of TE buffer for 5 min (*see* **Note 19**).
   (c) Document the result by photographing the membrane (*see* **Note 20**).
3. For chemiluminescent detection:
   (a) Put the membrane with the RNA side facing up on a plastic sheet (i.e., cut out of a hybridization bag) and add 20 drops of CDP-Star, Ready-to-Use reagent.
   (b) Immediately cover the membrane with another sheet to evenly distribute the reagent without creating any bubbles.
   (c) Incubate for 5 min at room temperature.
   (d) Squeeze out excess reagent and seal the bag.
   (e) Develop the membrane with an X-ray film in a darkroom (*see* **Note 20**).

4 Notes

1. The desired DNA template should be a plasmid containing a promoter for in vitro transcription (i.e., T7 or T3) and a target sequence whose 5′ end is placed as close as possible to the 3′ end of the promoter. We usually use pBluescript SK(+) and transcribe the target sequence using T7 polymerase. Minimize any unnecessary addition of non-lncRNA sequence into the plasmid to avoid impropriate RNA folding.
2. Northern blot analysis is a golden standard in RNA detection and analysis. There are many protocols developed by laboratories specialized in RNA research or companies. The protocol described here is adapted from the NorthernMax procedure from Invitrogen and DIG application manual for filter hybridization from Roche. In our hand, this protocol is time efficient and gives satisfying results without using radioactivity.
3. Always cast the gel in a fume hood as the denaturing solution contains formaldehyde. Solidified gels can be wrapped up and stored at 4 °C for overnight.

4. Do not let gel soaked in running buffer for more than 1 h before loading.
5. Load no more than 30 μg total RNA in each lane. As the binding capacity of the membrane is limited, more RNA loaded does not guarantee a stronger signal. Overloading can lead to the detection of minor degradation of targeted RNAs.
6. If the total volume of sample and dye exceeds the capacity of the wells, it is necessary to concentrate the RNA by precipitation and suspend the pellet in smaller volume of water before adding the loading dye.
7. Use a heat block instead of a water bath to avoid contaminating the samples with water.
8. The voltage is decided by the distance between the two electrodes (not the size of the gel). Usually, the run takes about 2 h. If the run is longer than 3 h, exchange the buffer at the two end chambers to avoid the pH gradient.
9. RNA gels that stained with ethidium bromide are not suitable for northern blot analysis. Therefore, if a visual examination or photograph of total RNA samples is needed as a reference for the northern blot, we suggest the researchers to either run the same set of samples on a separate gel or stain with ethidium bromide the gel just for visualization; or de-stain the gel before continuing northern blot analysis. If a gel will be subjected to northern analysis after UV visualization, avoid prolonged exposure of the gel to UV light.
10. It is essential to ensure that the only way for the transfer buffer to run from reservoir to the dry paper stack is through the gel. Therefore, extra care is needed to assemble the stack properly to avoid shortcut. The most common shortcut happens between the bridge and the paper beneath the gel. One can cover the edges of the gel with Parafilm to prevent this from happening.
11. Transfer longer than 4 h may cause small RNA hydrolysis and reduce yield.
12. The autocross-link Mode of Stratalinker® UV Crosslinker delivers a preset exposure of 1200 μJ to the membrane and takes about 40 s. Other methods of cross-linking RNA to membrane are available and can be used at this step as well.
13. Once the membrane is wet during prehybridization, it is important to avoid it getting dry during the hybridization and detection process. Dried membrane will have high background. Only if the membrane will not be stripped and reprobed, it can be dried after the last high-stringency wash and stored at 4 °C for future analysis.

14. For most northern blot hybridization using DIG easy Hyb buffer, 68 °C is appropriate for both prehybridization and hybridization. In cases of more heterologous RNA probes being used, the prehybridization and hybridization temperature need to be optimized.
15. Prehybridization/hybridization can be performed in containers other than bags, as far as it can be tightly sealed. Sealing the hybridization container can prevent the release of NH_4, which changes the pH of the buffer, during incubation.
16. For RNA probe synthesized by in vitro transcription, it is recommended that the probe concentration should be 100 ng per mL Hyb buffer.
17. If the probe is less than 80 % homologous to the target RNA, the high-stringency wash should be performed at a lower temperature, which needs to be empirically determined.
18. The DIG probe-target RNA hybrids can be detected in two ways. One uses chemiluminescent method, whereas the other uses chromogenic method. The chemiluminescent method is sensitive and fast, but it requires the usage of the films and the accessibility of a darkroom. The chromogenic method requires no film or darkroom and different targets can be detected simultaneously using different colored substrate. However, the chromogenic method may not be sensitive enough for low-abundance targets.
19. At this step, if there are multiple membranes, process one at a time. Depending on the abundance of target RNAs, the band may appear as quickly as a few minutes after adding the chromogenic agents. The reaction can be stopped when the band reaches a desired intensity.
20. If reprobing is needed, photograph the result while the membrane is wet and proceed to stripping and reprobing. If no reprobing is needed, dry the membrane, document the result by photograph, and store the dried membrane in a clean bag at room temperature.

Acknowledgments

This work was supported, in whole or in part, by the Basser Research Center for BRCA, the NIH (R01CA142776, R01CA190415, P50CA083638, P50CA174523), the Ovarian Cancer Research Fund (X.H.), the Breast Cancer Alliance, Foundation for Women's Cancer (X.H.), and the Marsha Rivkin Center for Ovarian Cancer Research.

References

1. Rinn JL, Chang HY (2012) Genome regulation by long noncoding RNAs. Annu Rev Biochem 81:145–166
2. Lee JT (2012) Epigenetic regulation by long noncoding RNAs. Science 338:1435–1439
3. Prensner JR, Chinnaiyan AM (2011) The emergence of lncRNAs in cancer biology. Cancer Discov 1:391–407
4. Guttman M, Rinn JL (2012) Modular regulatory principles of large non-coding RNAs. Nature 482:339–346
5. Spizzo R, Almeida MI, Colombatti A, Calin GA (2012) Long non-coding RNAs and cancer: a new frontier of translational research? Oncogene 31:4577–4587
6. Djebali S, Davis CA, Merkel A, Dobin A, Lassmann T, Mortazavi A et al (2012) Landscape of transcription in human cells. Nature 489: 101–108
7. Pachnis V, Brannan CI, Tilghman SM (1988) The structure and expression of a novel gene activated in early mouse embryogenesis. EMBO J 7:673–681
8. Bartolomei MS, Zemel S, Tilghman SM (1991) Parental imprinting of the mouse H19 gene. Nature 351:153–155
9. Brown CJ, Ballabio A, Rupert JL, Lafreniere RG, Grompe M, Tonlorenzi R et al (1991) A gene from the region of the human X inactivation centre is expressed exclusively from the inactive X chromosome. Nature 349:38–44
10. Kapranov P, Cawley SE, Drenkow J, Bekiranov S, Strausberg RL, Fodor SP et al (2002) Large-scale transcriptional activity in chromosomes 21 and 22. Science 296:916–919
11. Rinn JL, Euskirchen G, Bertone P, Martone R, Luscombe NM, Hartman S et al (2003) The transcriptional activity of human Chromosome 22. Genes Dev 17:529–540
12. Rinn JL, Kertesz M, Wang JK, Squazzo SL, Xu X, Brugmann SA et al (2007) Functional demarcation of active and silent chromatin domains in human HOX loci by noncoding RNAs. Cell 129:1311–1323
13. Gupta RA, Shah N, Wang KC, Kim J, Horlings HM, Wong DJ et al (2010) Long non-coding RNA HOTAIR reprograms chromatin state to promote cancer metastasis. Nature 464:1071–1076
14. Guttman M, Amit I, Garber M, French C, Lin MF, Feldser D et al (2009) Chromatin signature reveals over a thousand highly conserved large non-coding RNAs in mammals. Nature 458:223–227
15. Khalil AM, Guttman M, Huarte M, Garber M, Raj A, Rivea Morales D et al (2009) Many human large intergenic noncoding RNAs associate with chromatin-modifying complexes and affect gene expression. Proc Natl Acad Sci U S A 106:11667–11672
16. Maeda N, Kasukawa T, Oyama R, Gough J, Frith M, Engstrom PG et al (2006) Transcript annotation in FANTOM3: mouse gene catalog based on physical cDNAs. PLoS Genet 2:e62
17. Jia H, Osak M, Bogu GK, Stanton LW, Johnson R, Lipovich L (2010) Genome-wide computational identification and manual annotation of human long noncoding RNA genes. RNA 16:1478–1487
18. Orom UA, Derrien T, Beringer M, Gumireddy K, Gardini A, Bussotti G et al (2010) Long noncoding RNAs with enhancer-like function in human cells. Cell 143:46–58
19. Prensner JR, Iyer MK, Balbin OA, Dhanasekaran SM, Cao Q, Brenner JC et al (2011) Transcriptome sequencing across a prostate cancer cohort identifies PCAT-1, an unannotated lincRNA implicated in disease progression. Nat Biotechnol 29:742–749
20. Cabili MN, Trapnell C, Goff L, Koziol M, Tazon-Vega B, Regev A et al (2011) Integrative annotation of human large intergenic noncoding RNAs reveals global properties and specific subclasses. Genes Dev 25:1915–1927
21. Derrien T, Johnson R, Bussotti G, Tanzer A, Djebali S, Tilgner H et al (2012) The GENCODE v7 catalog of human long noncoding RNAs: analysis of their gene structure, evolution, and expression. Genome Res 22:1775–1789
22. Brown CJ, Hendrich BD, Rupert JL, Lafreniere RG, Xing Y, Lawrence J et al (1992) The human XIST gene: analysis of a 17 kb inactive X-specific RNA that contains conserved repeats and is highly localized within the nucleus. Cell 71:527–542
23. Zhao J, Sun BK, Erwin JA, Song JJ, Lee JT (2008) Polycomb proteins targeted by a short repeat RNA to the mouse X chromosome. Science 322:750–756
24. Jeon Y, Lee JT (2011) YY1 tethers Xist RNA to the inactive X nucleation center. Cell 146: 119–133
25. Tsai MC, Manor O, Wan Y, Mosammaparast N, Wang JK, Lan F et al (2010) Long noncoding RNA as modular scaffold of histone modification complexes. Science 329:689–693
26. Yap KL, Li S, Munoz-Cabello AM, Raguz S, Zeng L, Mujtaba S et al (2010) Molecular

interplay of the noncoding RNA ANRIL and methylated histone H3 lysine 27 by polycomb CBX7 in transcriptional silencing of INK4a. Mol Cell 38:662–674

27. Kotake Y, Nakagawa T, Kitagawa K, Suzuki S, Liu N, Kitagawa M et al (2011) Long noncoding RNA ANRIL is required for the PRC2 recruitment to and silencing of p15(INK4B) tumor suppressor gene. Oncogene 30:1956–1962
28. Huarte M, Guttman M, Feldser D, Garber M, Koziol MJ, Kenzelmann-Broz D et al (2010) A large intergenic noncoding RNA induced by p53 mediates global gene repression in the p53 response. Cell 142:409–419
29. Grote P, Wittler L, Hendrix D, Koch F, Wahrisch S, Beisaw A et al (2013) The tissue-specific lncRNA Fendrr is an essential regulator of heart and body wall development in the mouse. Dev Cell 24:206–214
30. Kino T, Hurt DE, Ichijo T, Nader N, Chrousos GP (2010) Noncoding RNA gas5 is a growth arrest- and starvation-associated repressor of the glucocorticoid receptor. Sci Signal 3:ra8
31. Hung T, Wang Y, Lin MF, Koegel AK, Kotake Y, Grant GD et al (2011) Extensive and coordinated transcription of noncoding RNAs within cell-cycle promoters. Nat Genet 43: 621–629
32. Tripathi V, Ellis JD, Shen Z, Song DY, Pan Q, Watt AT et al (2010) The nuclear-retained noncoding RNA MALAT1 regulates alternative splicing by modulating SR splicing factor phosphorylation. Mol Cell 39:925–938
33. Bernard D, Prasanth KV, Tripathi V, Colasse S, Nakamura T, Xuan Z et al (2010) A long nuclear-retained non-coding RNA regulates synaptogenesis by modulating gene expression. EMBO J 29:3082–3093
34. Tripathi V, Shen Z, Chakraborty A, Giri S, Freier SM, Wu X et al (2013) Long noncoding RNA MALAT1 controls cell cycle progression by regulating the expression of oncogenic transcription factor B-MYB. PLoS Genet 9:e1003368
35. Salmena L, Poliseno L, Tay Y, Kats L, Pandolfi PP (2011) A ceRNA hypothesis: the Rosetta Stone of a hidden RNA language? Cell 146:353–358
36. Kim TK, Hemberg M, Gray JM, Costa AM, Bear DM, Wu J et al (2010) Widespread transcription at neuronal activity-regulated enhancers. Nature 465:182–187
37. Wang D, Garcia-Bassets I, Benner C, Li W, Su X, Zhou Y et al (2011) Reprogramming transcription by distinct classes of enhancers functionally defined by eRNA. Nature 474: 390–394
38. Melo CA, Drost J, Wijchers PJ, van de Werken H, de Wit E, Oude Vrielink JA et al (2013) eRNAs are required for p53-dependent enhancer activity and gene transcription. Mol Cell 49:524–535
39. Lai F, Orom UA, Cesaroni M, Beringer M, Taatjes DJ, Blobel GA et al (2013) Activating RNAs associate with Mediator to enhance chromatin architecture and transcription. Nature 494:497–501

Chapter 15

Assessment of In Vivo siRNA Delivery in Cancer Mouse Models

Hiroto Hatakeyama, Sherry Y. Wu, Lingegowda S. Mangala, Gabriel Lopez-Berestein, and Anil K. Sood

Abstract

RNA interference (RNAi) has rapidly become a powerful tool for target discovery and therapeutics. Small interfering RNAs (siRNAs) are highly effective in mediating sequence-specific gene silencing. However, the major obstacle for using siRNAs as cancer therapeutics is their systemic delivery from the administration site to target cells in vivo. This chapter describes approaches to deliver siRNA effectively for cancer treatment and discusses in detail the current methods to assess pharmacokinetics and biodistribution of siRNAs in vivo.

Key words siRNA, Ovarian cancer, Delivery, Cancer therapy, Stem-loop RT-PCR

1 Introduction

Classical analyses of gene function are performed by generating knockout (KO) mouse models and observing a phenotype [1]. Even though KO of genes offers powerful means to discover disease related-genes such as oncogenes in vivo, development of drugs such as small-molecule compounds or antibodies is required for clinically relevant therapeutic strategies. However, these approaches do have limitations [2]. Small-molecule inhibitors are frequently associated with undesirable toxicities and antibodies are only useful for targets accessible in the circulation or located on the surface of target cells. Since the discovery of RNA interference (RNAi) [3] and the application of small interfering RNA (siRNA) to silence desired target genes [4], siRNA has become an alternative technology to analyze gene function and discover drug targets. Since siRNAs can inhibit the expression of any gene of interest, we can utilize this technology for targeting previously undruggable genes. Hence, the use of siRNA is attractive for cancer therapy.

Despite the promise, several hurdles must be overcome for successful use of siRNA in the clinic. SiRNA is easily degraded in the

Yi Feng and Lin Zhang (eds.), *Long Non-Coding RNAs: Methods and Protocols*, Methods in Molecular Biology, vol. 1402, DOI 10.1007/978-1-4939-3378-5_15,

bloodstream by ribonucleases (RNase), eliminated by renal excretion, and cannot pass through a cellular membrane readily because of its large molecular weight, high hydrophilicity, and negative charge [5]. Thus, effective siRNA delivery systems are needed for this approach to be successful. Many groups are developing siRNA delivery systems for cancer using a variety of formulations, such as liposomes, polymers, or micelles [5–7]. The physical properties of delivery systems such as size, shape, and surface charge are critical factors for delivery of nanoparticles to tumors after systemic administration. It is well established that long-circulating nanoparticles with an average diameter of ≤100 nm accumulate efficiently in tumor tissues via the enhanced permeability and retention (EPR) effect based on the fact that tumor vessels are irregularly shaped, defective, and leaky and have varying widths compared with normal capillaries [8, 9]. Intratumoral mobility of nanoparticles may be affected by higher interstitial fluid pressure and soluble factors in solid tumors, population of stromal cells, and density of extracellular matrix in tumor. For example, polymeric micelles of 30 nm in diameter showed penetration in stromal-rich pancreatic tumors but those of 70 nm showed no penetration [10]. After being taken up by target cells through endocytosis, siRNAs need to be released from endosomes into cytosol. These sequential steps from administration site to cytosol in target cells should be considered for development of siRNA delivery systems for cancer treatment [2, 5, 7]. Importantly, siRNAs need to be effectively delivered to tumors to exert therapeutic effect. Therefore, determination of pharmacokinetic profiles of administrated siRNA in the body is an important issue for the clinical development of siRNA medicine. Here, we describe in vivo siRNA delivery in orthotopic ovarian cancer (OvCa) models using chitosan/siRNA nanoparticles [11, 12], and quantification of siRNAs by stem-loop quantitative reverse transcribed (qRT)-PCR and fluorescence-based assays [13].

2 Materials

2.1 Commercial Reagents

1. Chitosan (CH), low molecular weight (Sigma-Aldrich).
2. Sodium tripolyphosphate (TPP; Sigma-Aldrich).
3. Glacial acetic acid (Thermo Scientific).
4. siRNAs (Sigma-Aldrich).
5. Human ovarian cancer cell lines, SKOV3, HeyA8 and A2780 (ATCC).
6. Hanks' Balanced Salt Solution (HBSS) (Mediatech, Inc).
7. Trizol (Life Technologies).
8. Direct-zol RNA Kit (Zymo Research).

9. Verso cDNA Synthesis Kit (Thermo Scientific).
10. TaqMan miRNA assays (Applied Biosystems).
11. 2× Fast SYBR Green Master Mix (Applied Biosystems).

2.2 Equipment

1. Centrifuges, 5417R and 5810R (Eppendorf).
2. A homogenizer, TISSUE MASTER 125 homogenizer (OMNI International, Kennesaw GA, USA) for homogenization of tissue samples.
3. A spectrophotometer, NanoDrop 2000c (Thermo Scientific) for RNA quantification.
4. A thermal cycler, Mastercycler pro (Eppendorf) for RT-PCR.
5. A real-time thermal cycler, 7500 Fast Real-Time PCR System (Applied Biosystmes) for real-time PCR.
6. In vivo imaging system, IVIS 200 system (Xenogen) for ex vivo imaging for siRNAs.

3 Methods

3.1 Preparation of siRNA/Chitosan (siRNA/CH) Nanoparticle

Chitosan (CH) is a linear polysaccharide composed of randomly distributed β-linked d-glucosamine and *N*-acetyl-d-glucosamine. CH is biodegradable, biocompatible, less immunogenic, and less toxic, which makes it a very attractive tool for clinical and biological applications [14–16]. Due to the presence of protonated amino groups, negatively charged nucleic acids can be loaded in CH, and siRNA/CH nanoparticles can effectively interact with cell membranes. Therefore, we developed CH nanoparticles to deliver siRNA into tumors [11, 12]. siRNA/CH nanoparticles are prepared based on ionic gelation of anionic TPP and siRNA with cationic CH.

1. 0.25 % Acetic acid is prepared by dissolving 0.25 ml glacial acetic acid in 99.75 ml of water.
2. CH solution is obtained by dissolving CH (2 mg/ml) in 0.25 % acetic acid.
3. TPP is prepared by dissolving 0.25 g of TPP in 100 ml of water.
4. Nanoparticles are spontaneously generated by the addition of TPP (0.25 % w/v) and siRNA (1 μg/μl) to CH solution under constant stirring at room temperature.
5. After incubating at 4 °C for 40 min, siRNA/CH nanoparticles are collected by centrifugation at 11,000 ×*g* for 40 min at 4 °C.
6. The pellet is washed three times to remove unbound chemicals or siRNA and siRNA/CH nanoparticles are stored at 4 °C until use.

3.2 Development of Orthotopic In Vivo Models of Ovarian Cancer

1. Female athymic nude mice (8–12 weeks old) are obtained from the National Cancer Institute.
2. Human ovarian cancer cells such as SKOV3ip1, HeyA8, or A2780 are cultured in RPMI1640 supplemented with 15 % fetal bovine serum in 10 or less passage prior to injection into mice.
3. Cells are detached with trypsin and complete media is added. Cells are then pelleted at $300 \times g$ for 5 min at 4 °C.
4. Cells are then washed twice with PBS.
5. Resuspend cells using HBSS at a concentration of 5×10^6 cells per ml.
6. Cells (1×10^6 cells per 200 μl per mouse) are injected into the peritoneal cavity using 30 G needles.
7. After tumors have been established, siRNA/CH nanoparticles are injected intraperitoneally into tumor-bearing mice at a dose of 1.25–5.0 μg siRNA per mouse.

3.3 RNA Isolation from Blood, Plasma, and Tissue

1 Sample preparation from blood:

(a) At different time points, mice are put under anesthesia using isoflurane and blood is collected from abdominal vena cava or by cardiac puncture into RNase-free cryotubes using 25-G needles.

(b) Blood (typically 200 μl) is mixed with three times volume of Trizol (600 μL). Vortex the mixture thoroughly (*see* **Note 1**).

(c) The mixture is centrifuged at $12{,}000 \times g$ for 10 min at 4 °C to obtain supernatants. The resulting supernatant is processed for total RNA isolation.

2 Sample preparation from plasma:

(a) Blood is to be stored in RNase-free tubes with each tube containing 84.3 μL K_2EDTA per mL of blood.

(b) Mix blood and anticoagulant thoroughly by inverting tube immediately ten times (*see* **Note 2**).

(c) Centrifuge the mixture at $12{,}000 \times g$ for 10 min at 4 °C. This will give three layers: (from top to bottom) plasma, leucocytes (buffy coat), and erythrocytes. Carefully aspirate the supernatant (plasma) to a tube. Prior to use, the plasma can be stored at −80 °C.

(d) Plasma (typically 200 μl) is mixed with three times volume of Trizol (600 μL).

(e) The mixture is centrifuged at $12{,}000 \times g$ for 10 min at 4 °C to obtain the supernatants. The resulting supernatant is processed for total RNA isolation.

3 Sample preparation from tissues:

   (a) Tissues, such as tumor, brain, lung, heart, liver, spleen, and kidney, are collected in RNase-free cryotubes using sharp scissors and forceps, and snap freeze them in liquid nitrogen. The organs can be stored at –80 °C until use.

   (b) Just prior to use, thaw the tissues on ice. Weighed tissue (typically 50–100 mg) is transferred into 5 ml polystyrene round-bottom tubes with 750 μL of Trizol (*see* **Note 3**).

   (c) A tissue is homogenized with a homogenizer (*see* **Note 4**).

   (d) The resulting tissue homogenate is centrifuged at 12,000 × *g* for 10 min at 4 °C to obtain the supernatants. The resulting supernatant is processed for total RNA isolation.

4 Total RNA is isolated from the supernatant of blood, plasma, or tissue using Direct-zol RNA Kit according to the manufacturer's protocol.

5 RNA concentration is quantified using a spectrophotometer.

3.4 siRNA Quantification in Blood and Tissue by Stem-Loop qRT-PCR

Stem-loop qRT-PCR has been utilized to quantify small RNA fragments (e.g., miRNA) [17]. First, a miRNA-specific stem-loop RT primer is hybridized to the miRNA and then reverse transcribed. Next, the RT product, cDNA, is amplified by regular real-time PCR using a miRNA-specific forward primer and the universal reverse primer. Stem-loop qRT-PCR method can also be applied for quantification of siRNA, which gives high sensitivity, selectivity, and wide dynamic range of detection of siRNA as compared with other means such as enzyme-linked immunosorbent assay (ELISA) or high-performance liquid chromatography (HPLC) [18]. Therefore, stem-loop qRT-PCR technique can be used for the quantification of administered siRNA in blood and tissues. The primer for siRNA in stem-loop PCR is designed for each sequence. The forward primer in PCR amplification is designed based on the siRNA sequence and a universal reverse primer (5′-GACCTGTCCGATCAC GACGAG-3′) is used.

1. Standard siRNA is prepared by serial twofold dilutions of siRNA with RNase-free water.
2. Standard blood sample is prepared by directly adding 5 ng siRNA to 200 μl of naïve blood or plasma obtained from non-treated mouse, and the siRNA/blood or plasma mixture is subjected to RNA isolation as mentioned before.
3. Isolated total RNA (1–10 μg) from blood and tissues, and standard siRNAs (~2.5 pg) are subjected to stem-loop PCR using TaqMan miRNA assays according to the manufacturer's protocol. 10 μl of total RNA/siRNA is combined with 5 μl of master mix of stem-loop PCR and 2 pmol stem-loop PCR

primer, and then stem-loop PCR is carried using Mastercycler pro. Condition for the stem-loop PCR reaction is as follows: 16 °C for 30 min, 42 °C for 30 min, 85 °C for 5 min, then hold at 4 °C.

4. 1.3 μl of cDNA is added into the PCR amplification reaction mix (10 μl 2× Fast SYBR Green Master Mix and 20 pmol of forward and reverse primer sets at a volume of 20 μl).
5. PCR amplification is carried out using the 7500 Fast Real-Time PCR System. Condition for the RT reaction is as follows: 95 °C for 15-min enzyme activation, then 40 cycles of 95 °C for 15-s denaturation, and 60 °C for 1-min annealing/extension.
6. siRNA amount in blood sample is calculated using the standard curve (Fig. 1a).
7. Blood concentration of siRNA at collected time points, $C(t)$, is expressed as the amount of siRNA per ml of blood.

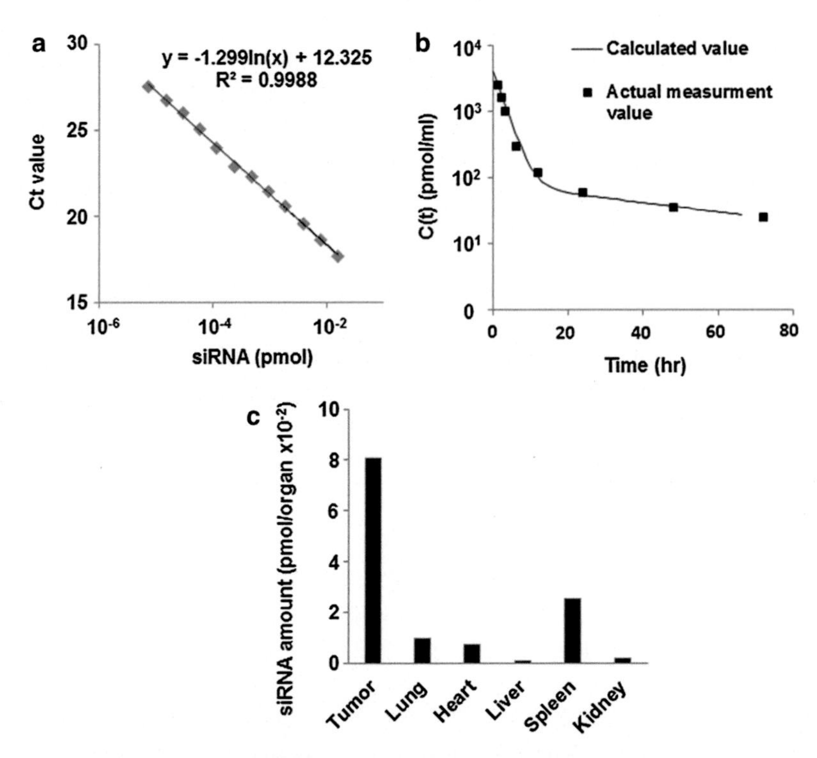

Fig. 1 Stem-loop PCR for siRNA quantification. (**a**) Standard curve showing a plot of log amount of siRNA vs. Ct values quantified using qRT-PCR method. The equation derived from this plot is used for calculating the absolute amount of siRNA. (**b**) Time profile of plasma siRNA concentration. *Closed squares and line* represent actual concentration of siRNA in blood and calculated curve of siRNA concentration by fitting the data using two-compartment model using MULTI program, respectively. (**c**) siRNA amount in organs is calculated using the siRNA standard curve shown in (**a**)

8. Pharmacokinetic analysis is performed as described below:

   C(t) is fitted by an appropriate equation for one- or two-compartment models using software such as Graphpad prism and MULTI, or other appropriate programs (Fig. 1b) (*see* **Note 5**).

   One compartment: $C(t) = Ae^{-\alpha t}$, two compartment: $C(t) = Ae^{-\alpha t} + Be^{-\beta t}$

   Area under the curve (AUC_{0-t}) of blood concentration is calculated by integration of C(t) up to a given time point:

   $$AUC = \int_0^t C(t)\,dt$$

9. Levels of siRNAs in various organs are also calculated using standard siRNA curve (Fig. 1c).

3.5 Fluorescence-Based Biodistribution Study

In vivo imaging has become an important tool for the development of drug delivery systems. The near-infrared (NIR) fluorescence provides simultaneous acquisition of full-color white light imaging with NIR images, deeper penetration of NIR signal, and decreased tissue autofluorescence compared to visible light [19]. Therefore, nanoparticle labeled with NIR fluorophore such as Cy5.5 or DiR allows determination of amount in organs ex vivo and noninvasive evaluation of biodistribution in vivo after their administration.

1. Cy5.5-labeled siRNA-loaded nanoparticle is administered i.p. into tumor-bearing mice at a dose of 2.5 μg siRNA.
2. At time points (typically 24 or 48 h), tumor and organs are excised.
3. Excised organs are washed with cold PBS and put on 6-well plate.
4. Fluorescence image in excised organs are captured using the Xenogen IVIS 200 system with Cy5.5 fluorophore excitation (678 nm) and emission (703 nm) filter.
5. Fluorescent images are analyzed using Living image 2.5 software. Regions of interest are drawn for each organ and total radiant efficiency ps^{-1} μW^{-1} cm^2 is measured (Fig. 2).

3.6 Quantifying Levels of Gene Knockdown Using Quantitative Reverse Transcription PCR (qRT-PCR)

1. Isolated RNA (500–1000 ng) from tumor tissue is reverse transcribed using a Verso cDNA Synthesis Kit as per the manufacturer's instructions using a Mastercycler pro.
2. 2 μl Diluted cDNA (typically two- to tenfold dilution) is then subjected to PCR amplification with 10 μl 2× Fast SYBR Green Master Mix and 20 pmol of forward and reverse primer sets at a volume of 20 μl.
3. PCR is performed using the 7500 Fast Real-Time PCR System. Each cycle consists of 15 s of denaturation at 95 °C and 1 min of annealing and extension at 60 °C (40 cycles).
4. Relative levels of gene expression are quantified using the ΔΔCt method.

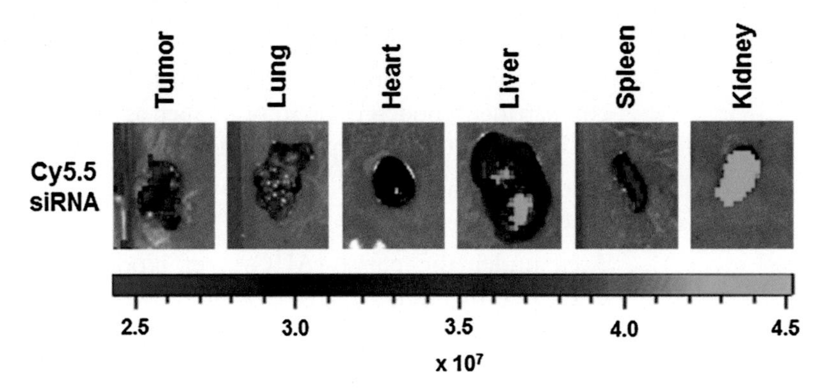

Fig. 2 Fluorescence-based images of Cy5.5-labeled siRNA in organs. The levels of Cy5.5-siRNA were measured in various organs at 48 h post-administration of nanoparticles in tumor-bearing mice. The scale bar represents the fluorescence intensity in ps^{-1} mW^{-1} cm^{2}

4 Notes

1. Blood samples should be vortexed immediately after mixing with Trizol, or blood solidifies in Trizol. If storage is necessary prior to use, blood needs to be collected in a tube with anticoagulant agent and stored at −80 °C.
2. Process samples as soon as possible. If storage is necessary prior to use, store the blood at room temperature, shielded from light.
3. Hard tissues should be cut into small pieces with scissors before adding Trizol, which results in efficient homogenization.
4. Output power of homogenizer should be adjusted depending on the softness of organ. Soft organs such as brain and liver are homogenized at low power to avoid bubble; hard organs such as spleen and heart are homogenized at high power. Homogenization should be done with tubes immersed in ice-cold water to avoid generating heat.
5. Five or more time points are required for curve fitting of the time profile of siRNA concentration in blood or plasma.

Acknowledgments

H.H. is supported by JSPS Postdoctoral Fellowships for Research Abroad. S.Y.W. is supported by Ovarian Cancer Research Fund, Inc., Foundation for Women's Cancer, and Cancer Prevention Research Institute of Texas training grants (RP101502 and RP101489). Portions of this work were supported by NIH grants (P50CA083639, CA109298, P50CA098258, U54CA151668, UH2TR000943, CA016672, U54CA96300, and U54CA96297),

CPRIT (RP110595 and RP120214), an Ovarian Cancer Research Fund Program Project Development Grant, the Betty Ann Asche Murray Distinguished Professorship, the RGK Foundation, the Gilder Foundation, the Judi A. Rees Ovarian Cancer Research Fund, the Chapman Foundation, and the Meyer and Ida Gordon Foundation. This research was also supported, in part, by the Blanton-Davis Ovarian Cancer Research Program.

References

1. Gondo Y (2008) Trends in large-scale mouse mutagenesis: from genetics to functional genomics. Nat Rev Genet 9:803–810
2. Pecot CV, Calin GA, Coleman RL, Lopez-Berestein G, Sood AK (2011) RNA interference in the clinic: challenges and future directions. Nat Rev Cancer 11:59–67
3. Fire A, Xu S, Montgomery MK, Kostas SA, Driver SE, Mello CC (1998) Potent and specific genetic interference by double-stranded RNA in Caenorhabditis elegans. Nature 391: 806–811
4. Elbashir SM, Harborth J, Lendeckel W, Yalcin A, Weber K, Tuschl T (2001) Duplexes of 21-nucleotide RNAs mediate RNA interference in cultured mammalian cells. Nature 411: 494–498
5. Hatakeyama H, Akita H, Harashima H (2011) A multifunctional envelope type nano device (MEND) for gene delivery to tumours based on the EPR effect: a strategy for overcoming the PEG dilemma. Adv Drug Deliv Rev 63: 152–160
6. Kanasty R, Dorkin JR, Vegas A, Anderson D (2013) Delivery materials for siRNA therapeutics. Nat Mater 12:967–977
7. Wu SY, Lopez-Berestein G, Calin GA, Sood AK (2014) RNAi therapies: drugging the undruggable. Sci Transl Med 6:240ps7
8. Matsumura Y, Maeda H (1986) A new concept for macromolecular therapeutics in cancer chemotherapy: mechanism of tumoritropic accumulation of proteins and the antitumor agent smancs. Cancer Res 46:6387–6392
9. McDonald DM, Choyke PL (2003) Imaging of angiogenesis: from microscope to clinic. Nat Med 9:713–725
10. Cabral H, Matsumoto Y, Mizuno K, Chen Q, Murakami M, Kimura M et al (2011) Accumulation of sub-100 nm polymeric micelles in poorly permeable tumours depends on size. Nat Nanotechnol 6:815–823
11. Kim HS, Han HD, Armaiz-Pena GN, Stone RL, Nam EJ, Lee JW et al (2011) Functional roles of Src and Fgr in ovarian carcinoma. Clin Cancer Res 17:1713–1721
12. Gharpure KM, Chu KS, Bowerman CJ, Miyake T, Pradeep S, Mangala SL et al (2014) Metronomic docetaxel in PRINT nanoparticles and EZH2 silencing have synergistic antitumor effect in ovarian cancer. Mol Cancer Ther 13:1750–1757
13. Wu SY, Yang X, Gharpure KM, Hatakeyama H, Egli M, McGuire MH et al (2014) 2′-OMe-phosphorodithioate-modified siRNAs show increased loading into the RISC complex and enhanced anti-tumour activity. Nat Commun 5:3459
14. Han HD, Song CK, Park YS, Noh KH, Kim JH, Hwang T et al (2008) A chitosan hydrogel-based cancer drug delivery system exhibits synergistic antitumor effects by combining with a vaccinia viral vaccine. Int J Pharm 350:27–34
15. Zhang HM, Chen SR, Cai YQ, Richardson TE, Driver LC, Lopez-Berestein G et al (2009) Signaling mechanisms mediating muscarinic enhancement of GABAergic synaptic transmission in the spinal cord. Neuroscience 158: 1577–1588
16. Katas H, Alpar HO (2006) Development and characterisation of chitosan nanoparticles for siRNA delivery. J Control Release 115: 216–225
17. Chen C, Ridzon DA, Broomer AJ, Zhou Z, Lee DH, Nguyen JT et al (2005) Real-time quantification of microRNAs by stem-loop RT-PCR. Nucleic Acids Res 33:e179
18. Cheng A, Li M, Liang Y, Wang Y, Wong L, Chen C et al (2009) Stem-loop RT-PCR quantification of siRNAs in vitro and in vivo. Oligonucleotides 19:203–208
19. Frangioni JV (2003) In vivo near-infrared fluorescence imaging. Curr Opin Chem Biol 7: 626–634

Chapter 16

Targeting Long Noncoding RNA with Antisense Oligonucleotide Technology as Cancer Therapeutics

Tianyuan Zhou, Youngsoo Kim, and A. Robert MacLeod

Abstract

Recent annotation of the human transcriptome revealed that only 2 % of the genome encodes proteins while the majority of human genome is transcribed into noncoding RNAs. Although we are just beginning to understand the diverse roles long noncoding RNAs (lncRNAs) play in molecular and cellular processes, they have potentially important roles in human development and pathophysiology. However, targeting of RNA by traditional structure-based design of small molecule inhibitors has been difficult, due to a lack of understanding of the dynamic tertiary structures most RNA molecules adopt. Antisense oligonucleotides (ASOs) are capable of targeting specific genes or transcripts directly through Watson–Crick base pairing and thus can be designed based on sequence information alone. These agents have made possible specific targeting of "non-druggable targets" including RNA molecules. Here we describe how ASOs can be applied in preclinical studies to reduce levels of lncRNAs of interest.

Key words Antisense oligonucleotide, Control ASO, Non-druggable targets, Off-target, RNA therapeutics, Target reduction, RNase H, qRT-PCR, Transfection, Free uptake

1 Introduction

Targeted drug discovery efforts have mostly focused on proteins, particularly enzymes, secretary factors, and G-protein-coupled receptors [1]. Many proteins are considered "non-druggable" targets because closely related protein family members exist, making specificity difficult [2]. The "non-druggable" category of proteins includes transcription factors, structural proteins, and RNAs [3]. Indeed, direct targeting of RNAs, including both protein-encoding genes and noncoding transcripts, would potentially allow modulation of all transcriptional products, such as specific splice variant forms [4], eRNAs [5], long noncoding RNAs (lncRNA), and all protein coding RNAs [6, 7]. Antisense oligonucleotides is a technology that enables the direct targeting of RNA and greatly expands the freedom of drug target selection for the treatment of human diseases.

Yi Feng and Lin Zhang (eds.), *Long Non-Coding RNAs: Methods and Protocols*, Methods in Molecular Biology, vol. 1402, DOI 10.1007/978-1-4939-3378-5_16, © Springer Science+Business Media New York 2016

LncRNAs, arbitrarily defined as RNA transcripts longer than 200 nucleotides that do not encode proteins, have been proposed to modulate diverse biological functions. Although the functional roles and mechanisms of actions for the majority of lncRNAs remain unknown, recent studies have revealed that lncRNAs are involved in chromosome dosage compensation, modulation of chromatin status, and cell differentiation among other cellular processes [8, 9]. Moreover, mutation or dysregulation of lncRNAs have been linked to many human diseases including diabetics, cardiovascular diseases, central nervous system disease, and cancer (reviewed in [10, 11]). Thus, the selective depletion of specific lncRNA will allow us to both experimentally explore lncRNA functions and to pursue the most attractive of these as therapeutic targets to the diseases. The selective depletion of lncRNAs has posed a common challenge in lncRNA research [9]. Knocking down lncRNA by RNAi is a well-established approach. However, the presence and activities of RNAi machinery in nucleus is not thought to be robust and its existence in this compartment has been under intense debate. It is possibly for this reason that RNAi is limited in its ability to target nuclear-retained lncRNAs [12]. The difficulty of knocking down nuclear-retained lncRNA may be overcome by ASOs, another nucleic acid-based technology which enables specific targeting of any gene in human transcriptome. ASOs rely on RNAse H to cleave target RNAs irrespective of their subcellular localization due to RNase H's ubiquitous presence in both cytoplasm and the nucleus [13, 14]. Importantly, ASOs' efficacy in man has recently led to the FDA approval of Kynamro®, an ASO drug targeting *ApoB*, that lowers cholesterol in patients with homozygous familial hypercholesterolemia (HoFH) [15]. More than 30 ASO drugs are currently in preclinical or clinical testing [16, 17]. In this chapter we discuss the design of ASOs to target lncRNA and methods employed to evaluate ASOs in both cell-based assays and animals.

1.1 Design of ASOs

ASO, as we discuss here, refers to a synthetic molecule comprising a string of nucleotides or nucleotide analogs that bind to complementary RNA sequences with high specificity through Watson–Crick base pairing. ASOs can modulate levels of the targeted RNA through several mechanisms: (1) ASOs with properties of deoxyoligonucleotides may recruit RNase H to the DNA–RNA heteroduplex to degrade RNA [14]. (2) Binding of ASOs to target sequences may inhibit biogenesis or translation of the transcript of gene [3, 18]. Splicing [19], 3′ polyadenylation [20], RNA localization [7] are some examples where ASOs were demonstrated to achieve potential therapeutic goals. ASOs acting through RNAse H mechanism are typically 12–20 nucleotides in length, because approximately 12 nucleotides are required to recognize a unique sequence in the genome given the size of human genome.

Unlike siRNAs, which are duplexes, ASOs are single-stranded molecules. Compared to double-stranded nucleotide compounds including siRNAs that are rigid, hydrophilic and have average molecular weight of 13,300 Da, ASOs are on average 5000–8000 Da, amphiphilic in nature and are more flexible, which allows efficient binding to target RNA.

Through chemical alterations of the natural nucleotides, ASOs have been designed to have drug-like properties. Naturally occurring nucleic acids are composed of ribonucleotides or deoxynucleotides linked with phosphodiester bonds. One of the first modifications made to ASOs was the phosphorothioate modification of the linkage. This modification protects ASOs from degradation by nucleases and increases half-life in serum, while still supporting RNase H activities [21]. These so-called first-generation ASOs were typically 20 nucleotides in length and are composed solely of deoxy residues [14]. However, due to low metabolic stability and suboptimal target binding affinity, the application of these early generation ASOs was limited in clinics [22]. Second-generation ASOs contain a central region of 8–10 phosphorothioate DNA nucleotides flanked by nucleotides modified at the sugar; this is called "gapmer" design [23]. Over the years, numerous nucleotide modifications have been tested in attempts to enhance binding affinity [24]. The bulky 2′-*O*-methoxyethyl (2′-MOE modification) improved metabolic stability of ASOs and prevented nonspecific protein interactions and thus improved overall safety profile relative to the first-generation ASOs [25]. Kynamro® is a systematically delivered 2′-MOE-modified 20-mer [26]. A more recently developed ASO chemistry incorporates the next generation 2′, 4′-constrained ethyl (cEt modification) in the residues flanking the deoxy central region. Because of the enhanced affinity provided by the cEt modification relative to the 2nd-generation ASOs, the cEt ASOs can be shorter; this contributes to higher ASO potency as the smaller molecular weight ASOs are more efficiently released in a cell [27]. Furthermore, ASOs of the same chemical class all have very similar pharmacodynamics, pharmacokinetic, and tissue accumulation features, making the overall drug development process more predictable and efficient for a given ASO drug. STAT3-Rx (AZD9150) is the first Gen 2.5 cEt ASO to enter clinical trials. It has shown single agent efficacy in patients with diffuse large B-cell lymphoma at modest doses [28].

Like many other drug classes, ASOs potentially have both "on-target" and "off-target" effects. "Off-target" effects occur due to ASOs' binding to unintended sequences in non-target RNAs or through direct interactions with proteins independent of target hybridization [29, 30]. Binding to off-target transcripts can be avoided in large part with the use of computational algorithms that ensure that ASO sequences have little homology to genomic sequence other than that of the desired target.

Off target effects have been further minimized with careful choice of chemical modifications and by extensive screening *in vitro* and in animal models.

The ability of ASOs to inhibit target RNAs does not depend on the abundance of the transcript. Levels of both rare transcripts such as enhancer RNAs, a class of relatively short noncoding RNAs that function to enhance gene expression, as well as very abundant transcripts such as metastasis associated lung adenocarcinoma transcript 1 (MALAT1) can be reduced equally well by ASOs that activate RNase H [5, 31, 32]. However, certain RNAs have proven difficult to target efficiently with ASOs despite repeated efforts. We speculate that specific features of the RNA such as transcript half-life, transcript secondary structures, and rates of RNA processing and nuclear export may contribute to such difficulties. On the other hand, targeting repeated sequences unique to the RNA transcript has been shown to greatly increase ASO potency [33]. Importantly, ASOs are capable of distinguishing between transcripts that differ by a single nucleotide, allowing for allele-specific suppression of a mutant gene while sparing the wild-type form [34, 35]. Such exclusive specificity achieved by ASOs has the potential for therapeutic targeting of otherwise essential genes.

1.2 Applications of ASO Designs with Different Mechanisms of Actions

ASOs decrease gene expression or disrupt the action of a functional RNA through two general mechanisms: RNase H-mediated and occupancy-based [3, 18]. ASOs that direct RNA degradation by RNase H bind to pre-mRNA or processed RNA through Watson–Crick base pairing, followed by RNase H1 recruitment to initiate cleavage of the target RNA. RNase H1 is a ubiquitously expressed nuclease that cleaves the RNA strand of an RNA–DNA hybrid. The enzyme is found in the cell nucleus, mitochondria, and, to a lesser extent, the cytoplasm. This makes ASO action different from the nucleic acid-based siRNA technology. The siRNA activity is mediated by the actions of the RNA-induced silencing complex (RISC), which is mainly localized to the cytoplasm. ASOs that mediate RNase H cleavage efficiently reduce levels of transcripts localized exclusively in the nucleus [7, 31, 32].

ASOs have also been shown to act through occupancy-based mechanisms, such as those designed to alter RNA splicing events. These ASOs bind to splicing regulatory sequences and function by hindering access of the splicing machinery. For an example, an ASO alters splicing of mutant *SMN2* pre-mRNA to generate the exon-incorporating productive form in mammalian cells and in mouse models of spinal muscular atrophy (SMA) [19]; the splice-altering *SMN2* ASO is currently being evaluated in clinical trials in SMA patients [36]. Additionally, ASOs have been designed to induce nonsense-mediated decay [37], to affect 3′ polyadenylation [20], to alter RNA localization [7], and to affect other RNA processing events [18]. In this chapter, we focus on applications of

RNase H-mediated ASO and describe procedures to inhibit the expression of target RNAs.

2 Materials

1. ASOs targeting MALAT1 in various species ISIS399479, ISIS395240, ISIS556089.
   Ultraviolet–visible spectrophotometer.
2. Cell lines:
   4T1, a mouse mammary carcinoma cell line; LNCaP, a human prostate cancer cell line; THP-1, a human myeloid leukemia cell line. All cells were obtained from American Type Culture Collection (ATCC) and were maintained in RPMI1460 supplemented with 10 % fetal bovine serum and antibiotics.
   ASO solutions adjusted to 200 μM.
   RNAiMAX (Life Technologies).
   Opti-MEM (Life Technologies).
   96-well electroporation manipulator (BTX Harvard Apparatus).
   High-throughput electroporation plates (BTX Harvard Apparatus).
3. Male 6–8 weeks CD.1 mice (Charles River Laboratories, USA).

3 Methods for Validation of ASO Activity

In this chapter, we describe protocols to reduce levels of a lncRNA MALAT1 with Gen 2.0 ISIS399479, ISIS395240 or the more potent Gen 2.5 ISIS556089 ASOs. These protocols can be applied to additional Gen 2.0 and Gen 2.5 ASOs for *in vitro* and *in vivo* preclinical studies.

3.1 Preparation of ASOs

Dissolve ASOs directly in PBS to approximately 10 mg/ml (*see* **Note 1**). Like all nucleic acids, ASOs absorb ultraviolet light. The extinction coefficient of an oligonucleotide depends on base composition in the ASO sequence [38]. The extinction coefficients for the ASOs used in this protocol are listed in Table 1. ASO concentrations are determined by measuring the absorbance of the solution at 260 nM in a UV-visible light spectrometer after 1:1000 dilutions and calculating using the formula:

$$\text{Concentration}\,(\text{mM}) =$$
$$A_{260} \times \text{dilution factor} \,/\, \text{extinction coefficient}$$

ASOs can be stored in aqueous buffer at 4 °C for ~1 month or at −20 °C for years without loss of activities. Occasionally some ASO

Table 1
Properties of MALAT1 ASOs

Isis No	Extinction coefficient ($mM^{-1} \times cm^{-1}$)	Molecular weight (Da)	Length (nucleotides)	Species	Sequence	Chemistry
395240	197.16	7209.24	20	Chimpanzee, Human, Mouse, Rabbit, Rat	5′-TGCCTTTAGGATTCTAGACA-3′	5-10-5 MOE gapmer w/phosphorothioate backbone
399479	201.24	7285.26	20	Mouse	5′-CGGTGCAAGGCTTAGGAATT-3′	5-10-5 MOE gapmer w/phosphorothioate backbone
556089	159.3	5414.59	16	Chimpanzee, Dog, Human, Mouse, Pig, Rat	5′-GCATTCTAATAGCAGC-3′	3-10-3 (S)-cEt gapmer w/phosphorothioate backbone

solutions precipitate after extended storage at 4 °C. If precipitation is observed, filter the solution and recalculate the concentration.

3.2 Evaluation of ASO Activity In Vitro

Many cancer cell lines take up ASOs under physiological conditions without lipid-mediated transfection reagent (referred to as "free uptake" hereinafter) [31, 39]. This process is independent of clathrin or caveolin pathways but specific receptors have not been identified yet [39]. In our extensive efforts to test free uptake in cancer cell line panels, we have identified at least one cell line in each cancer cell origin that takes up ASO very efficiently without the need for lipid transfection. Primary and early passage cells used for patient-derived xenograft models have higher propensity to take up ASOs for a particular type of cancer. This observation implies that the loss of free uptake abilities may be an artifact during cell line establishment. We have observed that the ability of a given cell line to take up ASO correlates with the ASO pharmacodynamics in the tumor models established from the same cell line [31]. Thus when possible, we test ASOs in cell lines *in vitro* by free uptake.

3.2.1 Delivery of ASOs to Cells by Free Uptake

1. Log phase 4T1 cells are plated at $2–5 \times 10^3$ cells per well into 96-well plates and are incubated for 16 h in 95 μl of culture medium (*see* **Note 2**).
2. Pre-diluted ASOs (5 μl of appropriate concentration stock) are added to cells to the desired final concentrations. A typical dose response analysis involves testing of the final concentrations at 80 nM, 400 nM, 2 μM, and 10 μM (*see* **Note 3**).
3. Cells are harvested 24 h after addition of ASO, and RNA is prepared. ASO activity is examined by qRT-PCR using Taqman assays with the primers and probe sequences designed to amplify the RNA of interest, normalized to the expression of a housekeeping gene; the primers and probe used to amplify mouse MALAT1 are:

   Forward primer: 5′-AGGCGGGCAGCTAAGGA-3′;

   Reverse primer: 5′-CCCCACTGTAGCATCACATCA-3′;

   Probe: 5′-FAM-TTCCTCTGCCGGTCCCTCGAAAG-TAMRA-3′ (*see* **Note 4**). Primers and probe sequences for housekeeping gene mouse Cyclophilin A are:

   Forward primer: 5′-TCGCCGCTTGCTGCA-3′;

   Reverse primer: 5′-ATCGGCCGTGATGTCGA-3′;

   Probe: 5′-FAM-CCATGGTCAACCCCACCGTGTTC-TAMRA-3′ [26].

   Typical data are shown in Fig. 1. ISIS399479 reduces levels of MALAT1 RNA with an IC_{50} of ~70 nM in 4T1 cells (*see* **Note 5**).

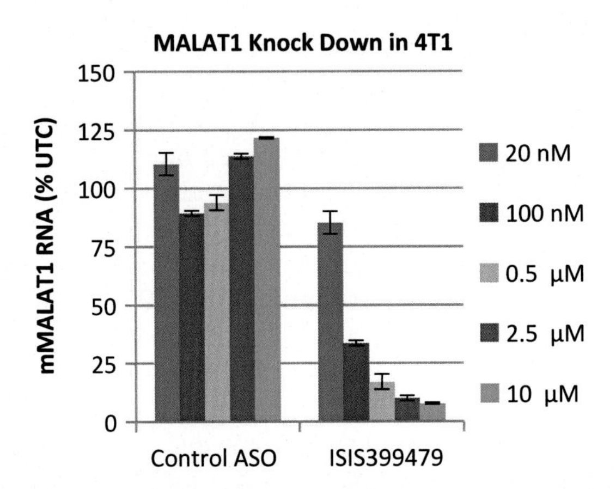

Fig. 1 Mouse mammary tumor 4T1 cells were treated with ISIS399479, a Gen 2.0 ASO targeting mouse MALAT1 along with a control ASO for 24 h. RNA was harvested, and target reduction was examined by qRT-PCR. ISIS399479 caused dose-dependent inhibition of target gene expression, whereas the control ASO had little effect. UTC: Untreated cells

3.2.2 Delivery of ASOs into Cells Using Transfection Reagents

It is not always possible to find a cell line model that is amenable to free uptake. To obtain proof of concept data *in vitro*, ASOs can be delivered to cells by lipid-mediated transfection. The following is a protocol for 96-well format. Reagents can be scaled up proportionally for other plate formats.

1. To examine the proliferation of LNCaP cells after ASO treatments, LNCaP are plated on 96-well plates at 2,000 cells per well in 100 μl 24 h prior to experiments.
2. For each well, 0.15 μl of RNAiMax is mixed with 12.5 μl of Opti-MEM by brief vortexing.
3. Five minutes later, 200 nM ASO diluted in 12.5 μl Opti-MEM is mixed with RNAiMAX by brief vortexing.
4. The 25-μl ASO-RNAiMAX solution is incubated at room temperature for 15 min and is subject to fivefold stepwise dilution in Opti-MEM to 20, 4, and 0.8 nM.
5. The 25-μl aliquot of ASO-RNAiMAX solution is added to each well containing cells to yield final concentrations of 20, 4, 0.8, and 0.16 nM.
6. Cells are harvested after 24 h for RNA analyses and 5–6 days later for cell proliferation assays (*see* **Note 6**). The primers and probe used to amplify human MALAT1 are:

   Forward primer: 5′-AAAGCAAGGTCTCCCCACAAG-3′;

   Reverse primer: 5′-TGAAGGGTCTGTGCTAGATCAAAA-3′;

   Probe: 5′-FAM-TGCCACATCGCCACCCCGT-TAMRA-3′.

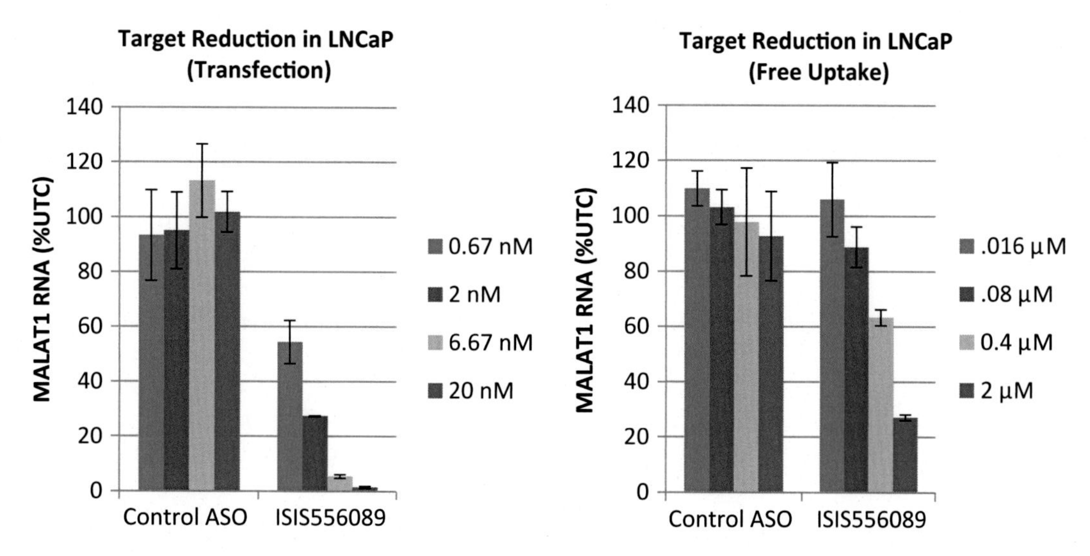

Fig. 2 Human prostate cancer LNCaP cells were treated with ISIS556089, a Gen 2.5 ASO targeting human MALAT1 along with a control ASO. (**a**) ASOs were transfected by transfection using RNAiMax reagents as described and RNA was harvested after 24 h. (**b**) ASOs were delivered to the cells by free uptake and RNA was collected after 48 h. Target knockdown was evaluated by qRT-PCR. UTC: Untreated cells

Human β-actin gene is used to normalize RNA amounts and sequences for primers and probe are:

Forward primer: 5′-CGGACTATGACTTAGTTGCGTTACA-3′;

Reverse primer: 5′-GCCATGCCAATCTCATCTTGT-3′;

Probe: 5′-FAM-CCTTTCTTGACAAAACCTAACTTGCGCAGA-TAMRA-3′. Representative data is shown in Fig. 2, where ASOs were introduced to cell by transfection (a) or by free uptake (b).

3.2.3 Delivery of ASOs to Cells by Electroporation

Some cell lines, including many suspension cells, are recalcitrant to lipid-mediated transfection. For analysis of ASO effects in these cells, we resort to electroporation in 96-well format.

1. THP-1 cells proliferating in log-phase are collected and resuspended at 1×10^7 cells per ml in complete growth medium.
2. Aliquots of 90 μl of cells are mixed with 10 μl ASOs at appropriate concentration; the solution is pipeted up and down (or vortexed gently) to mix (*see* **Note 7**). Samples are transferred to 96-well electroporation plate.
3. Cell mixtures are pulsed at desired voltage for electroporation, typically 130 V for 6 ms. Cells are then collected from each well and washed twice with 120 μl of fresh medium.
4. All cells are combined and plated at 50–100,000 cells per well for RNA extraction and 10,000 cells per well for analysis of proliferation.

3.3 Systemic Delivery of ASO in Mice

1. ASOs are formulated in PBS containing Ca^{2+} and Mg^{2+} at 5 mg/ml and filtered through 0.45 μm sterile filters before use.
2. ASOs can be administered via intraperitoneal (*see* **Note 8**), subcutaneous, or intravenous injection.
3. To examine whether ASO is tolerated in normal animals, CD.1 mice are treated with the target-specific ASO (in this case, ISIS395240 and ISIS399479 for mouse MALAT1) at 50 mg/kg, twice weekly for 6 weeks (*see* **Note 9**). Body weights are recorded after each dose is given. Twenty-four hours after the last dose, animals are sacrificed, and blood is collected by cardiac puncture. Liver, kidney, and spleen are weighed, and liver pieces are collected to prepare RNA (*see* **Note 10**).
4. Plasma is tested on a clinical analyzer for blood chemistry parameters, including alanine amino transferase (ALT), aspartate amino transferase (AST), total bilirubin, and blood urea nitrogen (*see* **Note 11**).
5. RNA is prepared from liver or other relevant organs or tissues, and target reduction is evaluated by qRT-PCR. Typical data is shown in Fig. 3 (*see* **Note 12**).

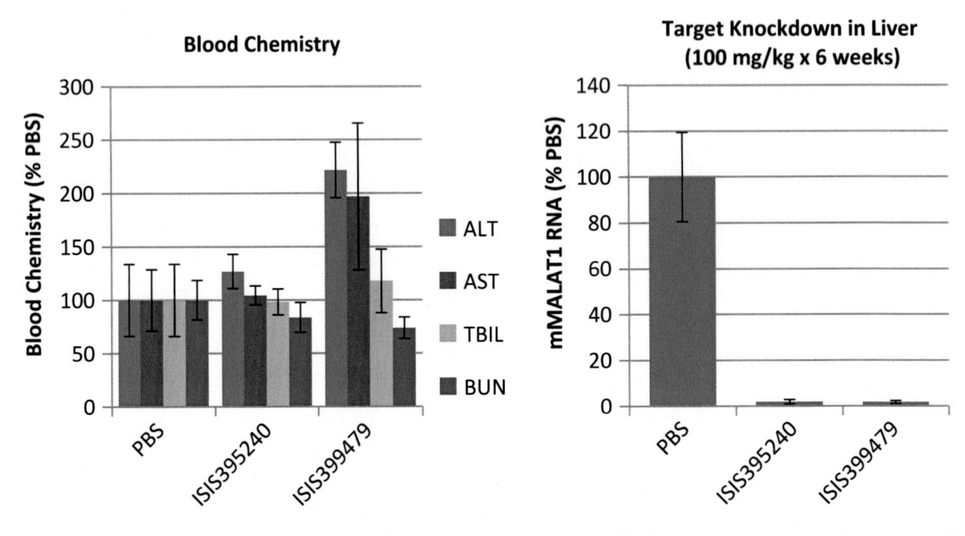

Fig. 3 Male CD.1 mice were treated intraperitoneally with indicated ASOs designed to target mouse MALAT1. (**a**) Blood chemistry markers were evaluated in plasma collected by cardiac puncture. (**b**) Mouse MALAT1 levels were measured in mouse livers using qRT-PCR. There were no notable changes in the blood chemistry from animals treated with two MALAT1 ASOs

4 Notes

1. The chemically modified ASOs described here are typically soluble up to 50 mg/ml in water. Sometimes at high concentrations (>20 mg/ml), some compounds show slight yellow or green tint.
2. ASO treatment affects the attachment of some cell lines, and thus we recommended that cells are incubated for at least 8 h after plating prior to ASO treatment. Cell plating densities between 10,000 and 100,000 cells per ml do not affect cells' free uptake ability.
3. Control ASOs designed to have no matches in human and mouse genome should also be included at the same concentrations. Some cell lines are especially sensitive to high concentrations of ASOs (>10 μM). In these lines, cell growth can be inhibited by ASOs in a sequence-independent manner. These ASO class effects can be better distinguished from on-target events when the control ASOs are included in the experiments in parallel.
4. It is important to design qRT-PCR assays outside the ASO-hybridizing sequences. ASOs remain in RNAs purified from ASO-treated samples and would interfere with RT-PCR reactions if ASOs hybridize to the PCR products defined by the PCR primers, generating extremely low "false" signals, and misleadingly high degree of target knockdown.
5. Typically cells are incubated with ASO-containing culture medium continuously for 24–96 h before cells are harvested for RNA analyses. Maximal RNA knockdown is observed in some non-dividing cells after 7–10 days. Incubation with MALAT1 ASO for merely 3 h is sufficient to initiate the necessary events leading to MALAT1 RNA downregulation 48 h later (Liang XH et al., manuscript in preparation). We observe that the incubation time required for the onset of ASO activities is cell line-specific. A careful time course study is necessary to reveal the dynamics of target RNA expression and inhibition for each ASO compound and each cell type.
6. A control ASO that has no matching sequence in the human and mouse genome should always be included in transfection experiments to determine whether cell growth inhibition is due to the inhibition of specific target or general class effects of ASOs. Some cell lines are highly sensitive to lipid-mediated transfection and as little as 10 nM ASO leads to the inhibition of cell growth; in these cells, observed growth inhibition is not target-related as similar effects are typically observed with both targeted and control ASOs.

7. Optimal experimental conditions are dependent on each cell line and electroporation apparatus used. We typically mix 200 μM of ASO with cell aliquots ranging from 0.5×10^6 to 2×10^6 cells. Tests are run with electroporation voltages ranging from 120 to 170 V. Cell viability is checked by trypan blue exclusion assay after electroporation. In order to ensure reliable data, we make sure at least 80 % cells survive the electric pulse. Efficiency of target reduction can be evaluated by comparing target RNA levels to levels in cells treated with control ASO and in mock electroporated cells. We always use the lowest voltage where >50 % target reduction is achieved.
8. Intraperitoneal injection of ASOs results in greater target reduction in peritoneal macrophages than does subcutaneous dosing, presumably because intraperitoneal dosing allows direct access of ASOs to peritoneal monocyte/macrophage cells.
9. The Gen 2.0 and Gen 2.5 ASOs discussed here demonstrate very similar tissue distribution profiles irrespective of their sequences. Target reduction in liver is observed 24 h after a single systemic administration with a peak in inhibition observed after 48–72 h. Generally a repeated dosing scheme is employed and we observe target inhibition in tumor cells in the 4T1 mouse model of breast cancer between 24–72 h after the last dose.
10. ASOs distribute widely into tissues within 2 h after systemic administration [40, 41]. Organs of high ASO accumulation include kidney, liver, and spleen. Efficient downregulation of target RNA in fat, muscle, and small intestines has also been reported despite low concentrations of ASOs [32]. ASOs remain efficacious for 2–4 weeks in liver and more than 6 months in muscle [41]. ASOs do not cross the blood–brain barrier, thus need to be administered directly to cerebral spinal fluid to reduce target RNA levels in the central nervous system [19]. ASOs are carried by serum proteins in plasma and either taken up by tissues or gradually degraded by various nucleases and cleared in urine [41].
11. ASOs are considered "well-tolerated" if mice treated with the ASOs show no significant changes in organ and body weights, no significant elevations in liver transaminases (ALT and AST) in serum, and no obvious signs of sickness. Different mouse strains may have different susceptibilities to ASOs' non-target related toxicities. Therefore ASOs should be tested for tolerability under the same condition as the intended animal model. Toxicities in mice caused by off-target effects of ASOs can complicate interpretation of experimental results. Off-target effects may be distinguished from on-target pharmacology by dose–response experiments using two or more ASOs targeting

the same target gene in the relevant animal models. We encourage the use of a second ASO designed to hybridize to a different region of the target RNA to confirm that observed pharmacology is not limited to one ASO compound. The relative potency of the two ASOs should be the same both in vitro and in vivo: the more potent ASO with a greater target reduction is predicted to demonstrate better efficacies in animals

12. To ensure efficient target reductions in tumor-bearing mice, various ASO dosing schemes should be tested for each animal model. Depending on mouse strain, ASOs can be tolerated at 100–1000 mg/kg/week.

4.1 Conclusions

Antisense oligonucleotide drugs can be used to specifically and efficiently reduce levels of any RNA of interest, including many lncRNAs, both in cultured cells and in animals without the need for formulation with delivery vehicles. Antisense technology is a promising, versatile modality in preclinical studies for target validation, and for the therapeutic targeting of previously non-druggable targets to treat human diseases.

Acknowledgements

We thank XueHai Liang for discussion and sharing unpublished data. We are grateful to Lauren Elder for editorial assistance.

References

1. Rask-Andersen M, Almen MS, Schioth HB (2011) Trends in the exploitation of novel drug targets. Nat Rev Drug Discov 10:579–590
2. Overington JP, Al-Lazikani B, Hopkins AL (2006) How many drug targets are there? Nat Rev Drug Discov 5:993–996
3. Bennett CF, Swayze EE (2010) RNA targeting therapeutics: molecular mechanisms of antisense oligonucleotides as a therapeutic platform. Annu Rev Pharmacol Toxicol 50:259–293
4. Yamamoto Y, Loriot Y, Beraldi E, Zhang F, Wyatt AW, Al Nakouzi N et al (2015) Generation 2.5 antisense oligonucleotides targeting the androgen receptor and its splice variants suppress enzalutamide resistant prostate cancer cell growth. Clin Cancer Res 21:1675–1687
5. Lam MT, Cho H, Lesch HP, Gosselin D, Heinz S, Tanaka-Oishi Y et al (2013) Rev-Erbs repress macrophage gene expression by inhibiting enhancer-directed transcription. Nature 498:511–515
6. Meng L, Ward AJ, Chun S, Bennett CF, Beaudet AL, Rigo F (2014) Towards a therapy for Angelman syndrome by targeting a long non-coding RNA. Nature 518(7539):409–12
7. Wheeler TM, Leger AJ, Pandey SK, MacLeod AR, Nakamori M, Cheng SH et al (2012) Targeting nuclear RNA for in vivo correction of myotonic dystrophy. Nature 488:111–115
8. Sauvageau M, Goff LA, Lodato S, Bonev B, Groff AF, Gerhardinger C et al (2013) Multiple knockout mouse models reveal lincRNAs are required for life and brain development. ELife 2:e01749
9. Li L, Chang HY (2014) Physiological roles of long noncoding RNAs: insight from knockout mice. Trends Cell Biol 24:594–602
10. Wapinski O, Chang HY (2011) Long noncoding RNAs and human disease. Trends Cell Biol 21:354–361
11. Li X, Wu Z, Fu X, Han W (2014) lncRNAs: insights into their function and mechanics in underlying disorders. Mutat Res Rev Mutat Res 762:1–21

12. Meister G (2013) Argonaute proteins: functional insights and emerging roles. Nat Rev Genet 14:447–459
13. Cerritelli SM, Frolova EG, Feng C, Grinberg A, Love PE, Crouch RJ (2003) Failure to produce mitochondrial DNA results in embryonic lethality in RNaseh1 null mice. Mol Cell 11:807–815
14. Wu H, Lima WF, Zhang H, Fan A, Sun H, Crooke ST (2004) Determination of the role of the human RNase H1 in the pharmacology of DNA-like antisense drugs. J Biol Chem 279:17181–17189
15. Lee RG, Crosby J, Baker BF, Graham MJ, Crooke RM (2013) Antisense technology: an emerging platform for cardiovascular disease therapeutics. J Cardiovasc Transl Res 6:969–980
16. Buller HR, Bethune C, Bhanot S, Gailani D, Monia BP, Raskob GE et al (2015) Factor XI antisense oligonucleotide for prevention of venous thrombosis. N Engl J Med 372:232–240
17. Gaudet D, Brisson D, Tremblay K, Alexander VJ, Singleton W, Hughes SG et al (2014) Targeting APOC3 in the familial chylomicronemia syndrome. N Engl J Med 371:2200–2206
18. Rigo F, Seth PP, Bennett CF (2014) Antisense oligonucleotide-based therapies for diseases caused by pre-mRNA processing defects. Adv Exp Med Biol 825:303–352
19. Rigo F, Chun SJ, Norris DA, Hung G, Lee S, Matson J et al (2014) Pharmacology of a central nervous system delivered 2′-O-methoxyethyl-modified survival of motor neuron splicing oligonucleotide in mice and nonhuman primates. J Pharmacol Exp Ther 350:46–55
20. Vickers TA, Wyatt JR, Burckin T, Bennett CF, Freier SM (2001) Fully modified 2′ MOE oligonucleotides redirect polyadenylation. Nucleic Acids Res 29:1293–1299
21. Baek MS, Yu RZ, Gaus H, Grundy JS, Geary RS (2010) In vitro metabolic stabilities and metabolism of 2′-O-(methoxyethyl) partially modified phosphorothioate antisense oligonucleotides in preincubated rat or human whole liver homogenates. Oligonucleotides 20:309–316
22. Henry SP, Geary RS, Yu R, Levin AA (2001) Drug properties of second-generation antisense oligonucleotides: how do they measure up to their predecessors? Current Opin Investig Drugs 2:1444–1449
23. Monia BP, Lesnik EA, Gonzalez C, Lima WF, McGee D, Guinosso CJ et al (1993) Evaluation of 2′-modified oligonucleotides containing 2′-deoxy gaps as antisense inhibitors of gene expression. J Biol Chem 268:14514–14522
24. Prakash TP (2011) An overview of sugar-modified oligonucleotides for antisense therapeutics. Chem Biodivers 8:1616–1641
25. Geary RS, Watanabe TA, Truong L, Freier S, Lesnik EA, Sioufi NB et al (2001) Pharmacokinetic properties of 2′-O-(2-methoxyethyl)-modified oligonucleotide analogs in rats. J Pharmacol Exp Ther 296:890–897
26. Lee RG, Fu W, Graham MJ, Mullick AE, Sipe D, Gattis D et al (2013) Comparison of the pharmacological profiles of murine antisense oligonucleotides targeting apolipoprotein B and microsomal triglyceride transfer protein. J Lipid Res 54:602–614
27. Seth PP, Vasquez G, Allerson CA, Berdeja A, Gaus H, Kinberger GA et al (2010) Synthesis and biophysical evaluation of 2′,4′-constrained 2′O-methoxyethyl and 2′,4′-constrained 2′O-ethyl nucleic acid analogues. J Org Chem 75:1569–1581
28. Hong DS, Kurzrock R, Kim Y, Woessner R, Younes A, Nemunaitis J, Fowler N, Zhou T, Schmidt J, Jo M, LeeSJ, Yamashita M, Hughes SG, Fayad L, Piha-Paul S, Nadella MVP, Mohseni M, Lawson D, Reimer C, Blakey DC, Xiao X, Hsu J, Monia BP, and MacLeod AR (2015). AZD9150, a nextgeneration antisense oligonucleotide Inhibitor of STAT3, with early evidence of clinical activity in lymphoma and lung cancer. Sci Transl Med (in press)
29. Senn JJ, Burel S, Henry SP (2005) Non-CpG-containing antisense 2′-methoxyethyl oligonucleotides activate a proinflammatory response independent of Toll-like receptor 9 or myeloid differentiation factor 88. J Pharmacol Exp Ther 314:972–979
30. Lima WF, Vickers TA, Nichols J, Li C, Crooke ST (2014) Defining the factors that contribute to on-target specificity of antisense oligonucleotides. PLoS One 9:e101752
31. Gutschner T, Hammerle M, Eissmann M, Hsu J, Kim Y, Hung G et al (2013) The noncoding RNA MALAT1 is a critical regulator of the metastasis phenotype of lung cancer cells. Cancer Res 73:1180–1189
32. Hung G, Xiao X, Peralta R, Bhattacharjee G, Murray S, Norris D et al (2013) Characterization of target mRNA reduction through in situ RNA hybridization in multiple organ systems following systemic antisense treatment in animals. Nucleic Acid Ther 23:369–378
33. Vickers TA, Freier SM, Bui HH, Watt A, Crooke ST (2014) Targeting of repeated sequences unique to a gene results in significant increases in antisense oligonucleotide potency. PLoS One 9:e110615
34. Skotte NH, Southwell AL, Ostergaard ME, Carroll JB, Warby SC, Doty CN et al (2014)

Allele-specific suppression of mutant huntingtin using antisense oligonucleotides: providing a therapeutic option for all Huntington disease patients. PLoS One 9:e107434

35. Ostergaard ME, Southwell AL, Kordasiewicz H, Watt AT, Skotte NH, Doty CN et al (2013) Rational design of antisense oligonucleotides targeting single nucleotide polymorphisms for potent and allele selective suppression of mutant Huntingtin in the CNS. Nucleic Acids Res 41:9634–9650
36. Castro D, Iannaccone ST (2014) Spinal muscular atrophy: therapeutic strategies. Curr Treat Options Neurol 16:316
37. Ward AJ, Norrbom M, Chun S, Bennett CF, Rigo F (2014) Nonsense-mediated decay as a terminating mechanism for antisense oligonucleotides. Nucleic Acids Res 42: 5871–5879
38. Cavaluzzi MJ, Borer PN (2004) Revised UV extinction coefficients for nucleoside-5′-monophosphates and unpaired DNA and RNA. Nucleic Acids Res 32:e13
39. Koller E, Vincent TM, Chappell A, De S, Manoharan M, Bennett CF (2011) Mechanisms of single-stranded phosphorothioate modified antisense oligonucleotide accumulation in hepatocytes. Nucleic Acids Res 39:4795–4807
40. Yu RZ, Grundy JS, Geary RS (2013) Clinical pharmacokinetics of second generation antisense oligonucleotides. Expert Opin Drug Metab Toxicol 9:169–182
41. Yu RZ, Lemonidis KM, Graham MJ, Matson JE, Crooke RM, Tribble DL et al (2009) Cross-species comparison of in vivo PK/PD relationships for second-generation antisense oligonucleotides targeting apolipoprotein B-100. Biochem Pharmacol 77:910–919

Chapter 17

Characterization of Circular RNAs

Yang Zhang, Li Yang, and Ling-Ling Chen

Abstract

Accumulated lines of evidence reveal that a large number of circular RNAs are produced in transcriptomes from fruit fly to mouse and human. Unlike linear RNAs shaped with 5′ cap and 3′ tail, circular RNAs are characterized by covalently closed loop structures without open terminals, thus requiring specific treatments for their identification and validation. Here, we describe a detailed pipeline for the characterization of circular RNAs. It has been successfully applied to the study of circular intronic RNAs derived from intron lariats (ciRNAs) and circular RNAs produced from back spliced exons (circRNAs) in human.

Key words Circular RNAs, ciRNAs, circRNAs, RNA fractionation, RNase R

1 Introduction

Single-stranded circular RNA molecules were firstly observed by electron microscopy in plant viroids [1], yeast mitochondrial RNAs [2], and hepatitis δ virus [3]. Later, a handful of circular RNAs processed from splicing were identified by Northern blots and/or RT-PCRs in higher eukaryotes, including human and mouse [4–6]. Due to their low abundance and scrambled order of exon–exon joining, these circular RNAs had long been considered as by-products of aberrant splicing, thus unlikely with important biological functions [6]. Nevertheless, circular RNAs produced from human *INK4a/ARF* or *CDR1* locus were reported to affect human atherosclerosis risk [7] or regulate gene expression [8–10], suggesting that some circular RNAs can be functional.

The application of next generation sequencing in transcriptome analyses (mRNA-seq) has provided unprecedented insight and quantitative measurements of gene expression, alternative splicing, etc. [11]. However, as formed covalently closed structures, circular RNAs lack canonical 3′ polyadenylation, thus falling below the radar of most canonical polyadenylated (linear) transcriptome profiling. Until recently, non-polyadenylated RNAs could be enriched from total RNAs by depleting both polyadenylated RNAs and

Yi Feng and Lin Zhang (eds.), *Long Non-Coding RNAs: Methods and Protocols*, Methods in Molecular Biology, vol. 1402, DOI 10.1007/978-1-4939-3378-5_17, © Springer Science+Business Media New York 2016

ribosomal RNAs for high throughput sequencing analyses [12], allowing the identification of new RNA transcripts. Non-polyadenylated RNA-seq signals were frequently identified in specific intron and exon regions [12], many of which were further proven to be circular RNAs produced from either introns [13] or exons [14, 15].

So far, two major groups of circular RNAs have been described [16]. One is produced from intronic regions (circular intronic RNAs, ciRNAs) [13] by escaping from debranching after splicing. The other group of circular RNAs (circRNAs) is generated from reversely back-spliced exons [10, 15, 17, 18]. Importantly, these two groups of circular RNAs have been successfully recapitulated with specific expression vectors [13, 15]. Currently, ten thousands of circular RNAs have been widely identified in metazoans from fruit fly to mouse and human [10, 15, 17–20], greatly expanding the complexity of transcriptomes and the diversity of noncoding RNAs [16]. To be noticed, special attention is required for circular RNAs enrichment and detection due to their intrinsic circular structures.

In this chapter, we describe an integrated pipeline for circular RNA characterization (Fig. 1), from the fractionation of non-polyadenylated RNAs from total RNAs, to the enrichment of circular RNAs with the RNase R digestion, and then to the validation of their existence by Northern blots and divergent PCRs. This pipeline provides a standard lab protocol to characterize circular RNAs, and has been successfully applied to the study of ciRNAs [13] and circRNAs [15] in human.

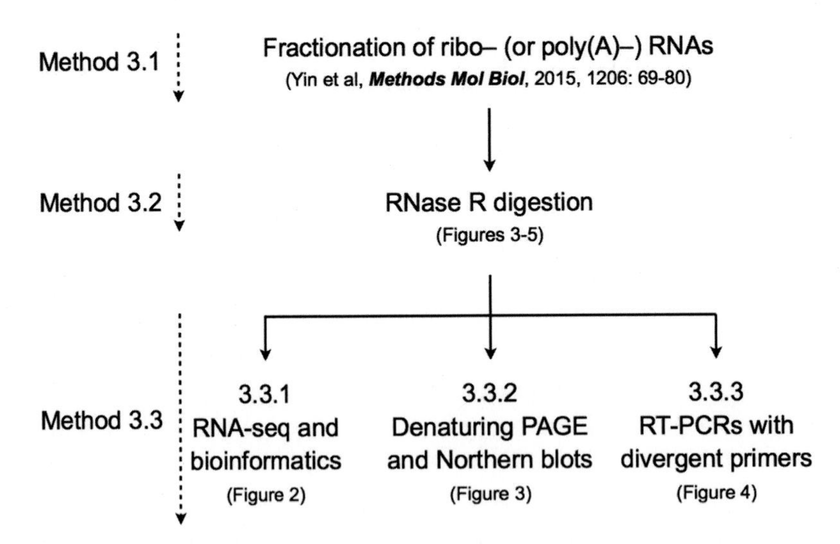

Fig. 1 A diagram of circular RNA fractionation, enrichment, and validation. See text for details

2 Materials

All solutions/recipes are prepared from analytical grade chemicals with deionized DEPC-treated water. All reagents are RNase-free. Sterilized reagents are aliquoted and stored at room temperature for immediate usage or −20 °C for long term storage. 1.5 ml RNase-free microcentrifuge tubes are purchased from Crystalgen (catalog number L-2507). 15 ml RNase-free centrifuge tubes are from Nest (catalog number 601052). 50 ml RNase-free centrifuge tubes are from Corning (catalog number 430829). Equipment required for polyacrylamide gel electrophoresis and electrophoretic transfer are from BIO-RED (catalog numbers 165-8001 and 170-3930, respectively).

2.1 RNA Fractionations

All reagents and recipes needed in this step have been described previously in great details [21].

2.2 RNase R Digestion

1. Ribonuclease R (RNase R): Epicentre, catalog number RNR07250, lot number RNR40620.
2. Phenol–chloroform–isoamyl alcohol (25:24:1, v/v): Life Technologies, catalog number 15593-031.
3. 4 M LiCl solution: weigh 3.3912 g LiCl (Sigma-Aldrich, catalog number L9650-500 G) and transfer to a 15 ml RNase-free centrifuge tube, add DEPC-treated water to 10 ml. Mix thoroughly and filter through a 0.22 μm Millex-GP Syringe Filter Unit (Millipore, catalog number SLGP05010). Split into small aliquots and store at −20 °C.
4. Glycogen, RNA grade: Thermo SCIENTIFIC, catalog number R0551.

2.3 Validation of Circular RNAs

2.3.1 RNA Deep-Sequencing and Genome-Wide Analysis of Circular RNAs

1. 75 % ethanol (v/v): transfer 30 ml absolute ethanol to a 50 ml RNase-free centrifuge tubes and add 10 ml DEPC-treated water. Mix well and store at −20 °C.

2.3.2 Denaturing Urea Polyacrylamide Gel Electrophoresis (Urea PAGE)

1. Urea: aMRESCO, catalog number 037-1 KG.
2. 30 % acrylamide: Sangon Biotech, catalog number SD6017.
3. 10× TBE Electrophoresis Buffer: weigh 108 g Tris base (Sigma-Aldrich, catalog number 15456-3), 55 g boric acid (Sigma-Aldrich, catalog number B6768-500 G), transfer to a beaker with 40 ml 0.5 M EDTA (pH 8.0), and add DEPC-treated water to 1 l. Filter through a 0.22 μm Millex-GP Syringe Filter Unit, and store at room temperature. 1× TBE buffer is diluted from 10× TBE buffer with DEPC-treated water.

4. 0.5 M EDTA, pH 8.0: weigh 186.1 g $Na_2EDTA \cdot 2H_2O$ (Sigma-Aldrich, catalog number E5134-250 G), transfer to a beaker with 500 ml DEPC-treated water, adjust pH to 8.0 with 10 M NaOH, and add DEPC-treated water to 1 l. Filter through a 0.22 μm Millex-GP Syringe Filter Unit, and store at room temperature.
5. 10 % (w/v) ammonium persulfate (APS): dissolute 0.1 g APS (Sigma-Aldrich, catalog number A3678-25 G) with DEPC-treated water to 1 ml. Store at 4 °C for few weeks.
6. TEMED ((*N*,*N*,*N*′,*N*′-tetramethylethylenediamine): BIO-RED, catalog number 161-0801.
7. 2× urea loading buffer: 8 M urea, 90 % formamide (Sigma-Aldrich, catalog number F9037-100 mL), 20 mM EDTA, 0.1 % (w/v) xylene cyanol, bromophenol blue. Store at 4 °C.
8. 50× Transfer Buffer: weigh 30.285 g Tris base, 17.02 g sodium acetate trihydrate (Sigma-Aldrich catalog number 236500-500 G), and 9.306 g EDTA, transfer to a beaker with 10 ml acetate, and add DEPC-treated water to 500 ml. Filter through a 0.22 μm Millex-GP Syringe Filter Unit, and store at room temperature. 1× transfer buffer is diluted from 50× transfer buffer with DEPC-treated water.
9. RiboMAX™ Large Scale RNA Production Systems: Promega, catalog number P1280 (SP6), P1300 (T7) for prepare RNA probe.
10. Dig RNA Labeling Mix: Roche, catalog number 11277073910.
11. DNA-free™ Kit, DNase Treatment and Removal Reagents: Ambion, catalog number AM1906.
12. Dig Easy Hyb Granules: Roche, catalog number 11796895001.
13. 20× SSC: weigh 175 g NaCl (Sigma-Aldrich, catalog number S9625-1 KG) and 88 g sodium citrate dihydrate (Sigma-Aldrich, catalog number W302600), transfer to a beaker with 500 ml DEPC-treated water, adjust pH to 7.0 with 1 M HCl, and add DEPC-treated water to 1 l. Filter through a 0.22 μm Millex-GP Syringe Filter Unit, and store at room temperature. 2× SSC and 0.2× SSC are diluted from 20× SSC with DEPC-treated water.
14. 10 % (w/v) SDS: dissolute 100 g SDS (Sigma-Aldrich, catalog number L3771-1KG) with DEPC-treated water to 1 l, filter through a 0.22 μm Millex-GP Syringe Filter Unit, store at room temperature. 0.1 % SDS dilute from 10 % SDS.
15. DIG Wash and Block Buffer Set: Roche, catalog number 11585762001.
16. Anti-Digoxigenin-AP, Fab fragments: Roche, catalog number 11093274910.

17. CDP-*Star*, ready-to-use solution: Roche, catalog number 12041677001.

2.3.3 Divergent RT-PCR

1. SuperScript™ III Reverse Transcriptase: Invitrogen, catalog number 18080.
2. Random hexamers: TaKaRa, catalog number RR037A.
3. dNTP mixture: TaKaRa, catalog number T4030.
4. Recombinant RNasin® Ribonuclease Inhibitor: Promega, catalog number N2511.

3 Methods

3.1 Fractionation of Non-polyadenylated RNAs from Mammalian Cells

Ribo-minus RNAs (ribo– RNAs) is enriched from DNase I treated total RNAs by depleting most redundant ribosomal RNAs with RiboMinus™ Human/Mouse Transcriptome Isolation Kit (Invitrogen™, catalog number K1550-01) as previously described [21]. In addition, fractionation of non-polyadenylated and ribo-minus RNAs (poly(A)–/ribo– RNAs, poly(A)– RNAs for short) from total RNAs by removal of poly(A)+RNA transcripts and ribosomal RNAs [21] shows cleaner background for circular RNA analysis (*see* **Note 1**).

3.2 RNase R Digestion (See Note 2)

1. Dissolve fractionated ribo– RNAs (or poly(A)– RNAs) from 20 μg total RNAs from Subheading 3.1 with 52 μl DEPC-treated water. Mix well.
2. Split RNAs to two aliquots in new 1.5 ml RNase-free microcentrifuge tubes: one for RNase R digestion and another for control with digestion buffer only.
3. For RNase R digestion, add 3 μl 10× RNase R Reaction Buffer and 1 μl RNase R (20 U/μl); for control, add 3 μl 10× RNase R Reaction Buffer and 1 μl DEPC-treated water. Mix well and quick spin the tubes for a few second. Incubate the samples at 37 °C for 1–2 h according to manufacturer's instructions (*see* **Note 3**).
4. After incubation, add 30 μl of phenol–chloroform–isoamyl alcohol to stop the exonuclease digestion. Vertex the tubes vigorously, and spin the tubes in a microcentrifuge at 13,000 ×*g* at 4 °C for 5 min.
5. Carefully remove the upper aqueous layers and transfer them to new 1.5 ml RNase-free microcentrifuge tubes, add 6 μl 4 M LiCl, 1 μl glycogen, 90 μl prechilled absolute ethanol (−20 °C). Mix well by inverting the tubes several times, and incubate the tubes at −80 °C for 1 h to precipitate RNAs. These RNA samples (RNase R treated or untreated) can be stored at −80 °C until use. This is a good stopping point in the process (*see* **Note 4**).

6. Precipitate RNAs by spinning at 13,000 ×*g* at 4 °C for 20 min. Remove the supernatants and wash the RNA pellet twice with 700 μl prechilled 75 % (v/v) ethanol and air-dry.
7. Resuspend RNA pellets with 20 μl DEPC-treated water.

3.3 Validation of Circular RNAs

3.3.1 High-Throughput Sequencing Analysis of Circular RNAs

1. Isolated ribo– RNAs (or poly(A)– RNAs) and/or their RNase R treatment can be directly used for RNA-seq library preparation according to the manufacturer's instructions, and then subjected to deep sequencing [21].
2. Specific pipelines can be used for genome-wide identification of ciRNAs [13] and/or circRNAs [10, 15, 17] by retrieving junction reads for circular RNAs.
3. Circular RNAs can be individually visualized at genome browser with uploaded track files, as indicated in Fig. 2 (Fig. 2(a) *ci-ankrd52* and (b) *circCAMSAP1*).

3.3.2 Northern Blots with Denaturing Polyacrylamide Gel Electrophoresis (PAGE)

Prepare the Gel

1. Assemble the gel plates according to the manufacturer's instructions, and fix the gel plates in the gel-casting apparatus.
2. Prepare the appropriate denaturing polyacrylamide gel solution by mixing 8 M urea, 10× TBE and 30 % acrylamide with DEPC-treated water (*see* **Note 5**). 10 ml gel solution is enough for a denaturing polyacrylamide gel of 10 cm × 7.5 cm × 1 mm. Dissolve the mixture by rotation, and filter the solution through 0.22 μl Millex-GP Syringe Filter Unit. Sequentially add 100 μl 10 % APS and 8 μl TEMED, and mix thoroughly. Immediately pour the gel solution between the gel plates and insert the comb. Wait for about 30 min to let the gel polymerize.

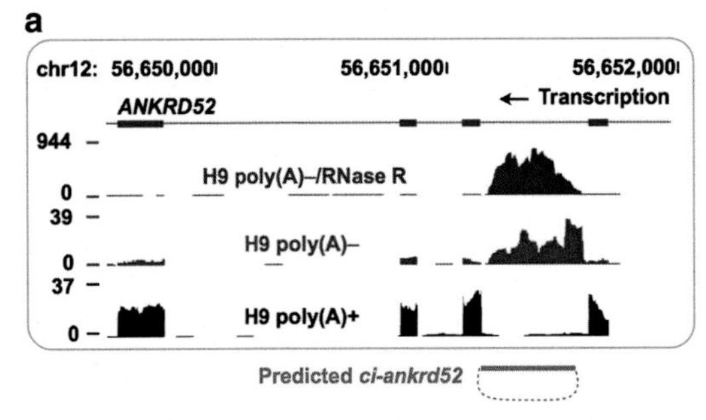

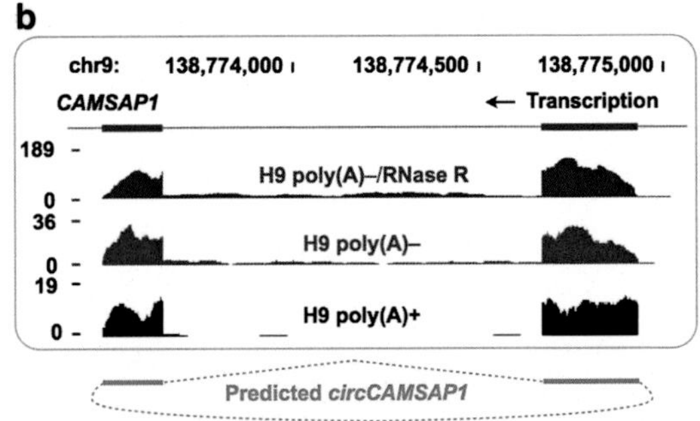

Fig. 2 Visualization of two types of circular RNAs with RNA-seq. (**a**) An example of ciRNAs (circular intronic RNAs), *ci-ankrd52*, from the second intron of *ANKRD52* gene. Deep sequencing signal from poly(A)+ (*black*), poly(A)– (*red*), poly(A)–/RNase R (*purple*) are shown. The predicted *ci-ankrd52* is indicated in *pink.*(**b**) An example of circRNAs from back spliced exons, *circCAMSAP1*, from the *CAMSAP1* loci. Deep sequencing signal from poly(A)+ (*black*), poly(A)– (*red*), poly(A)–/RNase R (*purple*) are shown. The predicted *circCAMSAP1* is indicated in *blue*

Prepare the Samples

3. Meanwhile, prepare RNA samples. Total RNAs (Subheading 3.1) were digested with RNase R (Subheading 3.2), cleaned up with phenol, precipitated by ethanol, and resuspended by DEPC-treated water. The final RNA concentration/amount was determined by measure UV absorption with a spectrophotometer. RNA concentration is determined by the OD reading at 260 nm. To get better signals for circular RNAs, similar amounts of RNAs with or without RNase R treatment were used for Northern blots (*see* **Note 6**).
4. Add 2× urea loading buffer to RNA samples (10 μl total volume is recommended to get sharp bands), for example RNAs with or without RNase R treatment as Subheading 3.2. Boil the RNA samples at 100 °C for 5 min, and chill on ice immediately.

Load the Samples and Run the Gel

5. Dismount the gel from the casting chamber, and assemble the gel electrophoresis apparatus according to the manufacturer's instructions.
6. Remove the comb, and fill the chamber with 1× TBE electrophoresis buffer (*see* **Note 7**).
7. Load denatured RNA samples and run the gel at 120 V for 3 h (*see* **Note 8**).

RNA Transfer and Fixation

8. Dissemble the gel electrophoresis apparatus, and assemble the gel transfer "sandwich" in a transfer blot. Put the bottom of the gel face to the nylon membrane, remove bubbles between the gel and nylon membrane. Transfer the RNA from gel to membrane in 1× transfer buffer with 100 V for 90 min.
9. Dissemble the gel transfer apparatus. Immobilize RNAs with UV cross-linking by an Ultravoliet Crosslinker (UVP, CL-1000, 254-nm wavelength) for the appropriate length of time (usually 180 mJ/cm^2 is used) (*see* **Note 9**).

Prepare RNA Probe

10. Prepare Dig (Digoxigenin)-labeled RNA probes according to the manufacturer's instructions (RiboMAX™ Large Scale RNA Production Systems, Promega). Briefly, mix DNA template (with T7 or SP6 promoter), 10× Dig labeling mixture, 5× transcription buffer (T7 or SP6) and enzyme mixture (T7 or SP6) with DEPC-treated water. Mix gently by pipetting, and incubate reaction at 37 °C for 2–3 h.
11. Add 1 μl DNase I and mix well. Incubate at 37 °C for 15 min to remove the DNA template.
12. To precipitate the RNA probe, sequentially add 4 μl 4 M LiCl, 100 μl DEPC-treated water and 300 μl prechilled absolute ethanol (−20 °C), mix well by inverting the tubes several times, and incubate the tubes at −20 °C for 1 h.

13. Spin the tubes at 13,000×g at 4 °C for 15 min. Remove the supernatants and wash the RNA pellet twice with 700 μl 75 % (v/v) cold ethanol (−20 °C) and air-dry.
14. Dissolve the RNA probes by adding 40 μl DEPC-treated water (usually 8–10 μg RNA probes can be obtained).

Hybridization

15. Pre-hybridize membrane with DIG Easy Hyb (3 ml/100 cm^2) for 30 min at 68 °C with gentle rotation.
16. Denature DIG-labeled RNA probe by heating at 100 °C for 5 min, and immediately place the tube on ice.
17. Discard prehybridization buffer, and add new prewarmed DIG Easy Hyb with denatured DIG-labeled RNA probe (100 ng/ml). Incubate overnight with gentle rotation.

Immunological Detection

The following steps followed protocol for DIG Wash and Block Buffer Set, and all performed at room temperature (unless indicated otherwise).

18. Wash the membrane twice with 2× SSC, 0.1 % SDS for 5 min with gentle rotation.
19. Wash the membrane twice with 0.2× SSC, 0.1 % SDS at 68 °C for 30 min with gentle rotation.
20. After stringency washes, briefly rinse the membrane in 1× Washing buffer.
21. Incubate the membrane in 1× Blocking solution for 30 min in an appropriate container with gentle agitation.
22. Incubate the membrane in Antibody Solution for 30 min with gentle agitation.
23. Wash the membrane three times with 1× Washing buffer for 20 min.
24. Incubate in Detection buffer for 5 min.
25. Place the membrane (RNA side-up) on a development folder, add CDP-*Star* ready-to-use solution (1 ml for 100 cm^2 membrane), and incubate for 2–5 min.
26. Expose to X-ray film for an appropriate time. Multiple exposures can be taken to achieve appropriate signals. Examples of Northern blots of circular RNAs migration in denaturing PAGE are shown in Fig. 3 (Fig. 3(a) *ci-ankrd52*, (b) *circCAMSAP1*).

3.3.3 RT-PCR with Divergent Primers

First Strand cDNA Synthesis

1. Before the first strand cDNA synthesis, ribo– RNAs (Subheading 3.1) were digested with RNase R (Subheading 3.2), cleaned up with phenol, precipitated by ethanol, and resuspended by DEPC-treated water. The final RNA concentration/amount was determined by measure UV absorption with

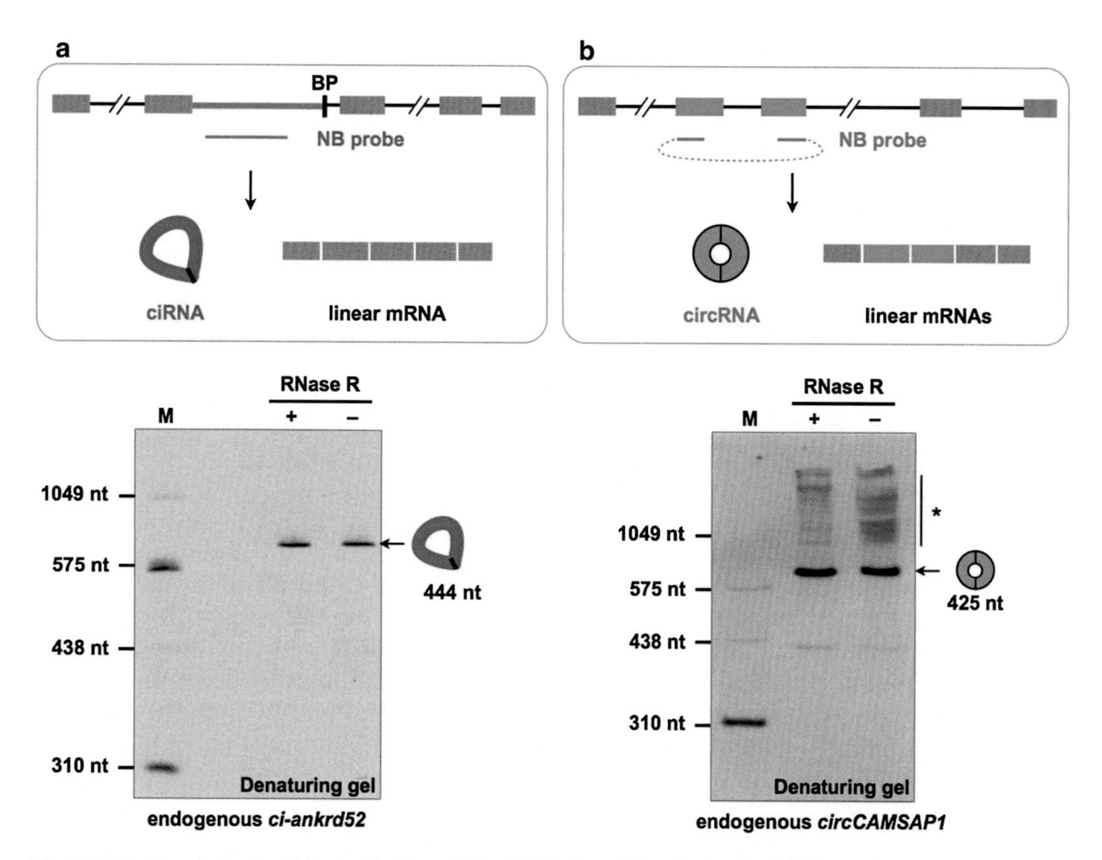

Fig. 3 Validation of circular RNAs by Northern blots. (***a***) Northern blot of *ci-ankrd52*. RNase R treated or untreated total RNAs from H9 cells are loaded on 5 % denaturing urea PAGE for Northern blot with a DIG-labeled antisense RNA probe (*pink bar*) as previously reported [13]. Note that *ci-ankrd52* remains stable after RNase R treatment and runs slowly on the denaturing urea PAGE. (***b***) Northern blot of *circCAMSAP1*. RNase R treated or untreated total RNAs from H9 cells are loaded on 5 % denaturing urea PAGE for Northern blot with a DIG-labeled antisense RNA probe (*blue bar*) as previously reported [15]. Note that *circCAMSAP1* remains stable after RNase R treatment and runs slowly on the denaturing urea PAGE. *Asterisk*, linear mRNAs

a spectrophotometer. RNA concentration is determined by the OD reading at 260 nm. To get better signals for circular RNAs, similar amounts of ribo– RNAs with or without RNase R treatment were used for RT-PCR validation (*see* **Note 6**). Prepare the first strand cDNA according to the manufacturer's instructions (SuperScript™ III Reverse Transcriptase, Invitrogen). Briefly, mix RNAs samples, random hexamers (100 μM), dNTP (2.5 mM each) with DEPC-treated water. Heat at 65 °C for 5 min, and immediately chill on ice for at least 1 min.

2. Briefly spin the tubes, add 5× First-Strand Buffer, 0.1 M DTT, RNasin, and SuperScript™ III RT. Pipette gently and incubate reaction at 25 °C for 5 min.
3. Incubate at 50 °C for 60 min, and inactivate the reaction by heating at 70 °C for 15 min.

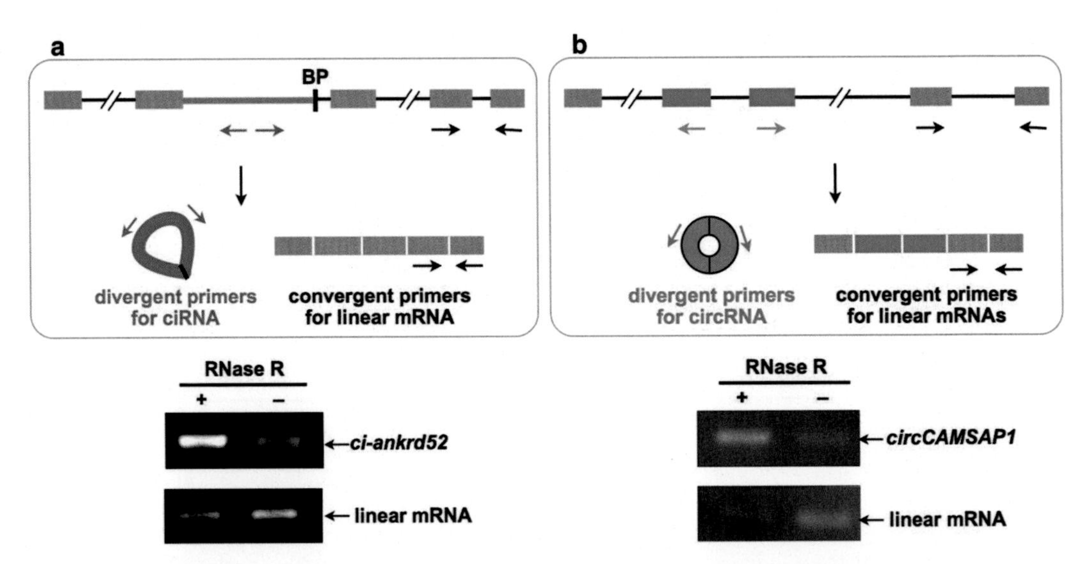

Fig. 4 Validation of circular RNAs by divergent PCRs. (**a**) PCR validation of *ci-ankrd52* with a divergent primer set (*pink arrows*) as previously reported [13]. Primers for linear mRNA are indicated as *black arrows*. Note that *ci-ankrd52* remains stable after RNase R treatment, while the linear mRNA is largely degraded with RNase R treatment. (**b**) PCR validation of *circCAMSAP1* with a divergent primer set (*blue arrows*) as previously reported [15]. Primers for linear mRNAs are indicated as *black arrows*. Note that *circCAMSAP1* remains stable after RNase R treatment, while the linear mRNA is largely degraded with RNase R treatment

Circular RNAs Divergent Primer PCR

1. Unlike linear RNA, circular RNAs can be amplified by divergent primers from both ribo– RNAs with or without RNase R treatment. Examples of circular RNAs semi-quantitative RT-PCR results are shown in Fig. 4 (Fig. 4(a) *ci-ankrd52*, (b) *circCAMSAP1*).

4 Notes

1. Ribo-minus RNAs (ribo– RNAs) is enriched from total RNAs by depleting ribosomal RNAs, which contain both poly(A)+ and poly(A)– RNAs. However, the fractionation of poly(A)– / ribo– RNAs (poly(A)– RNAs for short) removes both poly(A)+RNAs and ribosomal RNAs, which shows a much cleaner background for circular RNA analysis. We thus prefer to use poly(A)– RNAs (with or without further RNase R digestion) for RNA-seq analyses.
2. RNase R is a magnesium-dependent $3' \rightarrow 5'$ exribonuclease that digests linear RNAs and Y-structure RNAs, while preserving circular RNAs [22]. Besides RNase R, tobacco acid phosphatase and terminator exonuclease can also efficiently degrade linear RNAs, while leave circular RNAs intact [9].

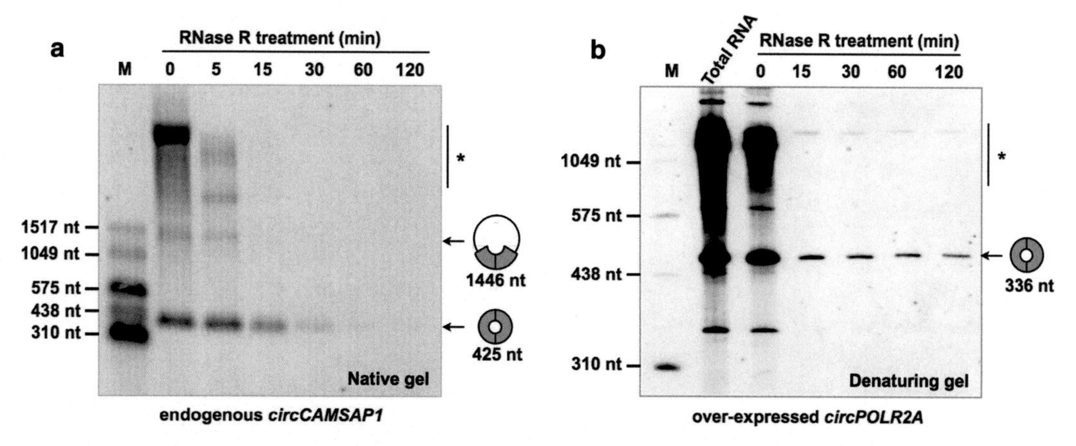

Fig. 5 Time course of the RNase R treatment. Ribo– RNAs from 10 μg total RNAs in either wild type (**a**) or transfected (**b**) cells are incubated with 1 μl RNase R at 37 °C for 0, 5, 15, 30, 60, 120 min, respectively, and then applied for Northern blots on either native agarose gel for endogenous *circCAMSAP1* (**a**) or denaturing PAGE for over-expressed *circPOLR2A* (**b**). The probes are the same ones as previously reported [15]. Note that linear RNAs (*asterisks*) are largely degraded with the short time of RNase R incubation, and circular RNAs are also reduced during the prolonged incubation

3. RNase R activity may vary from batch to batch. Thus, check the amount of enzyme and the incubation time for your particular experiment is highly recommended before carrying out further sequencing. For example, we have observed that circular RNAs can also be degraded with the RNase R incubation (Fig. 5). In this case, some RNase R-sensitive circular RNAs could be lost with long time incubation with RNase R. From the time course of the RNase R treatment (Fig. 5), we recommend that one-hour incubation of this batch of RNase R is good enough for circular RNA validation by Northern blots. Meanwhile, it is worthy to note that the short time course of RNase R digestion can lead to only partial digestion of linear RNAs, which may in turn result in unwanted linear RNA signals when such samples are applied to RNA-seq analyses (data not shown).
4. RNA samples are recommended to be used immediately or stored at –80 °C for later usage. RNA precipitation in ethanol can be stored at –80 °C for years.
5. Although we have applied the denaturing urea PAGE to detect some long circular RNAs [13], such PAGE is usually used to analyze RNAs less than 500 nt. In addition, due to the special covalently closed loop structure, circular RNAs migrate much more slowly than linear RNAs with the same molecular weight [13, 15]. On the contrast, the migration of circular RNAs on native agarose gels is similar with linear RNAs with similar molecular weights [13].

6. During the preparation of the RNase R treated RNA samples or RNA samples subjected to other treatments, RNAs will be lost due to different treatments and the subsequent RNA recovery. Thus, we generally use similar amounts of RNAs prior to any treatment (total RNA) and after the treatment of interest for Northern blots. For instance, about 20 μg total RNAs were first digested with RNase R (about 5–10 μg RNAs could be retrieved) and then applied for Northern blot; correspondingly, about 5–10 μg total RNAs were directly applied for Northern blot without RNase R digestion (Fig. 3). A similar strategy has been applied to RT-PCR analyses (Fig. 4).
7. Before loading the samples, rinse gel wells several times to remove dissolved urea, and then load the samples immediately.
8. The time of electrophorese depends on the molecular weights of interested circular RNAs. Usually, it requires about 3 h at 120 V to sufficiently resolve 400–nt circular RNAs in 5 % denaturing PAGE.
9. The cross-linked membrane can be immediately used for prehybridization or store at −80 °C for weeks after drying up.

Acknowledgements

This work was supported by 31271390, 31322018, 91440202 from NSFC to L.Y. and L.L.C.

References

1. Sanger HL et al (1976) Viroids are single-stranded covalently closed circular RNA molecules existing as highly base-paired rod-like structures. Proc Natl Acad Sci U S A 73:3852–3856
2. Arnberg AC et al (1980) Some yeast mitochondrial RNAs are circular. Cell 19:313–319
3. Kos A et al (1986) The hepatitis delta (delta) virus possesses a circular RNA. Nature 323:558–560
4. Nigro JM et al (1991) Scrambled exons. Cell 64:607–613
5. Cocquerelle C et al (1992) Splicing with inverted order of exons occurs proximal to large introns. EMBO J 11:1095–1098
6. Capel B et al (1993) Circular transcripts of the testis-determining gene Sry in adult mouse testis. Cell 73:1019–1030
7. Burd CE et al (2010) Expression of linear and novel circular forms of an INK4/ARF-associated non-coding RNA correlates with atherosclerosis risk. PLoS Genet 6:e1001233
8. Hansen TB et al (2011) miRNA-dependent gene silencing involving Ago2-mediated cleavage of a circular antisense RNA. EMBO J 30:4414–4422
9. Hansen TB et al (2013) Natural RNA circles function as efficient microRNA sponges. Nature 495:384–388
10. Memczak S et al (2013) Circular RNAs are a large class of animal RNAs with regulatory potency. Nature 495:333–338
11. Graveley BR et al (2008) Molecular biology: power sequencing. Nature 453:1197–1198
12. Yang L et al (2011) Genomewide characterization of non-polyadenylated RNAs. Genome Biol 12:R16
13. Zhang Y et al (2013) Circular intronic long noncoding RNAs. Mol Cell 51:792–806
14. Salzman J et al (2012) Circular RNAs are the predominant transcript isoform from hundreds

of human genes in diverse cell types. PLoS One 7:e30733

15. Zhang XO et al (2014) Complementary sequence-mediated exon circularization. Cell 159:134–147
16. Zhang Y, Yang L, Chen LL (2014) Life without A tail: new formats of long noncoding RNAs. Int J Biochem Cell Biol 54:338–349
17. Jeck WR et al (2013) Circular RNAs are abundant, conserved, and associated with ALU repeats. RNA 19:141–157
18. Salzman J et al (2013) Cell-type specific features of circular RNA expression. PLoS Genet 9:e1003777
19. Wang PL et al (2014) Circular RNA is expressed across the eukaryotic tree of life. PLoS One 9:e90859
20. Westholm JO et al (2014) Genome-wide analysis of drosophila circular RNAs reveals their structural and sequence properties and age-dependent neural accumulation. Cell Rep 9:1966–1980
21. Yin QF, Chen LL, Yang L (2015) Fractionation of non-polyadenylated and ribosomal-free RNAs from mammalian cells. Methods Mol Biol 1206:69–80
22. Suzuki H et al (2006) Characterization of RNase R-digested cellular RNA source that consists of lariat and circular RNAs from pre-mRNA splicing. Nucleic Acids Res 34:e63

Chapter 18

Methods for Characterization of Alternative RNA Splicing

Samuel E. Harvey and Chonghui Cheng

Abstract

Quantification of alternative splicing to detect the abundance of differentially spliced isoforms of a gene in total RNA can be accomplished via RT-PCR using both quantitative real-time and semi-quantitative PCR methods. These methods require careful PCR primer design to ensure specific detection of particular splice isoforms. We also describe analysis of alternative splicing using a splicing "minigene" in mammalian cell tissue culture to facilitate investigation of the regulation of alternative splicing of a particular exon of interest.

Key words Alternative splicing, RNA, RT-PCR, Variable exon, Minigene, Splicing factors, Splicing regulation

1 Introduction

Alternative splicing is a key cellular process whereby particular combinations of exons in a nascently transcribed pre-mRNA are either included or excluded to generate different mature mRNA splice isoform transcripts, resulting in multiple protein isoforms encoded by a single gene. It is estimated that nearly all protein-coding genes in the human genome are alternatively spliced, providing an essential source of protein diversity [1, 2]. Differentially spliced isoforms may exert distinct biological functions, therefore accurate quantification of the relative amounts of splice isoforms of a gene is essential to explore the role of alternative splicing in biological processes as well as disease [3]. The pre-mRNA of an alternative spliced gene contains both constitutive exons and variable exons interspaced by introns that are ultimately spliced out and excluded from the final mRNA transcript. In this chapter, we describe the characterization of exon-skipping, the most common form of alternative splicing, where one or more variable exons are either included or skipped from the final transcript to make up the array of spliced isoforms of a gene [4].

Yi Feng and Lin Zhang (eds.), *Long Non-Coding RNAs: Methods and Protocols*, Methods in Molecular Biology, vol. 1402, DOI 10.1007/978-1-4939-3378-5_18, © Springer Science+Business Media New York 2016

Reverse Transcription-Polymerase Chain Reaction (RT-PCR) is one of the most convenient methods for studying RNA transcripts and can generally be used with total RNA extracted from any biological source. Quantitative PCR (qPCR) following RT is often the method of choice given its extreme sensitivity of detection and accurate comparison of amplification of PCR products during the exponential phase of the PCR reaction [5]. Semi-quantitative methods are less useful for precise quantification of the relative abundance of splice isoforms, however, combined with electrophoresis, these methods allow for the direct visualization and comparison of splice isoform abundance based on size disparity between differentially spliced transcripts.

Quantification of levels of different splice isoforms is especially useful in studying the regulation of alternative splicing. Regulation of alternative splicing is accomplished by the recognition of cis-acting sequences in the pre-mRNA, usually located in the variable exonic or adjacent intronic sequences, by trans-acting RNA-binding proteins known as splicing factors [6]. In this protocol, we make use of a splicing "minigene" construct to study the regulation of alternative splicing of a variable exon within an alternatively spliced gene [7]. This minigene then offers a versatile tool for studying the regulation of splicing of a particular exon as it can be expressed in cell-culture along with potential splicing factors that could influence variable exon inclusion or skipping.

2 Materials

2.1 RNA Source Material

Alternative splicing can be examined using RNA from any source. In this protocol we use RNA extracted from mammalian cells grown in tissue-culture.

2.2 RNA Extraction and RT

1. E.Z.N.A.® Total RNA Isolation Kit (Omega Bio-Tek).
2. GoScript™ Reverse Transcriptase Reagents (Promega).

2.3 qPCR

1. GoTaq® Green Master Mix (Promega).
2. Primers for specific splice isoform detection and a housekeeping gene. As an example, primers for the detection of CD44 splice isoforms and housekeeping gene TBP are included in Table 1.

Table 1
qPCR Primers for the analysis of CD44 alternative splicing

Primer name	Forward (5′–3′)	Reverse (5′–3′)
CD44v with v5/v6 exons	GTAGACAGAAATGGCACCAC	CAGCTGTCCCTGTTGTCGAA
CD44s	TACTGATGATGACGTGAGCA	GAATGTGTCTTGGTCTCTGGT
Human TBP	GGAGAGTTCTGGGATTGTAC	CTTATCCTCATGATTACCGCAG

Table 2
Primers for the analysis of CD44 v8 minigene alternative splicing

Primer name	Forward (5′–3′)	Reverse (5′–3′)
qPCR v8 minigene inclusion	CAATGACAACGCTGGCACAA	CCAGCGGATAGAATGGCGCCG
qPCR v8 minigene skipping	GAGGGATCCGGTTCCTGCCCC	CAGTTGTGCCACTTGTGGGT
Semi-quantitative PCR v8	GAGGGATCCGGTTCCTGCCCC	CCAGCGGATAGAATGGCGCCG

2.4 Semi-quantitative PCR

1. HotStarTaq Plus DNA Polymerase Reagents (Qiagen).
2. Primers for detecting minigene splice isoforms. As an example, primers for the detection of a CD44 variable exon 8 (v8) minigene are included in Table 2.
3. 0.5× TBE buffer (45 mM Tris, 45 mM boric acid, 1 mM EDTA) for electrophoresis.
4. 1.5 % agarose gel prepared with 0.5× Tris–borate–EDTA (TBE) buffer and 0.5 μg/mL ethidium bromide.

2.5 Co-transfection of Splicing Minigene and Splicing Factors

1. Transfectable mammalian cell line, such as HEK293T (ATCC® CRL-3216™).
2. Gibco® Dulbecco's Modified Eagle Medium High Glucose (plain media as well as media supplemented with L-glutamine and 10 % fetal bovine serum).
3. Lipofectamine 2000 Reagent (Invitrogen).
4. Mammalian expression plasmids encoding splicing minigene and splicing factors of interest. As an example, we use a CD44 v8 minigene plasmid, splicing factor hnRNPM plasmid, and pcDNA3 control plasmid for co-transfection.

2.6 Equipment

1. Thermo Fisher Scientific NanoDrop 1000 Spectrophotometer for quantification of RNA.
2. A thermal cycler, such as the Bio-Rad DNA Engine Tetrad® 2, for Reverse Transcriptase reaction and semi-quantitative PCR.
3. A real-time thermal cycler such as the Roche Lightcycler® 480 II System with associated LightCycler software for qPCR.
4. Horizontal electrophoresis apparatus.
5. A UV transilluminator with camera such as the Bio-Rad Gel Doc XR System with Quantity One 1-D Analysis Software for visualization and densitometric analysis of ethidium-bromide stained DNA.
6. A tabletop centrifuge such as the Eppendorf Centrifuge 5424.

3 Methods

Prepare all reactions with molecular grade nuclease-free water to prevent RNA degradation.

3.1 RNA Extraction and Purification

1. Total RNA from mammalian cells in tissue culture is extracted using the E.Z.N.A.® Total RNA Isolation Kit (Omega Bio-Tek) by following the product manual (*see* **Note 1**). For RNA isolation via this kit, cells can be directly lysed using the provided lysis buffer.
2. As per the manufacturer protocol, first collect cells in 350 μL RNA lysis buffer.
3. Add 350 μL of 70 % ethanol to the lysate and mix thoroughly. Then transfer the sample to the RNA purification column.
4. Centrifuge at 10,000 × *g* for 1 min and discard flow-through.
5. Wash the column once with 500 μL of RNA wash buffer I and twice with RNA wash buffer II, centrifuging at 10,000 × *g* for 1 min between each wash and discarding flow-through.
6. Remove residual RNA wash buffer II from the column by centrifugation at maximum speed for 2 min.
7. Transfer the column into a new 1.5 mL tube, add 50 μL nuclease-free water at the center of the column matrix, and elute RNA by centrifugation at the maximum speed for 1 min.
8. Determine RNA quantity using a UV spectrometer (such as NanoDrop). High quality RNA should have a 260 nm/280 nm absorbance ratio of approximately 2.0. RNA should be stored at −80 °C for long-term storage.

3.2 RT Reaction

1. In a reaction totaling 20 μL, combine 250–1000 ng of total RNA with 0.05 μg random hexamer primers, 50 pmol $MgCl_2$, 10 pmol dNTPs, 2 μL 5× GoScript™ Buffer, and 1 μL of GoScript™ Reverse Transcriptase (*see* **Note 2**). Mix by vortex and briefly centrifuge.
2. Incubate reaction mixture 25 °C for 5 min for primer annealing, 42 °C for 60 min for the RT reaction, and 70 °C for 5 min to inactivate the RT enzyme. These incubations are best performed in a programmable thermocycler. Completed reactions may now be frozen at −20 °C for long-term storage.

3.3 Primer Design for qPCR

1. In this protocol we provide an example to detect the most common form of alternative splicing, exon skipping. In a hypothetical four-exon pre-mRNA (Fig. 1a), exon 3 is a variable exon that is either included in the mature mRNA to generate a longer isoform or skipped to generate a shorter isoform. Three primers are designed to differentiate one splice isoform from another.

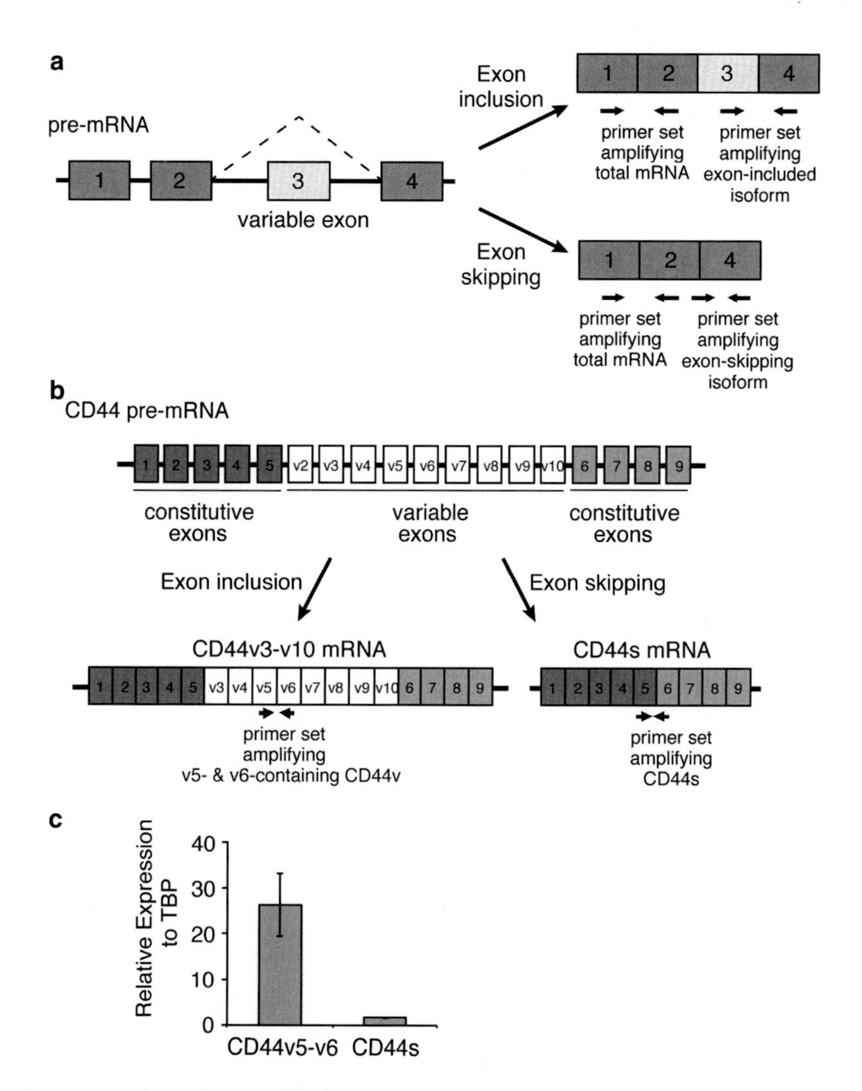

Fig. 1 Characterization of alternative splicing by qRT-PCR. (**a**) A schematic illustrating a hypothetical four-exon gene with a variable exon 3 that undergoes exon-inclusion or exon-skipping. Three primer sets for qPCR indicated by the paired arrows are designed to amplify the total mRNA, the variable exon 3 inclusion isoform, or the exon-skipping isoform. (**b**) A schematic depicting the exon structure of the transmembrane protein CD44, with the nine constitutive and nine variable exons labeled. CD44v3-v10 and CD44s are predominant splice isoforms and are depicted. *Paired arrows* indicate primer sets to amplify CD44v v5-v6 containing isoforms and the CD44s isoform which lacks all variable exons. (**c**) Representative qRT-PCR results using primers in Table 1 to show the relative expression of CD44v5/v6 and CD44s, normalized to TBP, in an immortalized human mammary epithelial cell line (HMLE)

A forward primer lies within the variable exon 3 with a reverse primer in the immediately downstream constitutive exon 4. These primers allow the selective amplification of the longer isoform with variable exon 3 included. To detect the shorter isoform with the variable exon 3 skipped, a forward primer is designed that spans the junction between exon 2 and 4.

The junction between exon 2 and 4 is disrupted with exon 3 inclusion. Thus, together with the corresponding reverse primer located in constitutive exon 4, this primer set only amplifies mRNA isoforms where exon 3 is skipped. When multiple consecutive variable exons exist in the gene of interest, primers can be designed in two different variable exons for detection of specific variable exon-containing isoforms (Fig. 1b).

2. It is useful to design qPCR primers that can detect all isoforms of the mRNA of interest. In this way the total expression of the gene as well as the expression of individual isoforms can be compared. In the hypothetical four-exon pre-mRNA (Fig. 1a), exons 1 and 2 are constitutive exons immediately adjacent to one another without any variable exons in between. By designing a forward primer in exon 1 and a reverse primer in exon 2, all mature mRNA isoforms generated from this pre-mRNA can be detected via PCR.
3. Primers should also be designed to amplify a reference "housekeeping" gene that is expected to be uniformly expressed in all samples. In this protocol we use primers amplifying a segment of human TATA-binding protein (TBP) (Table 1).
4. Primers should be designed with an ideal melting temperature (T_m) of approximately 55 °C. The primer length is generally 18–22 nucleotides. T_m of each primer can be estimated using online tools such as the oligoanalyzer of Integrative DNA Technologies (http://www.idtdna.com/analyzer/applications/oligoanalyzer). The amplicon for each primer pair should be between 80 and 150 base pairs.
5. As an example, we have depicted primer design to detect alternatively spliced isoforms of the CD44 gene that encodes a family of CD44 transmembrane proteins (Fig. 1b, Table 1). CD44 contains nine constitutive exons and nine variable exons, and in our example primers have been designed to detect isoforms containing CD44 variable exons 5–6 (CD44v5-6) and isoforms in which all variable exons have been skipped (CD44s).

3.4 Quantification of Splicing via qPCR

1. Prepare qPCR reactions in a total volume of 20 μL by combining 0.1–1.0 μL of cDNA generated from the RT reaction with 10 μL GoTaq® Green Master Mix and 5 pmol each of forward and reverse primers. Be sure to include reactions for the reference gene.
2. Perform triplicate qPCR reactions with 40–50 cycles of the following steps: 95 °C denaturation for 10 s, 58 °C annealing for 10 s, 60 °C extension for 30 s. Include a thermal denaturing step to generate dissociation curves that can be used to verify specific amplification of PCR products (*see* **Note 3**).

3. After the PCR is completed, the amplified signal from each reaction is presented as a threshold cycle (Ct) determined using an arbitrary detection threshold, usually set by the qPCR software to reside within the exponential range of PCR amplification. Relative quantification of splice isoforms is accomplished via comparison to expression of a reference gene such as TBP. Relative quantification assumes that the PCR amplification efficiency for each primer pair is perfect, resulting in doubling of PCR amplicons with every cycle. Since the reference gene expression remains constant between experimental samples, the quantity of mRNA for each splice isoform relative to the reference gene is calculated using the following formula: $2^{-\Delta Ct}$, where $\Delta Ct = (Ct$ splice isoform $- Ct$ reference mRNA). As an example, quantification of two splice isoforms of CD44 was conducted via qPCR using RNA extracted from the immortalized human mammary epithelial cell line HMLE (Fig. 1c). The primers used are in Table 1.
4. Relative levels of splice isoforms of the same gene can also be expressed as a ratio of one splice isoform to another, for example a variable exon inclusion/variable exon skipping ratio. In such a scenario, a reference gene is not required because splice isoforms of the same gene from the same sample are compared. The ratio of splice isoforms is calculated using the following formula: $2^{-\Delta Ct}$, where $\Delta Ct = (Ct$ inclusion splice isoform mRNA $- Ct$ skipping splice isoform mRNA in the same sample). Relative comparison of this ratio between different experimental samples is accomplished via the formula: $2^{-\Delta Ct(\text{experiment})}/2^{-\Delta Ct(\text{control})}$, where $\Delta Ct(\text{experiment}) = (Ct$ inclusion splice isoform mRNA in experimental sample $- Ct$ skipping splice isoform mRNA in experimental sample) and $\Delta Ct(\text{control}) = (Ct$ inclusion splice isoform mRNA in control sample $- Ct$ skipping splice isoform mRNA in control sample). In this way the fold change in inclusion/skipping ratio compared to the control sample is obtained.

3.5 Primer Design for Detection of Splice Isoforms via Semi-quantitative PCR

Through careful primer and PCR design, differences in splice isoform expression in cDNA generated from total RNA can be visualized using agarose gel electrophoresis. Returning to a hypothetical example of a four-exon pre-mRNA, a forward primer in constitutive exon 2 and a reverse primer in constitutive exon 4 will amplify differently sized PCR amplicons depending on whether or not the template includes variable exon 3 (Fig. 2). Using semi-quantitative PCR, these primers facilitate the amplification, separation via electrophoresis, and approximate quantification of different splice isoforms. Primers should be designed with an ideal T_m of 55 °C, and ideal amplicons should be no more than 300 base pairs long (*see* **Note 4**).

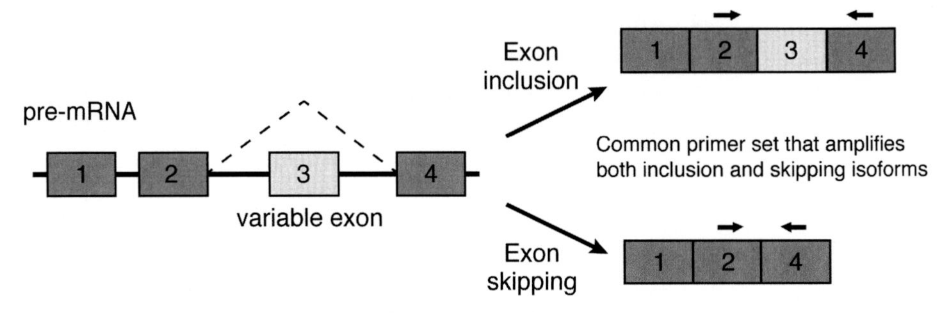

Fig. 2 Semi-quantitative PCR for examination of exon skipping. A schematic is shown depicting a hypothetical four-exon gene with variable exon 3 undergoing alternative splicing. The *paired arrows* indicate primer sets that flank the variable exon 3 and thus amplify both variable exon-inclusion and skipping isoforms from a cDNA template. The different splice isoforms can be distinguished by size using semi-quantitative PCR and electrophoresis

3.6 Performing and Analyzing Semi-quantitative PCR

1. In a 20 μL reaction, combine 0.1–1.0 μL cDNA with 2 μL 10× PCR Buffer (Qiagen), 4 μL 5× Q-Solution (Qiagen), 0.5 μL HotStarTaq DNA Polymerase (Qiagen), and 5 pmol each of forward and reverse primers.
2. Complete PCR reaction using the following "hot start" program: 95 °C denaturation for 5 min, approximately 30 cycles of 95 °C for 30 s, 55 °C for 30 s, 72 °C for 30 s, and lastly 72 °C for 10 min. Ensure that PCR is completed in the exponential range (*see* **Note 5**).
3. Visualize PCR reactions via horizontal gel electrophoresis using a 1.5 % agarose-TBE gel prepared with 0.5 μg/mL ethidium bromide (*see* **Note 6**). Detection of two distinct bands differing in size by exactly the length of the variable exon indicates a successful PCR amplification. The approximate quantity of each splice isoform present in the original RNA sample can be inferred using densitometric analysis by comparing the fluorescence intensity of the inclusion and skipping PCR amplicons using a UV transilluminator with camera and image intensity quantitation software (such as the Bio-Rad Gel Doc XR system with Quantity One 1-D Analysis Software). Brighter intensity of the band is equivalent to higher expression of the splice isoform in the original RNA.

3.7 Generating a Splicing Minigene to Analyze Regulation of Alternative Splicing of a Variable Exon

Splicing minigenes are useful tools to interrogate the regulation of splicing of a variable exon of interest. A splicing minigene is an expression plasmid consisting of a variable exon of interest flanked by two constitutive exons. It is important to include the immediately upstream and downstream intronic sequences flanking the variable exon where potential cis-acting sequences important for determining the inclusion or skipping of the exon may be located [7] (*see* **Note 7**). As an example, we have included a CD44 v8 minigene with variable exon 8 (v8) of CD44 bounded by the adjacent 250 base pairs of its upstream and downstream introns. This sequence is flanked by two constitutive exons from the human insulin gene with functioning 5′ and 3′ splice sites. Primers for detection of exon inclusion and exon skipping were designed in accordance with the principles mentioned in Subheadings 3.3 and 3.5 (Fig. 3a, Table 2).

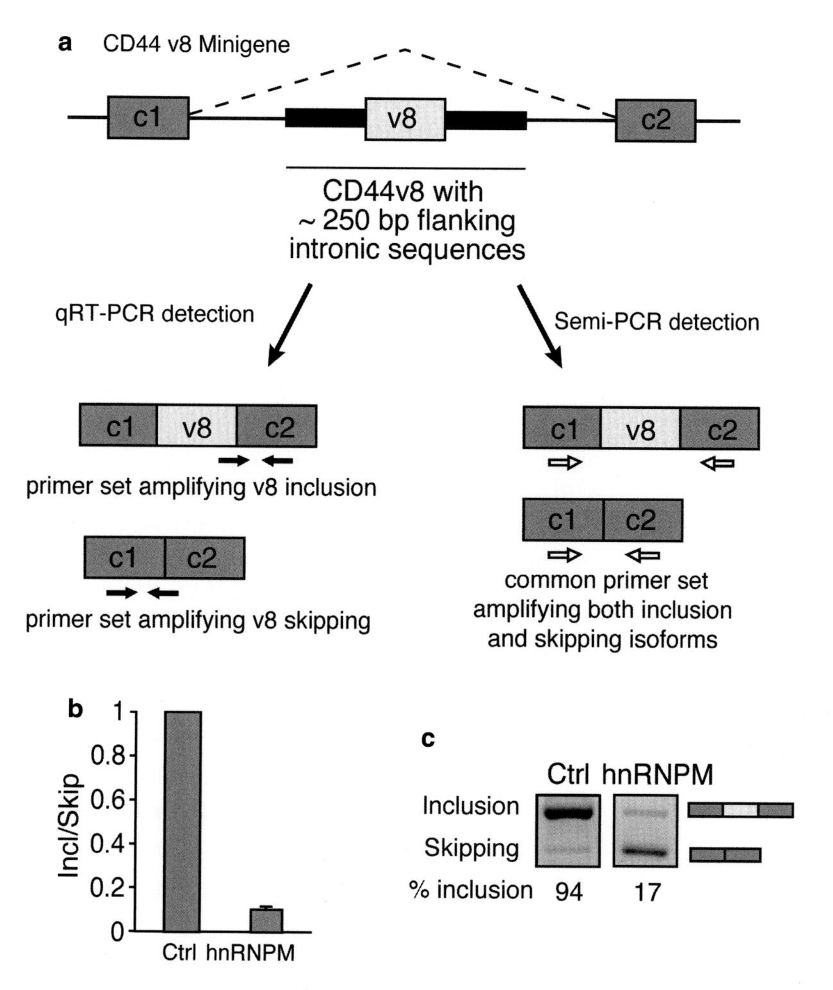

Fig. 3 A splicing minigene assay for characterization of exon skipping. (**a**) A schematic of the CD44 v8 minigene is shown. The CD44 v8 variable exon flanked by approximately 250 bp of upstream and downstream intronic sequence is cloned between constitutive exons c1 and c2. Exons are shown as *boxes* and introns as *lines*. Primer sets in *black* are used for qRT-PCR reactions. The primer set in *white* is designed for semi-quantitative methods. (**b**) qRT-PCR data after co-transfection of 100 ng CD44 v8 minigene with 400 ng of splicing factor hnRNPM using v8 inclusion and skipping primers in Table 2. A relative Inclusion/Skipping ratio (Incl/Skip) is plotted indicating that the Incl/Skip ratio decreases when hnRNPM is transfected. (**c**) Semi-quantitative PCR image showing decreased percent inclusion (% inclusion) after transfection with 400 ng hnRNPM using semi-quantitative v8 primers in Table 2. Both images were from the same gel with uniform exposure time. % inclusion was calculated using densitometric image analysis software to divide the pixel intensity of the inclusion band by the combined pixel intensity of both inclusion and exclusion bands. Figure 3b and c was modified from Fig. 4 in our previous publication [8] under Creative Commons License CC-BY-NC 4.0

3.8 Investigating the Regulation of Alternative Splicing via Co-transfection of a Splicing Minigene and Splicing Factors

1. As the splicing minigene contains native cis-acting splicing regulatory sequences within or near the variable exon, trans-acting RNA binding proteins are capable of modulating inclusion or skipping of the variable exon in the minigene in a transfectable cell model. As an example, we describe co-transfection of the CD44 v8 minigene plasmid and a plasmid expressing splicing factor hnRNPM in the HEK293T mammalian cell line. In this case, hnRNPM promotes exon-skipping of the CD44 v8 minigene [8] (Fig. 3b, c).
2. In a tissue culture hood using aseptic technique, plate 2.25×10^5 HEK293T cells in 24-well tissue culture plates using DMEM containing 10 % FBS (*see* Subheading 2).
3. 24 h after seeding cells, at which time the cells are approximately 90 % confluent, conduct transfections using Lipofectamine 2000 (Invitrogen) (*see* **Note 8**). Per well of a 24-well plate, combine 1.5 μL Lipofectamine 2000 with 50 μL plain DMEM (*see* **Note 9**) and incubate for 5 min at room temperature.
4. Combine Lipofectamine and media with transfecting plasmids diluted in 50 μL plain DMEM and mix well. As an example, transfect all wells with 100 ng CD44 v8 minigene, one well with a control plasmid such as pcDNA3, and other wells with splicing factors of interest such as hnRNPM. Each well should be transfected with a total of 800 ng of plasmid, so add in the remaining DNA with a control vector such as pcDNA3 (for example, a well could be transfected with 100 ng CD44 v8 minigene, 400 ng hnRNPM, and 300 ng pcDNA3). Incubate mixture for 20 min at room temperature.
5. During the above incubation, replace media on HEK293T cells with fresh DMEM containing 10 % FBS.
6. Add lipofectamine–transfectant mixture dropwise slowly over cells and incubate at 37 °C for 24 h (*see* **Note 10**).
7. Collect cells using RNA extraction protocol detailed in Subheading 3.1.
8. As an example, qRT-PCR data (Fig. 3b) and semi-quantitative RT-PCR data (Fig. 3c) from the CD44 v8 splicing minigene experiments were collected using primers in Table 2 [8] (*see* **Note 11**).

4 Notes

1. Using the E.Z.N.A.® Total RNA Isolation Kit (Omega Bio-Tek), genomic DNA contamination is minimal. In addition, as described in Subheading 3, the primers are designed in different exons. This would eliminate PCR products amplified

from genomic DNA which contains long intronic sequences. Therefore, the RNA may be used directly for RT-PCR analysis. If genomic DNA contamination is a concern, the manufacturer provides instructions for an on-column DNase treatment protocol. Also, during the elution step, we elute with RNAse-free H_2O. If DEPC-treated H_2O is used as an eluent, it can interfere with quantification of RNA via UV spectrometry [9].

2. Depending on the expression level of the gene of interest and the amount of different splice isoforms in the mRNA, the total input RNA into the Reverse Transcriptase reaction can be adjusted. Usually 250 ng of total RNA is sufficient, however up to 1 μg of RNA may be added to the reaction to increase the cDNA concentration without saturating the Reverse Transcriptase during cDNA synthesis.
3. The thermal denaturing step in qPCR is critical to determine the specific amplification of the target PCR product. The thermal denaturation curve for each pair of primers per reaction should show a single isolated peak with a uniform melting temperature. If multiple peaks with different melting temperatures are observed, then nonspecific amplification is occurring and the PCR reaction must be optimized before qPCR results can be analyzed [5]. Redesigning primers and adjusting the annealing temperature of the reaction may be necessary. We also recommend electrophoresing the PCR products on an agarose gel and visualizing the product using ethidium-bromide under UV. If amplification is specific, a single band of the appropriate size should be observed.
4. Designing primers so that the PCR amplicons for each splice isoform are relatively small (i.e., 300 base pairs or less) ensures that both the large and small amplicons are amplified with similar efficiency during the PCR reaction.
5. Accuracy of semi-quantitative PCR for visualizing the relative amounts of splice isoforms is dependent on the number of PCR cycles. The relative quantities of splice isoforms in a sample of mRNA remain proportional during the exponential phase of a PCR reaction, but this proportional relationship is less accurate in later PCR cycles once the reaction reaches the linear and plateau phases as PCR reagents and the polymerase are exhausted [5]. This is the reason qPCR offers more accurate quantification of expression compared to semi-quantitative PCR. It is important to perform a series of PCR cycles to identify the range of cycles where the proportional intensity of the different splice isoform bands remains constant [10]. This range also depends on the expression level of the gene in the original RNA sample. It is also recommended to conduct ^{32}P-labeled PCR by adding ^{32}P-labeled dCTP to the PCR reaction to increase the sensitivity of detecting bands on a gel.

In ^{32}P-labeled PCR, the total PCR cycle number can be reduced to approximately 20 cycles to ensure that results are captured during the exponential phase of the reaction.

6. To improve the resolution of PCR product bands, it may be necessary to increase to the concentration of the agarose gel to 2 %. Polyacrylamide gel electrophoresis (6 %) may also be used to resolve even smaller PCR products or resolve splice isoform bands that differ by a small size.
7. Splicing minigenes using constitutive exons not derived from the gene of interest offer a convenient tool for cloning different variable exons of interest into the same minigene construct. However, the actual exons flanking the variable exon may also be included in the minigene to better approximate the genomic cis-acting sequences present for native splicing of the gene. These cis-elements are located in both the variable exon of interest as well as adjacent introns [6].
8. When using the transfection reagent Lipofectamine 2000 (Invitrogen), it is essential that cells be at a high confluency of at least 90 %. This is to ensure that cells are still dividing to maximize uptake of transfected DNA while minimizing the impact of cell death that can occur during transfection.
9. Do not use DMEM containing serum when preparing transfection mixtures with the transfection reagent and transfected DNA. Serum proteins can interfere with the formation of the DNA-lipid transfection complexes and reduce transfection efficiency.
10. Although 24 h is usually sufficient to observe efficient expression of transfected DNA, cells can be incubated for up to 48 h to obtain even higher transgene expression.
11. To confirm protein transgene expression, for example transfected splicing factor expression constructs, Western blot analysis using cell lysates and an antibody specific for the protein of interest can be conducted. Efficient expression should result in an intense band on Western blot compared to control.

Acknowledgement

This work was supported by research grants from the US National Institutes of Health (R01GM110146) and the American Cancer Society (RSG-09-252-01-RMC) to C.C.

References

1. Wang ET, Sandberg R, Luo S, Khrebtukova I, Zhang L, Mayr C, Kingsmore SF, Schroth GP, Burge CB (2008) Alternative isoform regulation in human tissue transcriptomes. Nature 456(7221):470–476. doi:10.1038/nature07509

2. Pan Q, Shai O, Lee LJ, Frey BJ, Blencowe BJ (2008) Deep surveying of alternative splicing complexity in the human transcriptome by high-throughput sequencing. Nat Genet 40(12):1413–1415. doi:10.1038/ng.259
3. Liu S, Cheng C (2013) Alternative RNA splicing and cancer. Wiley Interdiscip Rev RNA 4(5):547–566. doi:10.1002/wrna.1178
4. Keren H, Lev-Maor G, Ast G (2010) Alternative splicing and evolution: diversification, exon definition and function. Nat Rev Genet 11(5):345–355. doi:10.1038/nrg2776
5. Freeman WM, Walker SJ, Vrana KE (1999) Quantitative RT-PCR: pitfalls and potential. Biotechniques 26(1):112–122, 124-125
6. Wang Z, Burge CB (2008) Splicing regulation: from a parts list of regulatory elements to an integrated splicing code. RNA 14(5): 802–813. doi:10.1261/rna.876308
7. Cooper TA (2005) Use of minigene systems to dissect alternative splicing elements. Methods 37(4):331–340. doi:10.1016/j.ymeth.2005.07.015
8. Xu Y, Gao XD, Lee JH, Huang H, Tan H, Ahn J, Reinke LM, Peter ME, Feng Y, Gius D, Siziopikou KP, Peng J, Xiao X, Cheng C (2014) Cell type-restricted activity of hnRNPM promotes breast cancer metastasis via regulating alternative splicing. Genes Dev 28(11):1191–1203. doi:10.1101/gad.241968.114
9. Okamoto T, Okabe S (2000) Ultraviolet absorbance at 260 and 280 nm in RNA measurement is dependent on measurement solution. Int J Mol Med 5(6):657–659
10. Horner RM (2006) Relative RT-PCR: determining the linear range of amplification and optimizing the primers:competimers ratio. CSH Protoc 2006(1). doi:10.1101/pdb.prot4109

Chapter 19

NONCODEv4: Annotation of Noncoding RNAs with Emphasis on Long Noncoding RNAs

Yi Zhao, Jiao Yuan, and Runsheng Chen

Abstract

The rapid development of high-throughput sequencing technologies and bioinformatics algorithms now enables detection and profiling of a large number of noncoding transcripts. Long noncoding RNAs (lncRNAs), which are longer than 200 nucleotides, are accumulating with important roles involved in biological processes and tissue physiology. In this chapter, we describe the use of NONCODEv4, a database that provide a comprehensive catalog of noncoding RNAs with particularly detailed annotations for lncRNAs. NONCODEv4 stores more than half million transcripts, of which more than 200,000 are lncRNAs. NONCODEv4 raises the concept of lncRNA genes and explores their expression and functions based on public transcriptome data. NONCODEv4 also integrated a series of online tools and have a web interface easy to use. NONCODEv4 is available at http://www.noncode.org/ http://www.bioinfo.org/noncode.

Key words Sequencing, lncRNA, lncRNA gene, Expression, Function

1 Introduction

Noncoding RNAs (ncRNAs) participate in many biological processes such as translation [1], RNA splicing [2], and DNA replication [3]. It has been recognized that the number of ncRNAs is much larger than expected. Although long noncoding RNAs (lncRNAs) which refer to ncRNAs longer than 200 nucleotides [4] form a group, they have diverse functions and mechanisms [4]. The continuously developed high-throughput sequencing technology has resulted in an explosion of transcriptome data [5, 6]. Through exploring these data resource, NONCODEv4 identified a large number of lncRNAs from both qualitative and quantitative perspective [7].

In the situation that more and more interest has been focused on lncRNAs, NONCODEv4 provides a platform that would felicitate and benefit researches into lncRNAs for both traditional biologists and bioinformaticians. In order to provide a reliable list of

Yi Feng and Lin Zhang (eds.), *Long Non-Coding RNAs: Methods and Protocols*, Methods in Molecular Biology, vol. 1402,
DOI 10.1007/978-1-4939-3378-5_19,

ncRNAs with detailed annotation, NONCODEv4 integrated information following a pipeline described below:

1. Collecting ncRNA sequences from literature, the latest version of specialized databases [8, 9] and the third version of NONCODE [10] following redundancy elimination, exclusion of protein coding transcripts, and mapping to various genome. Currently, NONCODEv4 contains more than half million ncRNA sequences from more than 1000 organisms. Of them, the majority comes from model organism: human, mouse, nematode, and fly. The genomic location for some short sequences that could not be mapped to unique position is not shown.
2. Defining lncRNA genes and classifying lncRNA genes into four categories according to their position relationship with protein coding genes. A lncRNA gene is region of genome which encodes overlapping lncRNAs. Currently, NONCODEv4 defines lncRNA genes for only human and mouse. 56,018 and 46,475 lncRNA genes of human and mouse are further classified into four categories: intergenic, antisense, sense non-exonic, and sense exonic [11].
3. Constructing expression pattern of lncRNA transcripts and lncRNA genes respectively based on public RNA-Seq data from different tissues of human and mouse. Expression profile is represented as a bar graph.
4. Predicting functions of lncRNA genes based on expression correlation between lncRNA genes and protein coding genes inferred from public RNA-Seq data.

All data curated in NONCODEv4 are available for browse and download. There is also an online pipeline named "iLncRNA" incorporated in order to provide an access to deal with their own raw RNA-Seq data for users to identify novel lncRNAs. NONCODEv4 also encourages users to submit their own discoveries of novel lncRNAs to facilitate future updates. Other tools have been incorporated to make NONCODEv4 more friendly to use, such as Genome Browser by which users could check neighboring genes and isoforms of the lncRNA they are interested in and an ID conversion tool which quickly convert accessions of NONCODE to those of other databases.

2 Materials

NONCODEv4 is available online via the URL http://www.noncode.org/. A Web browser is needed by a workstation of UNIX, Windows, or Macintosh with an Internet connection. In addition, decompression software is required since files provided for download is compressed.

3 Methods

The methods presented in this chapter describes how to use the NONCODEv4 Web interface to obtain information for a specific lncRNA (Subheading 3.1), how to navigate the record of a lncRNA of interest (Subheading 3.2), and how to submit data to NONCODEv4 server (Subheading 3.3).

3.1 Browsing Information for a Specific lncRNA

Access to the data content of NONCODEv4 may be performed by browsing the list of all lncRNAs which provides a quick overview about the dataset. The following steps describe how to browse the list of all lncRNAs and the detailed information:

1. Click "Browse DB" on the home page to open the browse page. A user who is interested in specific organism might first select it from the list of species (Fig. 1). By default, lncRNA transcripts are listed. It could be switched to the list of lncRNA genes by selection through the check box followed by clicking "Display". An accession designated by NONCODEv4, genomic coordinates, exon number, length and CNCI score suggesting coding potential are all listed in the shown table.

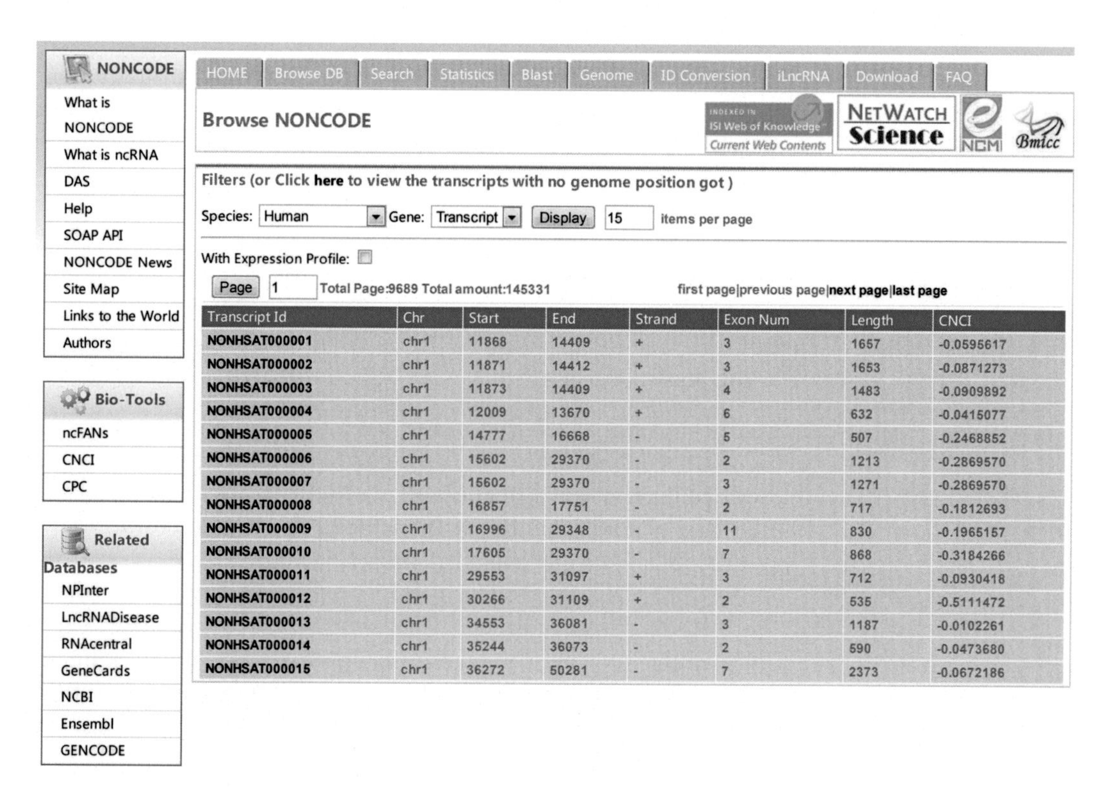

Fig. 1 A screenshot of browse page. The species of interest might be selected by *check box*. It is optional to browse lncRNA transcript list or lncRNA gene list. ncRNAs have no exact genomic coordinates might be browsed by clicking "here" *above* the *check box*

2. Click on the accession listed on the first column of the table to launch a detailed page providing further information on that lncRNA transcript or gene (Figs. 2 and 3).
3. The detailed information for a specific lncRNA transcript includes five sections (Fig. 2):
   (a) In the section for general information, clicking the "NONCODE Gene ID" will link to the webpage of the lncRNA gene encoding the lncRNA transcript described at present page.
   (b) In the second section, the full length of sequence is provided in a fasta format [12].
   (c) In the third section, the expression pattern of the lncRNA transcript across different tissues is given by both numerical value and bar graph.
   (d) In the section for isoforms, the accessions of other lncRNA transcripts encoded by the same lncRNA gene are listed.
   (e) In the section for data resources, accessions of Ensembl, RefSeq, or NONCODE v3 might be listed to indicate the origin of the lncRNA transcript.
4. The detailed information contained in the page of lncRNA gene (Fig. 3) includes four sections. The first three sections are similar to those of lncRNA transcript. Notice that the category of the lncRNA gene is classified into (intergenic, antisense, sense non-exonic, or sense exonic) is provided in the section for general information. The last section listed predicted function of the lncRNA gene by the software ncFANs [13].

3.2 Navigating to the Record for a Specific lncRNA

From the browse page, it might not be easy to quick navigate to the right webpage for the lncRNA of interest. Navigating to the exact webpage for a specific lncRNA might be performed via three options summarized below.

1. Click "Search" on the home page to open the search page, which enables users to search ncRNAs by keywords or accessions. Single word or multiple words separated by whitespace might be entered into the blank box, "HOTAIR" for example (Fig. 4). Multiple types of terms, including accessions from NONCODEv4, NONCODEv3, RefSeq, and Ensembl, lncRNA name, and other keywords, are supported. Click "Search" to view the search result. Selecting and clicking an accessions from the list in the result page would direct a new page for browsing annotations for the lncRNA transcript.
2. For a ncRNA of which none information except sequence is known, search based on keywords might not work. Sequence alignment [14] is a solution for this situation. Click "Blast" on the home page to open the blast page, which enables users to

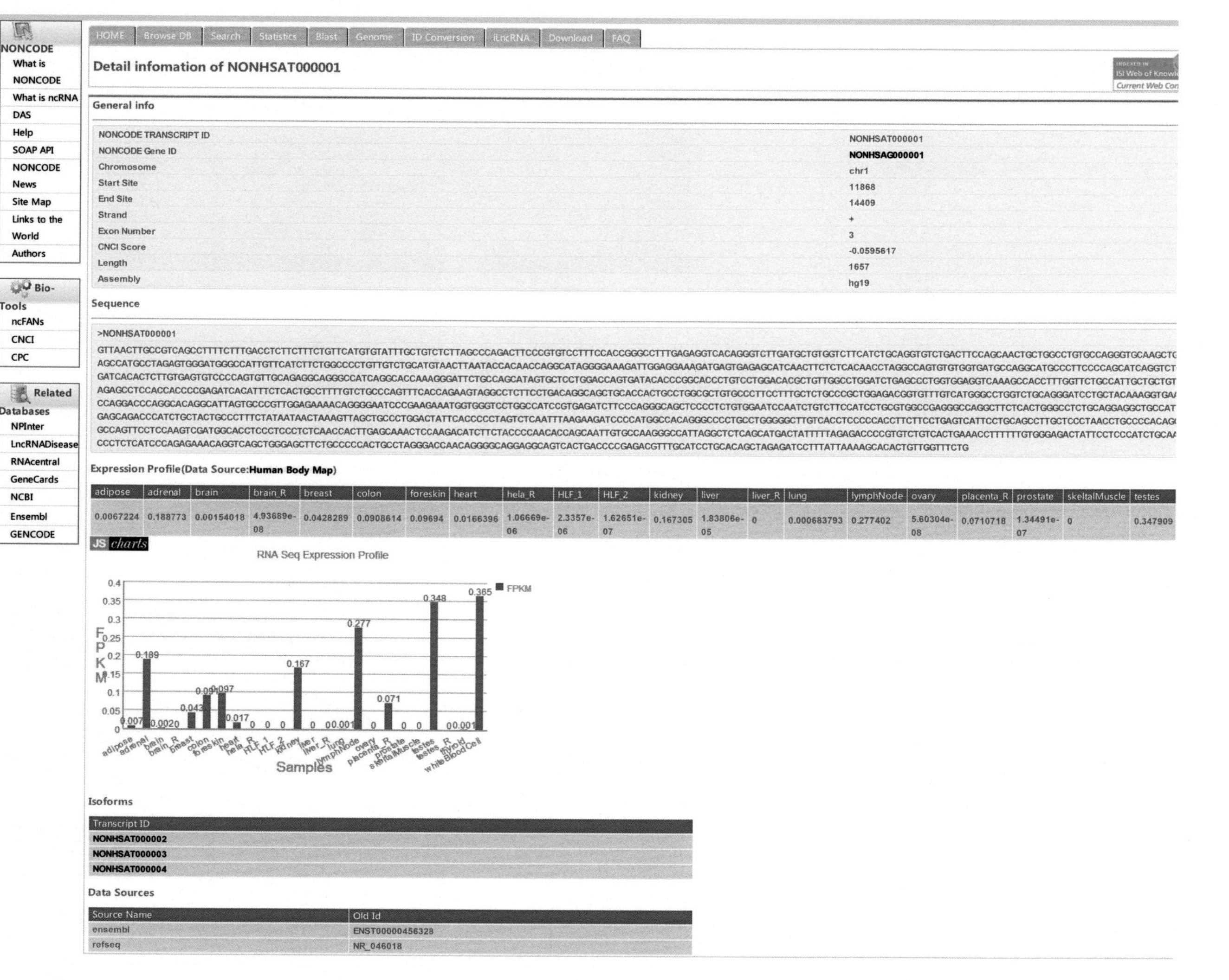

Fig. 2 A sample screenshot of detail information of a lncRNA transcript. Annotation includes five sections: general information, sequence, expression profile, isoforms, and data sources. A link to the webpage for annotation of the lncRNA gene encoding the lncRNA transcript describe at present webpage is included in the section of general information

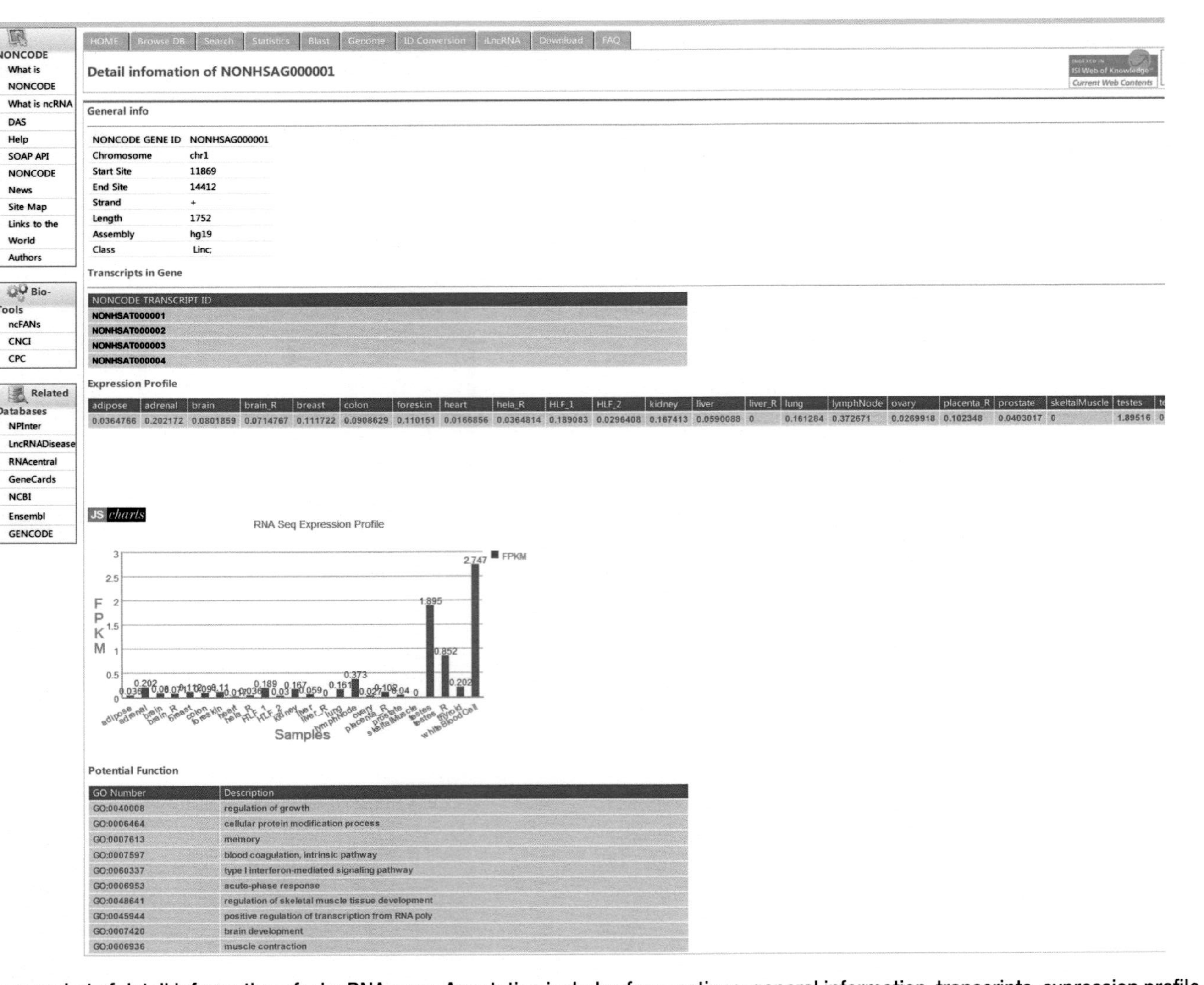

Fig. 3 A sample screenshot of detail information of a lncRNA gene. Annotation includes four sections: general information, transcripts, expression profile, and potential function. The category into which the lncRNA gene is classified is shown in the section of general information

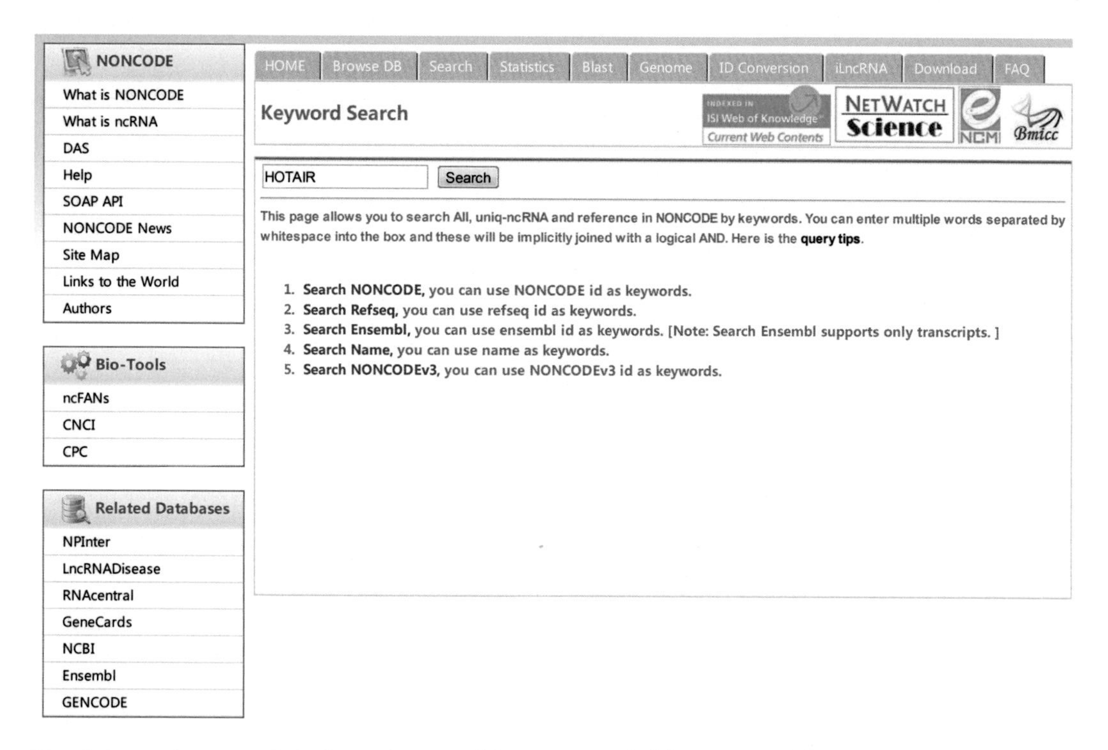

Fig. 4 A sample screenshot of search page. It is recommended for users to follow the query tips listed to determine their keywords

search ncRNAs by sequences (Fig. 5). Enter the sequences in fasta format into the blank box or upload the file containing sequences in fasta format from local disk. Click "Search" button to start the search. Parameters for BLAST might be adjusted through check boxes below. In the result page of BLAST output, the top NONCODEv4 accessions with highest score might match the input sequence. If there are no match records with score high enough according to the length of input sequence, then the sequence might not be curated in NONCODEv4.

3. Another option is navigating to the record for a specific lncRNA based on its genomic coordinates. NONCODEv4 provides Genome Browser, a visualization tool, to quickly navigate to a specific genome region. Click "Genome" on the home page to open the Genome Browser page. Select the genome of interest as described above, and then type the genomic coordinate of the lncRNA of interest into the "genomic position" text box. Click "submit" button. In the display page (Fig. 6), all lncRNA transcripts and genes overlapping the submitted genomic coordinates are shown, as well as annotation from other tracks within the region. Clicking a track item within the browser launches a detailed page providing further information on that item. The width of the displayed coordinate range could be

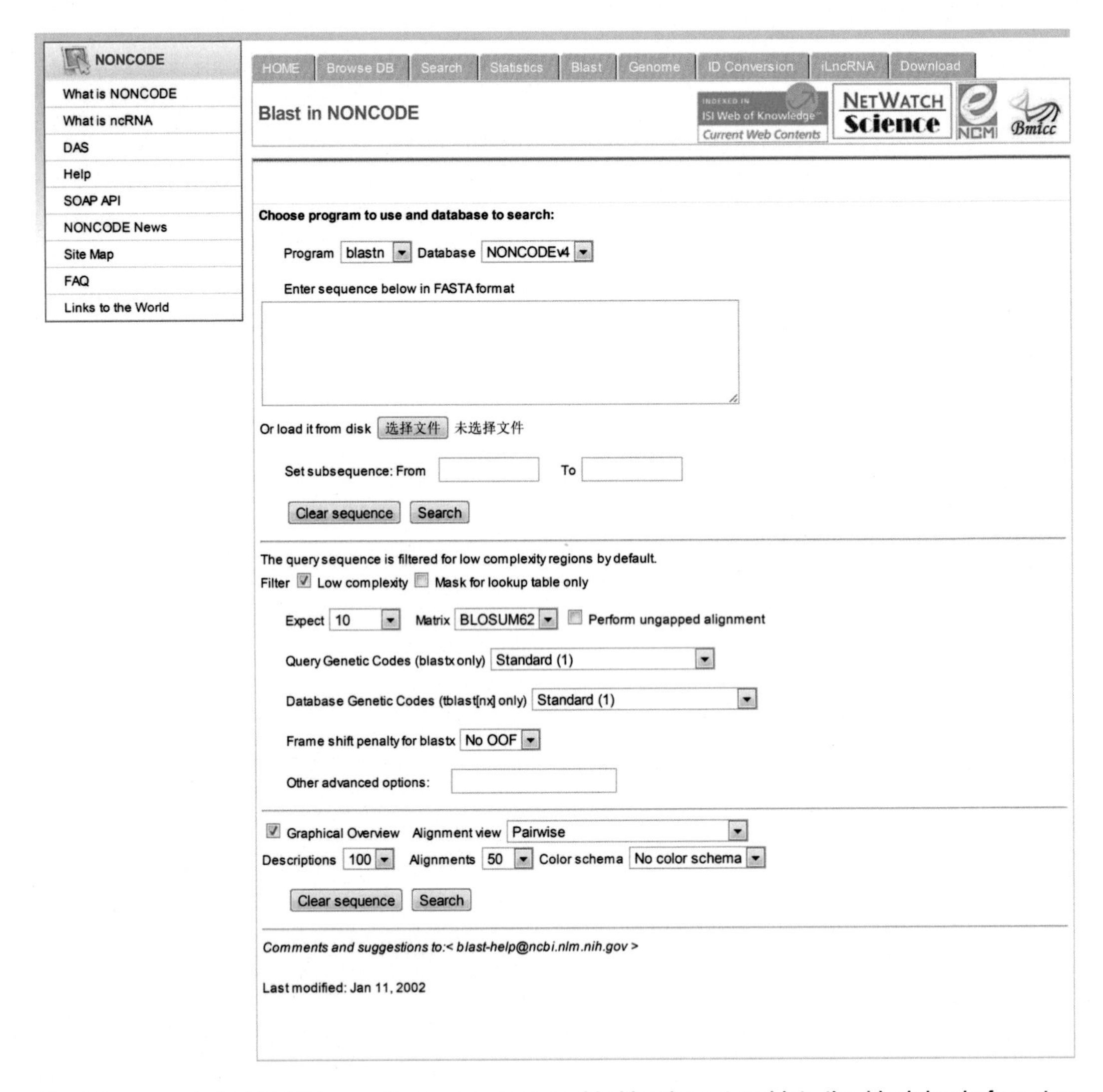

Fig. 5 A screenshot of BLAST page. Query sequence could either be entered into the *blank box* in format or uploaded in a file from local disk. Parameters for blast could be adjusted

adjusted by clicking the "zoom out" or "zoom in" button. The genomic coordinates could also be shifted to the left or right by clicking "move left" or "move right" button. Custom tracks might be uploaded to compare it with tracks on NONCODEv4 server. Optionally, it could be custom determined which track should be shown of hidden through the track option track.

3.3 Submitting Raw Data to NONCODEv4 Server

There is continuing emergence of high-throughput sequencing data. NONCODEv4 provides an online pipeline named "iLncRNA" to help users with identification of novel lncRNAs based on their own deep-sequencing data.

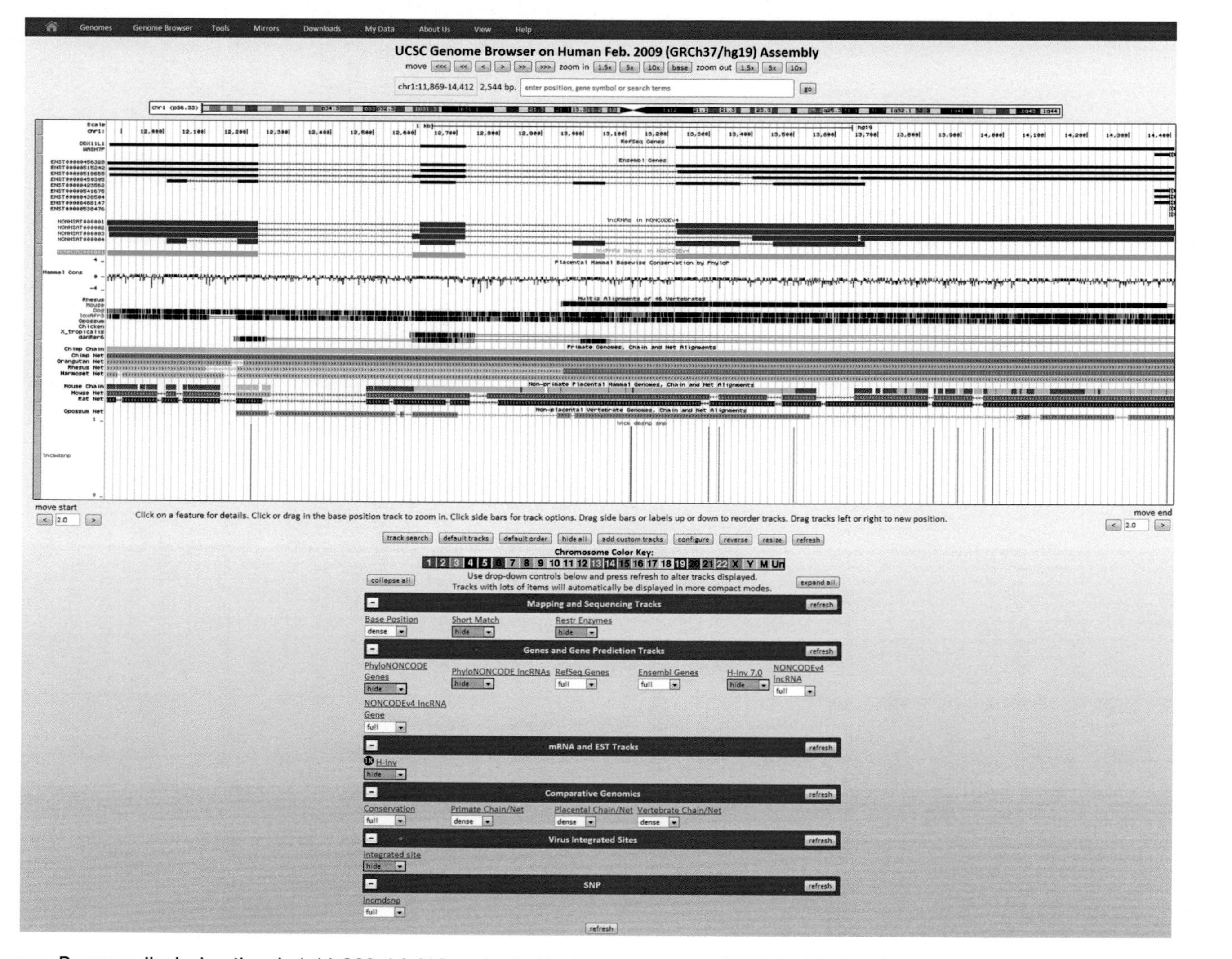

Fig. 6 The Genome Browser displaying the chr1:11,869-14,412 region in the human genome (UCSC hg19). The Genome Browser provides an integrated view by integrating annotations from NONCODEv4, RefSeq, Ensembl, and UCSC. The display region might be adjusted by the navigation buttons at the *top* of the image, either shifted or scaled

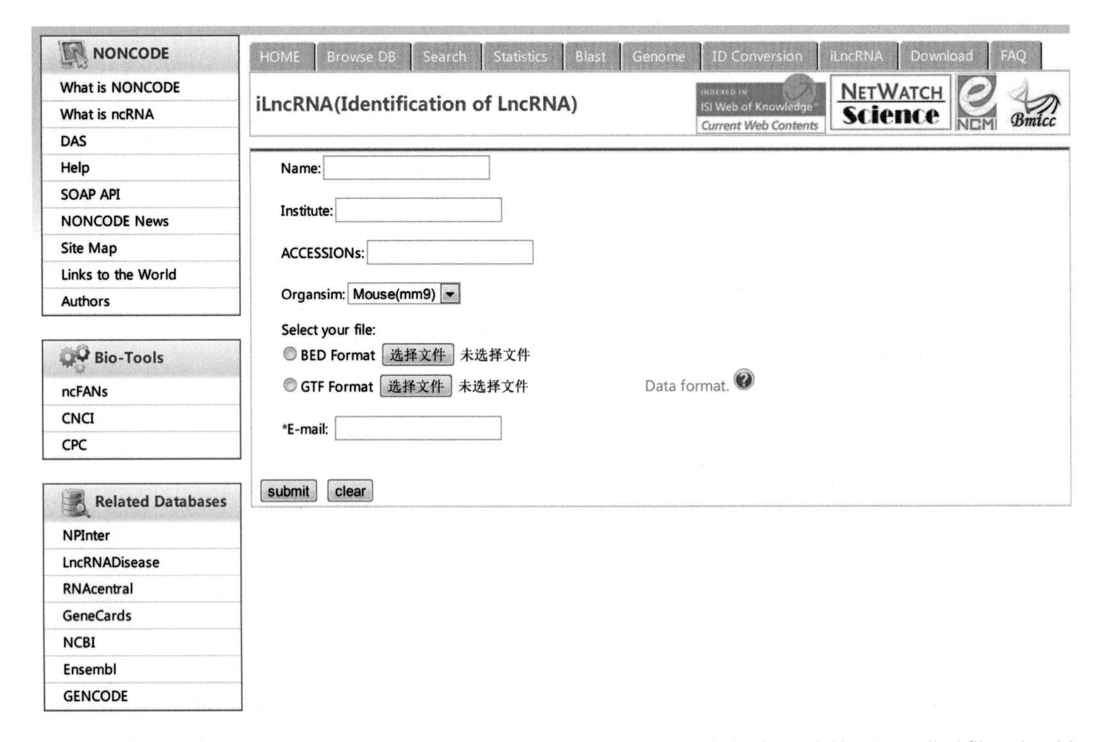

Fig. 7 A screenshot of iLncRNA webpage. The species of interest should be selected. User supplied files should be either bed format or gtf format. An e-mail address is needed for receiving message when analysis is done

1. Click “iLncRNA” on the home page to open the webpage for iLncRNA pipeline (Fig. 7).
2. Enter information including name, institute and accessions of the user into the corresponding boxes. Select organism (human or mouse) in the check box and upload the file in bed format or gtf format generated by assembly software or other tools. User’s e-mail is necessary for receiving feedback when the analysis is done.
3. After reading user supplied files, the iLncRNA server would first extract sequences with length more than 200 nucleotides. Then transcripts completely matching known protein coding transcripts from RefSeq or pseudogenes from Ensembl would be discarded. CNCI [15] would be used to judge whether the sequences which have passed the previous filtration have coding potential. Sequences with no coding potential are kept for further annotation by Cuffcompare [16]. Sequences which do not completely match lncRNA transcripts curated in NONCODEv4 would be classified as novel lncRNAs. The result would be sent to the email address which user filled in the check box.

4 Notes

1. NONCODEv4 is under periodically update. There would be statement of the latest modification highlighted in red on the home page. In turn, it is encouraged for users to report their problems regarding to usage of NONCODEv4 or interpreting annotations made by NONCODEv4 to the group working for NONCODEv4 by e-mail.
2. It is supported to search accessions from Ensembl or RefSeq on the search webpage (*see* Subheading 3.1). Besides, NONCODEv4 provided an online tool named "ID Conversion". It enables quick conversion between accessions of NONCODEv4 and other resources including Ensembl, RefSeq, and NONCODEv3. Batch conversion is supported.
3. All data of NONCODEv4 are stored in relational tables of MySQL database. Accessions of ncRNAs in NONCODEv4 are designated systematically. Take human as an example, lncRNA transcripts of human are designated with accessions from NONHSAT000001 to NONHSAT148172. The prefix of "NON" stands for "noncoding". The following "HSA" stands for "Homo sapiens". Similarly, it should be replaced by other letters for other organisms, such as "MMU" for "Mus musculus". The next letter "T" stands for "transcript", which should be replaced by "G" in accessions of lncRNA genes. By default, the numeric string with which NONCODEv4 accessions end is according to the order of transcripts or genes sorted by chromosome. Additionally, it is a little different for ncRNAs with unclear genomic coordinates, with "NOBED" in the middle. For example, NONHSANOBEDT000001 denotes a ncRNA sequence form human which could not be uniquely mapped to human genome.

Acknowledgment

This work was supported by National High-tech Research and Development Projects 863 [2012AA020402, 2012AA022501], National Key Basic Research and Development Program 973 [2009CB825401], Training Program of the Major Research plan of the National Natural Science Foundation of China [91229120], and National Natural Science Foundation of China [31371320]. Funding for open access charge: Training Program of the Major Research plan of the National Natural Science Foundation of China [91229120].

References

1. Himeno H, Kurita D, Muto A (2014) tmRNA-mediated trans-translation as the major ribosome rescue system in a bacterial cell. Front Genet 5:66
2. Jones TA, Otto W, Marz M, Eddy SR, Stadler PF (2009) A survey of nematode SmY RNAs. RNA Biol 6:5–8
3. Christov CP, Gardiner TJ, Szuts D, Krude T (2006) Functional requirement of noncoding Y RNAs for human chromosomal DNA replication. Mol Cell Biol 26:6993–7004
4. Mercer TR, Dinger ME, Mattick JS (2009) Long non-coding RNAs: insights into functions. Nat Rev Genet 10:155–159
5. Croucher NJ, Thomson NR (2010) Studying bacterial transcriptomes using RNA-seq. Curr Opin Microbiol 13:619–624
6. Mortazavi A, Williams BA, McCue K, Schaeffer L, Wold B (2008) Mapping and quantifying mammalian transcriptomes by RNA-seq. Nat Methods 5:621–628
7. Xie C, Yuan J, Li H, Li M, Zhao G, Bu D, Zhu W, Wu W, Chen R, Zhao Y (2014) NONCODEv4: exploring the world of long non-coding RNA genes. Nucleic Acids Res 42:D98–D103
8. Flicek P, Ahmed I, Amode MR, Barrell D, Beal K, Brent S, Carvalho-Silva D, Clapham P, Coates G, Fairley S et al (2013) Ensembl 2013. Nucleic Acids Res 41:D48–D55
9. Pruitt KD, Tatusova T, Brown GR, Maglott DR (2012) NCBI Reference Sequences (RefSeq): current status, new features and genome annotation policy. Nucleic Acids Res 40:D130–D135
10. Bu D, Yu K, Sun S, Xie C, Skogerbo G, Miao R, Xiao H, Liao Q, Luo H, Zhao G et al (2012) NONCODE v3.0: integrative annotation of long noncoding RNAs. Nucleic Acids Res 40:D210–D215
11. Picardi E, D'Erchia AM, Gallo A, Montalvo A, Pesole G (2014) Uncovering RNA editing sites in long non-coding RNAs. Front Bioeng Biotechnol 2:64
12. Lipman DJ, Pearson WR (1985) Rapid and sensitive protein similarity searches. Science 227:1435–1441
13. Liao Q, Xiao H, Bu D, Xie C, Miao R, Luo H, Zhao G, Yu K, Zhao H, Skogerbo G et al (2011) ncFANs: a web server for functional annotation of long non-coding RNAs. Nucleic Acids Res 39:W118–W124
14. Altschul SF, Gish W, Miller W, Myers EW, Lipman DJ (1990) Basic local alignment search tool. J Mol Biol 215:403–410
15. Sun L, Luo H, Bu D, Zhao G, Yu K, Zhang C, Liu Y, Chen R, Zhao Y (2013) Utilizing sequence intrinsic composition to classify protein-coding and long non-coding transcripts. Nucleic Acids Res 41(17):e166
16. Trapnell C, Roberts A, Goff L, Pertea G, Kim D, Kelley DR, Pimentel H, Salzberg SL, Rinn JL, Pachter L (2012) Differential gene and transcript expression analysis of RNA-seq experiments with TopHat and Cufflinks. Nat Protoc 7:562–578

Chapter 20

Computational Analysis of LncRNA from cDNA Sequences

Susan Boerner and Karen M. McGinnis

Abstract

Based on recent findings, long noncoding (lnc) RNAs represent a potential class of functional molecules within the cell. In this chapter we describe a computational scheme to identify and classify lncRNAs within maize from full-length cDNA sequences to designate subsets of lncRNAs for which biogenesis and regulatory mechanisms may be verified at the bench. We make use of the Coding Potential Calculator and specific Python scripts in our approach.

Key words Long noncoding RNA, miRNA, siRNA, NATs, Maize

1 Introduction

Single sequence (alignment-free) de novo identification of long noncoding (lnc)RNAs has traditionally been based on three characteristics: base pair length (>200 bp), open reading frame (ORF) length (variable, but generally < 100 aa), and homology of the amino acid sequence of the open reading frame to known proteins [reviewed in ref. 1]. Classification based on this scheme is problematic, however. For example, there are mRNA molecules that have a dual role, and both code for a short peptide and serve a function as a noncoding RNA. In addition, lncRNAs can have long ORFs that do not code for a functional protein, or very short ORFs that code for a functional peptide [reviewed in ref. 1]. In an attempt to refine this approach, computational tools have been developed to increase the accuracy of coding vs. noncoding prediction by taking advantage of support vector machines (SVM) and logistic regression models [2–4]. The Coding Potential Calculator (CPC) was one of the first tools developed to address coding vs noncoding prediction using an SVM, and takes into account six features of the sequence associated with the ORF and the BLASTX results of the ORF to the UniRef90 database of nonredundant protein sequences. Although more recent, and faster, alignment-free computational tools have been developed to classify transcripts as

Yi Feng and Lin Zhang (eds.), *Long Non-Coding RNAs: Methods and Protocols*, Methods in Molecular Biology, vol. 1402,
DOI 10.1007/978-1-4939-3378-5_20,

coding or noncoding [3, 4], we use CPC for the identification of potential lncRNAs in maize herein due to the accessibility of the webserver.

As lncRNAs are emerging as an important class of molecules within the cell with a range of potential functional roles [reviewed in ref. 5], their identification and functional classification is of increasing interest. Towards this end, we have paired the identification of potential lncRNAs in maize using CPC with further classification schemes based on location and small RNA content, to create small subsets of potential lncRNAs that can be verified at the bench. Our starting point is a set of strand-specific full-length cDNA sequences [6]. After selecting sequences greater than 200 bp, and passing these through CPC to distinguish potential noncoding sequences, we classify them further based on small RNA content and location within the genome (either genic or intergenic). Sequences that share homology with (presumably transcribed from) either verified or predicted gene models in maize are further localized and oriented (sense vs. antisense) within their respective gene model (Fig. 1). This secondary classification beyond CPC is

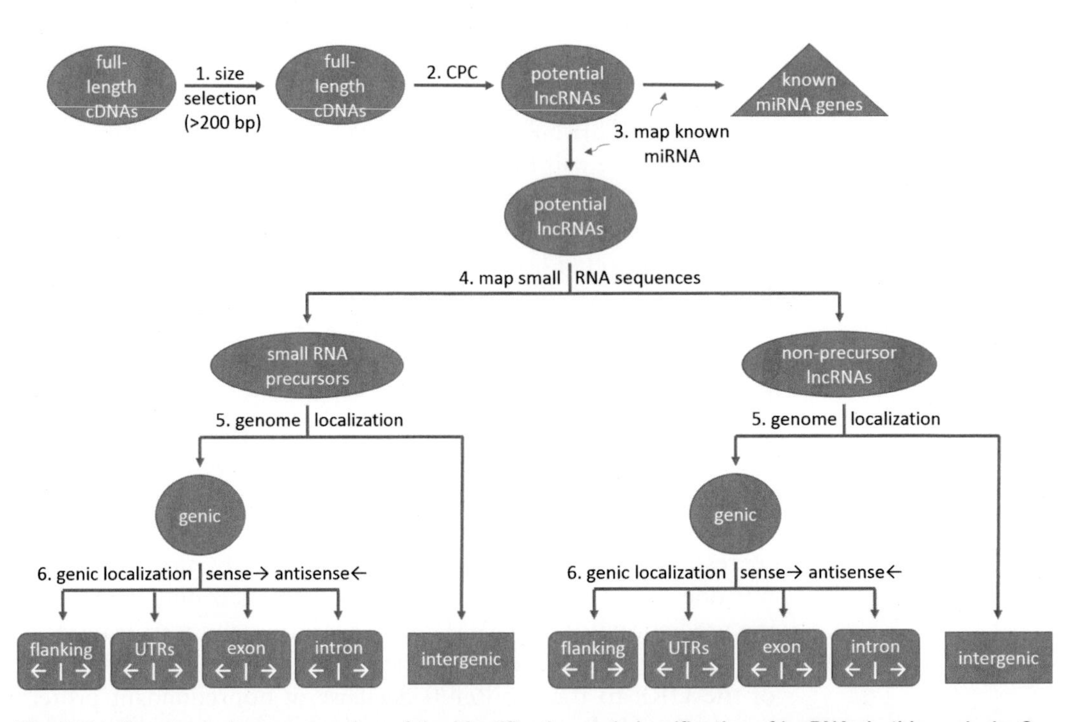

Fig. 1 This flowchart gives an overview of the identification and classification of lncRNAs in this analysis. Our starting point is a set of full-length cDNA sequences. We retain sequences greater than 200 bp in length (step 1) and run these through the Coding Potential Calculator (CPC) (step 2). We then filter sequences containing known miRNAs (step 3), and map sequenced small RNAs to the remaining set (step 4). Next, we localize each set (those with small RNA sequences and those without) within the genome (either genic or intergenic) (step 5), and finally localize genic lncRNAs within their respective gene model (step 6)

achieved with a set of simple Python scripts that take advantage of a Biopython wrapper for the BLAST algorithm.

This analysis in its current form is specific for maize, but can be adapted to other species. The subgroups of potential lncRNAs generated are very useful if the goal is to narrow a search for a specific type of lncRNA to verify at the bench. This analysis is by no means exhaustive, however, and multiple tools are available to further classify predicted lncRNAs. These could include miRNA prediction tools [7], RNA structural prediction tools [8], and homology searches against noncoding RNA databases [9] (*see* **Note 1**).

2 Materials

2.1 Hardware Requirements

Personal computer, preferably with a multi-core processor.

2.2 Software Requirements

1. Python 2.7. The website https://wiki.python.org/moin/BeginnersGuide has everything you need to get started with Python.
2. Biopython. The latest version with prerequisite software and installation instructions can be found here: http://biopython.org/wiki/Download. Also install NCBI Standalone BLAST, listed under Optional Software.
3. NCBI BLAST+ executables. The following website contains links to manuals on how to download, install, and run BLAST on your machine: http://www.ncbi.nlm.nih.gov/books/NBK1762/.

3 Methods

3.1 Size Selection of Long Noncoding RNAs

1. *Downloading full-length cDNA sequences*
   The full-length cDNAs we use for this analysis can be downloaded from the Maize full-length cDNA project website: http://www.maizecdna.org/download/ [6]. Download the five individual FASTA files listed under "Finished sequences" and combine them into one FASTA file (*see* **Notes 2** and **3**). Rename this file: maize_flcDNA.fasta.
2. *Size selection of cDNA sequences*
   The following Python script (*see* **Notes 4** and **5**) will select cDNA sequences greater than 200 base pairs in length and write them to a new FASTA file. The input and output file names are in bold text in the Python script, and can be changed if desired.

```
#!/usr/bin/env python
from Bio import SeqIO
input1 = open("maize_flcDNA.fasta", "rU")
```

```
L = list(SeqIO.parse(input1, "fasta"))
l=[]
for record in L:
    if len(record.seq) > 200:
        l.append(record)
print "number of starting cDNA sequences:", len(L)
print "number of cDNAs over 200 bp:", len(l)
output=open("maize_flcDNA_over_200bp.fasta", "w")
SeqIO.write(l, output, "fasta")
output.close()
input1.close()
```

3.2 Identification of Potential lncRNAs Using the Coding Potential Calculator

1. *Running the Coding Potential Calculator*
   Go to the Coding Potential Calculator website: cpc.cbi.pku.edu.cn. Click on "Run CPC" from either the top or side menu. Under "file upload" click "Browse" and select the FASTA file that you created in Subheading 3.1, **step 2** (maize_flcDNA_over_200bp.fasta), and click "run" (*see* **Note 6**). Depending on how many cDNA sequences you have, this analysis can take several minutes or several hours to complete. After the run begins, you are given a Task ID. Save this ID so you can retrieve your results at a later time (if needed) by clicking on "Get Results" and entering your Task ID.
2. *Extracting noncoding cDNA sequence IDs and creating a FASTA file of potential lncRNAs*
   If you saved your Task ID, you can return to the CPC website and click on "Get Results" from the side menu, enter your Task ID and click "view". At the top of the page, click on "Raw Data" (red text), and then click on "CPC Results". Save the results as a text file (e.g., right click and select "save as"), and name the file "cpc.txt". Run the following Python script to extract the necessary data from this file and create a new FASTA file containing the cDNA sequences identified as noncoding by CPC (*see* **Note 7**). Again, the input and output files are in bold text.

```
#!/usr/bin/env python
from Bio import SeqIO
input1 = open("cpc.txt", "rU")
report = file("noncoding.fasta", 'w')
for line in input1:
    line = line.rstrip()
    if line[14] == "n":
        if int(line[25]) >= 1:
            print >> report, ">" + line [0:8]
report.close()
input2 = open("maize_flcDNA_over_200bp.fasta", "rU")
```

```
input3 = open("noncoding.fasta", "rU")
a = list(SeqIO.parse(input2, "fasta"))
b = list(SeqIO.parse(input3, "fasta"))
noncoding = []
for record in a:
    for rec in b:
        if record.id == rec.id:
            noncoding.append(record)
print "number of potential lncRNAs identified by CPC:", len(noncoding)
output = open("lncRNAs.fasta", "w")
SeqIO.write(noncoding, output, "fasta")
output.close()
input1.close()
input2.close()
input3.close()
```

You now have a new FASTA file of potential lncRNAs, in this case named "lncRNAs.fasta".

3.3 Filtering Known miRNA Precursors

1. *Download known miRNA sequences from ensembl*
   Go to http://ensembl.gramene.org/Zea_mays/Info/Index (*see* **Note 8**). Under "Gene annotation", click on the FASTA link next to "Download genes, cDNAs, ncRNA, proteins". Click on the "ncrna" folder. Click on Zea_mays.AGPv3.23.ncrna.fa.gz to download the file. Unzip the file, which contains known miRNA sequences in *Zea mays*, and rename it "known_miRNA.fasta".
2. *Run Python script to filter known miRNA genes from lncRNAs*
   Before running the following Python script, you need to create a BLAST database of the lncRNAs generated in Subheading 3.2 (*see* **Note 9**). At the command line, pass the following:

```
makeblastdb -in lncRNAs.fasta -dbtype nucl -title lncRNA_db -out lncRNA_db
```

The script creates a text file that lists the lncRNA ID with its associated known miRNA ID, the hsp score, and hsp identities for their alignment. The script determines lncRNA precursor potential based on a 98 % identity match between the entire known miRNA sequence and the lncRNA (without gaps in the alignment). This stringency can of course be adjusted if desired. The percent identity, the name of the report file, input and output files, and the BLAST database name are in bold text in the script.

```
#!/usr/bin/env python
from Bio import SeqIO
from Bio.Blast.Applications import NcbiblastnCommandline
from Bio.Blast import NCBIXML
```

```
input1 = open("known_miRNA.fasta", "rU")
RNAs = list(SeqIO.parse(input1, "fasta"))
report = file("known_miRNA_in_lncRNAs.txt", 'w')
lncRNA_list = file("lncRNAs with RNA list.fasta", 'w')
for record in RNAs:
      sequence = open("fasta1.fasta", "w")
      SeqIO.write(record, sequence, "fasta")
      sequence.close()
      blast_cline = NcbiblastnCommandline(db="lncRNA_db",
      ungapped="yes", num_alignments=1, num_threads=4,
      query="fasta1.fasta", outfmt=5, out="out.xml",
      max_target_seqs=1)
      blast_cline()
      handle = open("out.xml", "rU")
      blast_record = NCBIXML.read(handle)
      for alignment in blast_record.alignments:
            for hsp in alignment.hsps:
              if hsp.identities >= (len(record.seq)*0.98):
                 print >> report, "RNA:", record.id
                 print >> report, "length(bp):", len(record.
                 seq)
                 print >> report, "lncRNA:", alignment.
                 title
                 print >> report, "length(bp):", align-
                 ment.length
                 print >> report, "hsp score:", hsp.score
                 print >> report, "hsp identities:", hsp.
                 identities
                 print >> report, "---------"
                 print >> lncRNA_list, ">" +
                 alignment.title[alignment.title.
                 find("BT"):(alignment.title.
                 find("BT"))+10]
      handle.close()
lncRNA_list.close()
input2 = open("lncRNAs.fasta", "rU")
input3 = open("lncRNAs with RNA list.fasta", "rU")
lncRNA = list(SeqIO.parse(input2, "fasta"))
lncRNAlist = list(SeqIO.parse(input3, "fasta"))
lncRNARNA = []
for record in lncRNA:
    for rec in lncRNAlist:
        if record.id == rec.id:
            lncRNARNA.append(record)
lncRNARNA = list(set(lncRNARNA))
print "number of lncRNAs with miRNA:", len(lncRNARNA)
```

```
for record in lncRNARNA:
      lncRNA.remove(record)
print "number of lncRNAs without miRNA:", len(lncRNA)
output = open("lncRNAs_with_known_miRNAs.fasta", "w")
SeqIO.write(lncRNARNA, output, "fasta")
output.close()
output = open("lncRNAs_without_known_miRNAs.fasta", "w")
SeqIO.write(lncRNA, output, "fasta")
output.close()
report.close()
input1.close()
input2.close()
input3.close()
```

We can now move forward with the new set of lncRNAs without known miRNA genes.

3.4 Small RNA Precursor Prediction via Mapping of Available small RNA Sequences

1. *Generation of a small RNA FASTA file*
   There are many possible sources of small RNA sequences (*see* **Note 10**). For this analysis, we use a dataset of maize siRNAs available on Gene Expression Omnibus via NCBI [10]. Visit http://www.ncbi.nlm.nih.gov/geo/ and enter the following accession number: GSE15286. From here, you can download four sets of small RNA sequences: shRNA.root.fa.gz, shRNA.shoot.fa.gz, siRNA.root.fa.gz, and siRNA.shoot.fa.gz (*see* **Note 11**). These files are quite large; *see* **Note 2** for the fastest way to combine them into one file, if desired.
2. *Mapping small RNA sequences to lncRNAs*
   Before running the Python script, you need to create a BLAST database of the lncRNAs generated in Subheading 3.3. At the command line, pass the following:

   ```
   makeblastdb -in lncRNAs_without_known_miRNAs.fasta -dbtype nucl -title lncRNAs_without_known_miRNAs_db -out lncRNAs_without_known_miRNAs_db
   ```

To map small RNA sequences to the set of lncRNAs use the Python script from Subheading 3.3, **step 2** (*see* **Note 12**). For an exact sequence alignment of small RNAs, change the bold text value "0.98" to "1". Be sure to change the file names and database names accordingly. We can now move forward with our two sets of lncRNAs, those with small RNA sequences and those without.

3.5 Localization of lncRNAs within the Genome, Genic versus Intergenic

1. *Downloading genome and gene model sequences*
   To download chromosomal sequences (*see* **Note 13**) for the maize genome, go to: http://ensembl.gramene.org/Zea_mays/Info/Index. Under "Genome assembly: AGPv3", click on "Download DNA sequence", and then click on "Zea_mays.

AGPv3.23.dna.chrmosome.1.fa.gz" to download the file. Then download the remaining chromosome FASTA files (through 10). Extract the files, and combine them into one FASTA file (*see* **Note 3**). Rename the file "maize_v3_chromosomes". Next, we use Biomart to extract gene model sequences. Go to: http://plants.ensembl.org/biomart/martview/7b4abd04a0d0c38a22a7c261cda9abdf. Select "Plant Mart" under the "CHOOSE DATABASE" drop down menu, and then "Zea mays genes (AGPv3 (5b))" under the "CHOOSE DATASET" drop down menu. Click on "Filters" on the left-side menu, and under "GENE" check the box for "Limit to genes" and select "with Affymetrix array Maize ID(s)". Click on "Attributes" on the left-side menu, then select "Sequences" at the top, and select "Unspliced (Gene)" under "SEQUENCES". Then check the two boxes corresponding to "Upstream flank" and "Downstream flank" below and enter a value for the number of base pairs of the flanking region to include in the gene model sequence file. For this analysis, we focus on 1500 base pairs of flanking sequence for both upstream and downstream regions of the gene models. Click on "Results" on the top left menu bar, and for "Export all results to" select "File" and "FASTA" and check the "Unique results only" box. When the FASTA file of gene models has finished downloading, rename it "maize_v3_genemodels_flank1500.fasta".

2. *Localize lncRNAs within genome*

   To run the following Python script to localize lncRNAs within the genome you will need to create two BLAST databases, one for the chromosome sequences, and one for the gene model sequences (maize_v3_chromosomes_db and maize_v3_genemodels_flank1500_db). The script prints IDs for lncRNAs that were not classified, and the total number of genic and intergenic lncRNAs that were classified.

```
#!/usr/bin/env python
from Bio.Blast.Applications import NcbiblastnCommandline
from Bio import SeqIO
input1 = open("lncRNAs_without_known_smallRNAs.fasta",
"rU")
L = list(SeqIO.parse(input1, "fasta"))
intergenic = []
for record in L:
        sequence = open("fasta1.fasta", "w")
        SeqIO.write(record, sequence, "fasta")
        sequence.close()
        blast_cline = NcbiblastnCommandline(db="maize_v3_
        genemodels_flank1500_db", num_alignments=1, num_
        threads=4,     query="fasta1.fasta",     outfmt=5,
        out="out1.xml", max_target_seqs=1)
        blast_cline()
```

```
        blast_cline = NcbiblastnCommandline(db="maize_
        v3_chromosomes_db", num_alignments=1, num_
        threads=4, query="fasta1.fasta", outfmt=5,
        out="out2.xml", max_target_seqs=1)
        blast_cline()
        handle1 = open("out1.xml", "rU")
        from Bio.Blast import NCBIXML
        blast_record1 = NCBIXML.read(handle1)
        handle2 = open("out2.xml", "rU")
        from Bio.Blast import NCBIXML
        blast_record2 = NCBIXML.read(handle2)
        x = []
        for alignment1 in blast_record1.alignments:
            for hsp1 in alignment1.hsps:
                x.append(hsp1.score)
        y = []
        for alignment2 in blast_record2.alignments:
            for hsp2 in alignment2.hsps:
                y.append(hsp2.score)
        if len(x) == 0 and len(y) == 0:
            print "lncRNA that doesn't have a hit in
            sequenced chromosomes", record.id
        elif len(x) == 0 and len(y) != 0:
            intergenic.append(record)
        elif len(x) != 0 and len(y) == 0:
            print "lncRNA that is somehow present in
            genemodel but not in sequenced chromosomes",
            record.id
        else:
            if len(x) != 0 and len(y) != 0:
                if y[0] > x[0]:
                    intergenic.append(record)
        handle1.close()
        handle2.close()
print "intergenic lncRNAs:", len(intergenic)
for record in intergenic:
        L.remove(record)
print "genic lncRNAs:", len(L)
output = open("lncRNA_without_smallRNA_GENIC.fasta", "w")
SeqIO.write(L, output, "fasta")
output.close()
output = open("lncRNA_without_smallRNA_INTERGENIC.
fasta", "w")
SeqIO.write(intergenic, output, "fasta")
output.close()
input1.close()
```

For this script, the FASTA file for lncRNAs without small RNA sequences was used. The lncRNAs that contain small RNA sequences can also be mapped within the genome using the same script; just change the file names accordingly.

3.6 Sub-localization of Genic lncRNAs within Gene Models

1. *Downloading gene model sequences for localization*
   Sequences within gene models are downloaded using Biomart as described in Subheading 3.5, **step 1**. For this analysis, we grab the sequences for: "Flank (Gene)", "5′ UTR", "3′ UTR", Unspliced (Transcript), and "Coding sequence". The "Flank (Gene)" option needs to be processed twice. Click on "Flank (Gene)", and then below check the box for "Upstream flank" and enter 1500, and get results. Do this again for "Downstream flank" (check box and enter 1500). Rename these files: "maize_v3_upflank1500.fasta", "maize_v3_downflank1500.fasta", "maize_v3_5UTR.fasta", "maize_v3_3UTR.fasta", "maize_v3_unspliced_trans.fasta", and "maize_v3_coding.fasta", respectively. The unspliced transcript and coding databases are used in Subheading 3.6, **step 3** below.
2. *Sub-localization of genic lncRNAs within flanking regions and UTRs*
   To run the following Python script to sub-localize lncRNAs within the flanking regions and UTRs, you will need to create BLAST databases of the FASTA files from Subheading 3.6, **step 1**. As the full-length cDNAs used in this analysis are strand specific, the script was written to classify the genic lncRNAs based on their orientation within their respective gene model (sense or antisense). The script determines lncRNA localization based on an 80 % identity match between the lncRNA sequence and the gene model region sequence. Again, this stringency can be adjusted if desired. This script also creates a text report file that lists the lncRNA ID, its orientation, the gene model ID, and the alignment scores. You can rerun the script for each database and for each set of genic lncRNAs separately. The script below uses the set of genic lncRNAs without small RNA sequences, and maps these to the upstream flanking regions of maize gene models (*see* **Note 14**).

```
#!/usr/bin/env python
from Bio import SeqIO
from Bio.Blast.Applications import NcbiblastnCommandline
from Bio.Blast import NCBIXML
input1 = open("lncRNA_without_smallRNA_GENIC.fasta",
"rU")
genic = list(SeqIO.parse(input1, "fasta"))
report = file("genic lncRNAs without small RNA in upstream
flanking region.txt", 'w')
sense = []
antisense = []
```

```
for record in genic:
    sequence = open("fasta1.fasta", "w")
    SeqIO.write(record, sequence, "fasta")
    sequence.close()
    blast_cline = NcbiblastnCommandline(db="maize_v3_
    upflank1500_db", num_alignments=1, num_threads=4,
    query="fasta1.fasta", outfmt=5, out="out.xml", max_
    target_seqs=1)
    blast_cline()
    handle = open("out.xml", "rU")
    blast_record = NCBIXML.read(handle)
    for alignment in blast_record.alignments:
        for hsp in alignment.hsps:
            if hsp.identities >= (len(record.seq)*0.8):
                if hsp.sbjct_start > hsp.sbjct_end:
                    antisense.append(record)
                    print >> report, "antisense"
                elif hsp.sbjct_start < hsp.sbjct_end:
                    sense.append(record)
                    print >> report, "sense"
                print >> report, "lncRNA:", record.id
                print >> report, "length(bp):",
                len(record.seq)
                print >> report, "gene model:", align-
                ment.title
                print >> report, "length(bp):", align-
                ment.length
                print >> report, "hsp.score:", hsp.score
                print >> report, "hsp.identities:", hsp.
                identities
                print >> report, "---------"
        handle.close()
print "number of lncRNAs in the upstream region in the
sense orientation:", len(sense)
print "number of lncRNAs in the upstream region in the
antisense orientation:", len(antisense)
output = open("genic_lncRNA_without_smRNA_upflank
1500_sense.fasta", "w")
SeqIO.write(sense, output, "fasta")
output.close()
output = open("genic_lncRNA_without_smRNA_upflank
1500_antisense.fasta", "w")
SeqIO.write(antisense, output, "fasta")
output.close()
input1.close()
```

3. *Sub-localization of genic lncRNAs, exons* vs. *introns*
   As intron sequences are not available for v3 gene model builds for maize at this time, a script similar to the genic vs. intergenic

localization one will be used to distinguish between intronic and exonic lncRNAs. The script compares the highest alignment scores for the lncRNA to the unspliced transcript sequence and the coding sequence. If the top hit of the lncRNA to the unspliced transcript is greater than the top hit of the lncRNA to a coding sequence, then at least part of the lncRNA is transcribed from an intronic region. In the following script, intronic lncRNAs were filtered as such if more than 50 % of the sequence is complimentary to an intron. This step is in bold text in the script, and can be adjusted if desired.

```
#!/usr/bin/env python
from Bio.Blast.Applications import Ncbiblastn Commandline
from Bio import SeqIO
input1 = open("lncRNA_without_smallRNA_GENIC.fasta", "rU")
L = list(SeqIO.parse(input1, "fasta"))
intronic = []
for record in L:
        sequence = open("fasta1.fasta", "w")
        SeqIO.write(record, sequence, "fasta")
        sequence.close()
        blast_cline = NcbiblastnCommandline (db="maize_v3_unspliced_trans_db", num_alignments=1, num_threads=4, query="fasta1.fasta", outfmt=5, out="out1.xml", max_target_seqs=1)
        blast_cline()
        blast_cline = NcbiblastnCommandline(db="maize_v3_coding_db", num_alignments=1, num_threads=4, query="fasta1.fasta", outfmt=5, out="out2.xml", max_target_seqs=1)
        blast_cline()
        handle1 = open("out1.xml", "rU")
        from Bio.Blast import NCBIXML
        blast_record1 = NCBIXML.read(handle1)
        handle2 = open("out2.xml", "rU")
        from Bio.Blast import NCBIXML
        blast_record2 = NCBIXML.read(handle2)
        x = []
        for alignment1 in blast_record1.alignments:
            for hsp1 in alignment1.hsps:
                x.append(hsp1.score)
        y = []
        for alignment2 in blast_record2.alignments:
            for hsp2 in alignment2.hsps:
                y.append(hsp2.score)
```

```
        if len(x) != 0 and len(y) == 0:
            intronic.append(record)
        elif len(x) != 0 and len(y) != 0:
            if x[0] > y[0]:
                if x[0] > (len(record.seq)*0.5):
                    intronic.append(record)
        handle1.close()
        handle2.close()
print "intronic lncRNAs:", len(intronic)
for record in intronic:
        L.remove(record)
print "exonic lncRNAs:", len(L)
output = open("lncRNA_without_smallRNA_GENIC_exonic.
fasta", "w")
SeqIO.write(L, output, "fasta")
output.close()
output = open("lncRNA_without_smallRNA_GENIC_intronic.
fasta", "w")
SeqIO.write(intronic, output, "fasta")
output.close()
input1.close()
```

To further filter the intronic and exonic lncRNAs based on strand specificity, the resulting FASTA files of this script can be run through the script from Subheading 3.6, **step 2** using the database "maize_v3_unspliced_trans_db" and the appropriate file names.

4 Notes

1. These are just a few examples of the computational tools available. We recommend the recent Methods in Molecular Biology book, "RNA Sequence, Structure, and Function: Computational and Bioinformatic Methods", for an in-depth view of what is currently available [11].
2. The file containing cDNA sequences needs to be in FASTA format. Several online tools exist to convert files to FASTA from other formats. A simple script in Biopython to convert to FASTA from GenBank follows. Replace the input and output file names as needed.

```
#!/usr/bin/env python
from Bio import SeqIO
input1 = open("cDNAs in genbank format.gb", "rU")
cDNAs = SeqIO.parse(input1, "genbank")
output = open("cDNAs in fasta format.fasta", "w")
```

```
SeqIO.write(cDNAs, output, "fasta")
output.close()
input1.close()
```

3. A very efficient way to combine large FASTA files is to use cat at the command line. If your two FASTA files are "shRNAshoot.fasta" and "siRNAshoot.fasta" and you want to combine them and rename the new file "smallRNA.fasta", then pass the following.

```
cat shRNAshoot.fasta siRNAshoot.fasta > smallRNA.fasta.
```

4. To ease the overall understanding of the analysis, sys.argv was not used to create command-line arguments for the Python scripts. Instead, we felt that using boldface for the text in the script which can/need be modified would enable the reader to better follow the process (especially considering the scripts are relatively short).
5. When copying the script into a text editor to run on your machine, be sure to retain the indentation.
6. There is an "additional options" section on the Run CPC webpage that includes a search against the UTRdb and RNAdb, as well as running the antisense strand of your submitted sequence through the Coding Potential Calculator. The database searches greatly increase the time it takes to complete the run, and are therefore not recommended for this analysis. If the cDNA sequences you have are not strand specific, you may want to include the antisense strand analysis, and only select sequences that are predicted to be noncoding in the sense and antisense direction.
7. CPC has four classifications of coding potential based on the output score: noncoding, noncoding (weak), coding (weak), and coding. Noncoding transcripts have a score < -1.0; the script here selects only those transcripts with this classification. If you wish to lower the stringency, you can change the highlighted value of "1" in the script to "0". This will include sequences classified as: noncoding (weak).
8. This analysis is specific for maize, but other databases of miRNAs exist for many other species, as well as databases for other small ncRNAs (piwi, sno, etc.).
9. Due to the attributes of the XML file generated for each BLAST alignment and the length of the known miRNA sequences (shorter than lncRNAs), the BLAST database must consist of the lncRNA sequences.
10. Several small RNA databases exist, and next-generation sequencing data of small RNAs can be produced or downloaded from the Gene Expression Omnibus (http://www.ncbi.nlm.nih.gov/gds/).

11. The classification of sh vs si RNA in this study was based on the likelihood of the genomic region surrounding the small RNA sequence to form a short hairpin (ref). For simplicity, these were grouped together as downstream analysis of potential lncRNAs will likely classify small RNA precursors based on this feature.
12. Depending on the power of your processor, this script may take a long time to run. Be sure to change the number of threads used for BLAST within the script to reflect the number of cores your processor has ("num_threads=4" is in bold text within the script, and assumes you have a quad-core processor to run each thread).
13. The Python script to distinguish between intergenic and genic lncRNAs is based on a comparison of the alignment scores of the lncRNA to complete genome sequences and gene model sequences. Assembled genome sequences are required for an alignment score of the lncRNA to this set to be comparable to the score against the gene model sequences. The chromosomal sequences are assembled. For the rare cases where the lncRNA falls outside of the assembled chromosomal sequences and therefore the gene model sequences as well, the script will print the corresponding lncRNA ID for further analysis, if desired.
14. Depending on how stringent your alignment is, you may have lncRNAs that localize to more than one region of a gene model.

References

1. Dinger ME, Pang KC, Mercer TR, Mattick JS (2008) Differentiating protein-coding and noncoding RNA: challenges and ambiguities. PLoS Comput Biol 4(11), e1000176
2. Kong L, Zhang Y, Ye ZQ, Liu XQ, Zhao SQ, Wei L, Gao G (2007) CPC: assess the protein-coding potential of transcripts using sequence features and support vector machine. Nucleic Acids Res 35:W345–W349
3. Wang L, Park HJ, Dasari S, Wang S, Kocher J-P, Li W (2013) CPAT: Coding-Potential Assessment Tool using an alignment-free logistic regression model. Nucleic Acids Res 41(6), e74
4. Li A, Zhang J, Zhou Z (2014) PLEK: a tool for predicting long non-coding RNAs and messenger RNAs based on an improved k-mer scheme. BMC Bioinformatics 15(1):311
5. Johnny TY, Kung DC, Lee JT (2013) Long noncoding RNAs: past, present, and future. Genetics 193(3):651–669
6. Soderlund C, Descour A, Kudrna D, Bomhoff M, Boyd L, Currie J, Angelova A, Collura K, Wissotski M, Ashley E, Morrow D, Fernandes J, Walbot V, Yu Y (2009) Sequencing, mapping, and analysis of 27,455 maize full-length cDNAs. PLoS Genet 5(11), e1000740
7. Xue C, Li F, He T, Liu G, Li Y, Zhang X (2005) Classification of real and pseudo microRNA precursors using local structure-sequence features and support vector machine. BMC Bioinformatics 6:310
8. Gruber AR, Lorenz R, Bernhart SH, Neuböck R, Hofacker IL (2008) The Vienna RNA Websuite. Nucleic Acids Res 36(Web Server issue):W70–W74
9. Amaral PP, Clark MB, Gascoigne DK, Dinger ME, Mattick JS (2011) lncRNAdb: a reference database for long noncoding RNAs. Nucleic Acids Res 39:D146–D151
10. Wang X, Elling AA, Li X, Li N, Peng Z, He G, Sun H, Qi Y, Liu XS, Deng XW (2009) Genome-wide and organ-specific landscapes of epigenetic modifications and their relationships to mRNA and small RNA transcriptomes in maize. Plant Cell 21:1053–1069
11. Jan G, Ruzzo WL (2014) RNA sequence, structure, and function: computational and bioinformatic methods. Methods Mol Biol 1097:437–456

Chapter 21

Analyzing MiRNA–LncRNA Interactions

Maria D. Paraskevopoulou and Artemis G. Hatzigeorgiou

Abstract

Long noncoding RNAs (lncRNAs) are noncoding transcripts usually longer than 200 nts that have recently emerged as one of the largest and significantly diverse RNA families. The biological role and functions of lncRNAs are still mostly uncharacterized. Their target-mimetic, sponge/decoy function on microRNAs was recently uncovered. miRNAs are a class of noncoding RNA species (~22 nts) that play a central role in posttranscriptional regulation of protein coding genes by mRNA cleavage, direct translational repression and/or mRNA destabilization. LncRNAs can act as miRNA sponges, reducing their regulatory effect on mRNAs. This function introduces an extra layer of complexity in the miRNA–target interaction network. This chapter focuses on the study of miRNA–lncRNA interactions with either in silico or experimentally supported analyses. The proposed methodologies can be appropriately adapted in order to become the backbone of advanced multistep functional miRNA analyses.

Key words microRNA, lncRNA, HITS-CLIP, PAR-CLIP, Sponge, In-silico predictions, Experimentally supported, Interaction

Abbreviations

CLIP	Crosslinking immunoprecipitation
miRNA	microRNA
lncRNA	Long noncoding RNA
MRE	miRNA recognition element

1 Introduction

ncRNA families are being vigorously researched for their physiological and pathological implications. Recent advances in next-generation sequencing (NGS) technologies allowed large-scale transcriptome studies [1] to unveil the number of noncoding transcripts and their regulatory roles in the cell.

miRNAs are small noncoding RNAs (~22 nts) and are considered central posttranscriptional gene regulators, acting through transcript degradation and/or translation suppression in the case of mRNAs

Yi Feng and Lin Zhang (eds.), *Long Non-Coding RNAs: Methods and Protocols*, Methods in Molecular Biology, vol. 1402,
DOI 10.1007/978-1-4939-3378-5_21,

[2]. Their number has increased in a super linear rate surpassing 2500 identified miRNAs for the human genome [3]. lncRNAs, noncoding transcripts usually longer than 200 nucleotides, exhibit numerous functions, many of which are under debate or remain to be uncovered [4]. They are spatially classified into four main categories (sense, antisense, intergenic, bidirectional) according to their loci of origin and transcription orientation as compared to protein coding genes.

Many lncRNAs are polyadenylated, 5′ capped and spliced. Their low abundance is probably connected with the underestimation of lncRNA transcript length and number of exons [5]. Their expression is characterized by high tissue, disease and developmental stage specificity. Recent studies also describe conserved function of lncRNAs despite their poor sequence conservation [6]. They perform different roles in all cell compartments, controlling gene expression in cis and/or trans by participating in almost every known level of regulation. lncRNAs promote chromatin modifications, mediating gene silencing; can act as guide molecules and scaffolds for proteins, contributing to the formation of cellular substructures; while they are also shown to control protein synthesis, RNA maturation and transport [7, 8].

Another important function of lncRNAs is that they may act as endo-siRNAs or encode small non coding RNAs. Recent studies suggest that lncRNAs could play a "sponge"/"decoy" role, competing other genes for miRNA binding and therefore reduce the regulatory effect of miRNAs on targeted mRNAs [9–11]. This recently proposed posttranscriptional mechanism alters the components of endogenous regulatory interaction networks: ncRNAs sharing MREs with mRNAs can act as miRNA decoys and participate in the elegantly described "competing endogenous RNA" [12] (ceRNA) activity (Fig. 1).

2 Materials

The essential prerequisites to perform the described miRNA-lncRNA-guided analyses are as follows: a Unix/Linux-based environment, a free registered access to the DIANA tools webserver (www.microrna.gr). The free to access online tools: DIANA-LncBase [13] database, DIANA-miRPath webserver [14], and UCSC Genome Browser (http://genome.ucsc.edu/) are utilized.

3 Methods

3.1 Identification of lncRNAs that Regulate miRNA Transcription

LncRNAs (such as Meg3, Dleu2, H19, and Ftx) can function as pri-miRNA host genes [11]. Genomic regions where miRNA transcripts and lncRNAs overlap have dual/multiple functionality. Different biological processes can either trigger lncRNA function

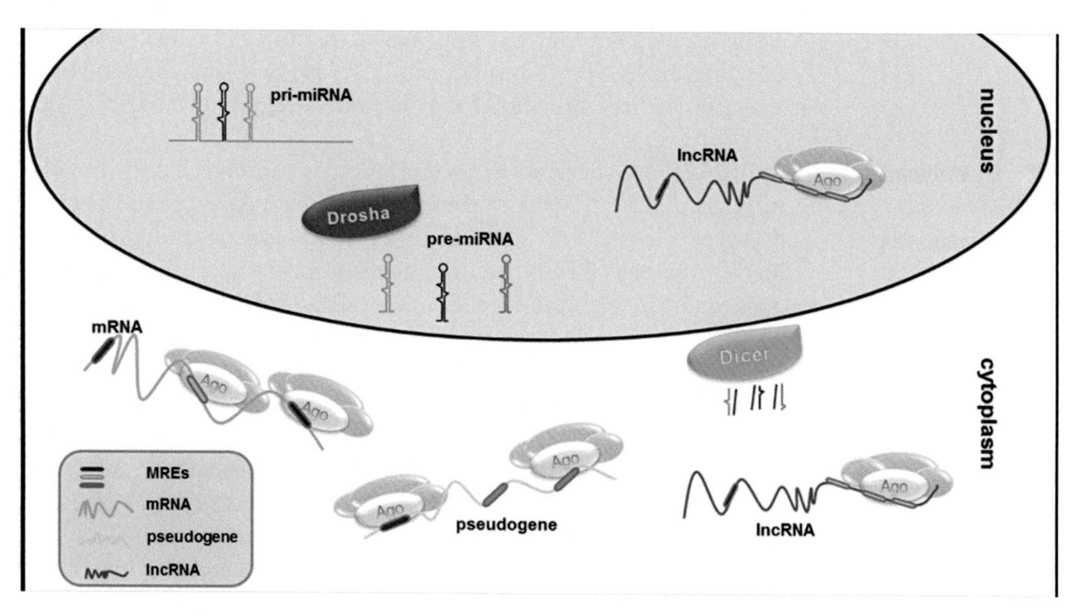

Fig. 1 Overview of the ceRNA activity in nucleus and cytoplasm. miRNAs loaded in the RISC complex post-transcriptionally regulate protein coding genes through mRNA cleavage, direct translational repression and/or mRNA destabilization in the cytoplasm. lncRNAs compete with mRNAs for miRNA binding by acting as "sponge" molecules in both cell compartments

or promote the activation of the miRNA biogenesis pathway. Several well-known polycistronic miRNA gene clusters, including members of let-7 family, derive form intergenic regions that also encode lncRNAs.

The characterization of pri-miRNA transcripts remains widely unknown and is hindered by practical obstacles [15]. The rapid cleavage of primary miRNA transcripts by Drosha enzyme in the nucleus does not allow complete transcript annotation with conventional approaches. microTSS [15] is a versatile computational framework that enables tissue specific identification of miRNA transcription start sites. Its current version requires RNA-Seq datasets in order to detect expressed regions upstream of miRNA precursors. The area around the 5′ of the RNA-Seq signal is scanned for H3K4me3, Pol2, and DNase enrichment, corresponding to putative regions for pri-miRNA transcription initiation. The candidate miRNA promoters are scored based on three distinct SVM models, trained on deep sequencing data.

The annotation of intergenic pri-miRNA transcripts with microTSS can enable the identification of overlapping lncRNAs, the revision of lncRNA and pri-miRNA annotation that in many cases is considered incomplete, the detection of common lncRNA–miRNA promoter regions as well as further support lncRNA-centered functional analyses. This machine learning approach outperforms any other similar existing methodologies and can be easily applied on any cell line/tissue of human or mouse species utilizing RNA-Seq, Chip-Seq and DNase-Seq data. microTSS is available for free download at www.microrna.gr/microTSS.

3.2 Experimentally Verified miRNA–lncRNA Interactions

LncRNAs have been shown to function as "sponges" coordinating miRNA function. Most of these interactions take place in the cytoplasm, while there are also examples of miRNA targeting lncRNAs in the nucleus. PTEN pseudogene competes its coding counterpart for miRNA binding; CDR1as/ciRS-7 circular antisense transcript acts as a sponge by harboring multiple miRNA binding sites, while it is also cleaved in the nucleus through a miRNA-AGO mediated mechanism; linc-MD1, a muscle-specific lncRNA, functions in the nucleus as a pri-miRNA host gene, while it also exports in the cytoplasm acting in a "target mimetic" fashion for two miRNAs. An extended collection of functional direct miRNA–lncRNA interactions are described in Table 1.

Several other lncRNA–miRNA indirect interactions have been identified by low throughput expression experiments that quantify miRNA effect on mRNA levels and vice versa [41, 42].

The complex network of miRNA–lncRNA–mRNA regulatory machinery is difficult to be determined by exploring individual pairs of interactions. To this end, CLIP-Seq (Crosslinking and Immunoprecipitation) techniques, a family of high-throughput Next Generation Sequencing methodologies, have been implemented and applied on different cell lines and tissues [18, 25, 43]. They are able to chart cross-linked miRNA: AGO binding sites in a transcriptome-wide scale and have revolutionized miRNA–gene interactions research. CLIP-Seq experiments are cell-based cross-linking protocols that identify binding sites of AGO RNA-binding protein and miRNA-containing ribonucleoprotein complex (miRNPs). HITS-CLIP (High-throughput sequencing of RNA isolated by crosslinking immunoprecipitation) [44] is considered as the first CLIP-Seq methodology. PAR-CLIP [45] (Photoactivatable-Ribonucleoside-Enhanced Crosslinking and Immunoprecipitation), a slightly different technique, provides a nucleotide resolution on the miRNA binding sites via the incorporation of T-to-C transition sites in the miRNPs. Recently introduced modified versions of HITS/PAR-CLIP methodologies, such as CLASH (crosslinking, ligation, and sequencing of hybrids) [46] experiment, also reveal a fraction of chimeric microRNA–target ligated pairs. Each of these techniques has its own merits and disadvantages [43].

The number of publications that describe in miRNA–lncRNA regulation is increasing. The collection of the related literature is already considered as a demanding and time consuming practice. The manual curation can be assisted by text-mining pipelines successfully applied for the inquiry of miRNA–gene interactions [43].

Table 1
A collection of direct miRNA-lncRNA experimentally verified interactions. lncRNA sponge/decoy function is characterized by disease, tissue specificity and is encountered in cytoplasm and/or nucleus. Certain interactions appear to be a conserved mechanism identified in more than one species

Publication	lncRNA	miRNA	Tissue	Organism	Compartment
Hansen et al. [16]	CDR1as/ ciRS-7	miR-671	Embryonic kidney	*H. sapiens*	Nucleus
Memczak et al. [17], Hansen et al. [18]	CDR1as/ ciRS-7	miR-7	Embryonic kidney, brain	*H. sapiens*, *M. musculus*, *C. elegans*, *Zebrafish*	Cytoplasm, nucleus
Wang et al. [9]	HULC	miR-372	Liver	*H. sapiens*	Cytoplasm, nucleus
Calin et al.[19]	T-UCRs	miR-155, miR-24, miR-29b	Bone marrow	*H. sapiens*	–
Faghihi et al. [20]	BACE1-AS	miR-485-5p	Brain	*H. sapiens*, *M. musculus*	Cytoplasm
Yoon et al. [21]	lincRNA-p21	let-7	Cervix, embryo	*H. sapiens*, *M. musculus*	Cytoplasm
Poliseno et al. [22]	PTENP1	sequesters miRNAs that target PTEN	Prostate	*H. sapiens*	Cytoplasm
Cesana et al. [23]	linc-MD1	miR-133	Muscle	*H. sapiens*	Cytoplasm
Cesana et al. [23]	linc-MD1	miR-135	Muscle	*H. sapiens*	Cytoplasm
Franco-Zorrilla et al. [24]	IPS1	miR-399	–	*A. thaliana*	–
Imig et al. [25]	H19	miR-106a	Myoblast	*H. sapiens*	Cytoplasm
Imig et al. [25]	H19	miR-17-5p	Myoblast	*H. sapiens*	Cytoplasm
Imig et al. [25]	H19	miR-20b	Myoblast	*H. sapiens*	Cytoplasm
Kallen et al. [26]	H19	let-7	Muscle	*H. sapiens*, *M. musculus*	Cytoplasm
Zhou et al. [27]	linc-RoR	miR-145	ESC	*H. sapiens*	Cytoplasm

(continued)

Table 1
(continued)

Publication	lncRNA	miRNA	Tissue	Organism	Compartment
Wang et al. [28]	CHRF	miR-489	Heart	*H. sapiens, M. musculus*	–
Wang et al. [29]	MALAT1	miR-101	ESCC	*H. sapiens*	–
Wang et al. [29]	MALAT1	miR-217	ESCC	*H. sapiens*	–
Leucci et al. [30]	MALAT1	miR-9	Brain	*H. sapiens*	Nucleus
Han et al. [31]	MALAT1	miR-125b	Bladder	*H. sapiens*	–
Wang et al. [32]	UCA1	miR-1	Urinary bladder	*H. sapiens*	–
Prensner et al. [33]	*PCAT-1*	miR-3667-3p	Prostate	*H. sapiens*	Cytoplasm, nucleus
Wang et al. [34]	*CASC2*	miR-21	Brain	*H. sapiens*	–
Wang et al. [35]	MDRL	miR-361	Cardiomyocytes	*M. musculus*	Nucleus, cytoplasm
Gao et al. [36]	HOST2	let-7b	Ovary	*H. sapiens*	–
Chiyomaru et al. [37]	HOTAIR	miR-34a	Prostate	*H. sapiens, M. musculus*	–
Ma et al. [38]	HOTAIR	miR-130a	Gallbladder	*H. sapiens*	–
Cao et al. [39]	UFC1	miR-34a	Liver	*H. sapiens, M. musculus*	Cytoplasm
Zhang et al. [40]	GAS5	miR-21	Breast	*H. sapiens, M. musculus*	Cytoplasm

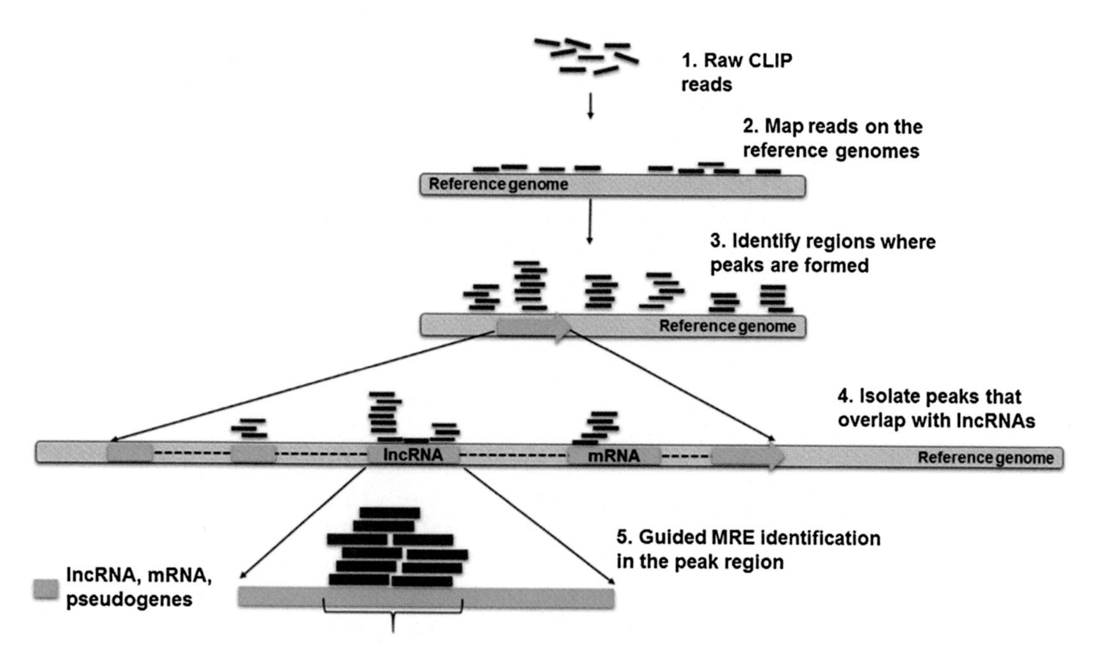

Fig. 2 Raw CLIP-seq data were initially processed for contaminant removal and reads were aligned against the reference genome. Enriched regions in CLIP-Seq signal are formed from overlapping reads. Peaks were annotated in transcript loci. A CLIP-peak-guided MRE search algorithm was utilized to compute interactions of expressed miRNAs

3.3 DIANA-LncBase

DIANA-LncBase (http://www.microrna.gr/LncBase/) [13] provides a comprehensive collection of >5000 experimentally supported and >10^6 in silico predicted miRNA–lncRNA interactions.

DIANA-LncBase is the first extensive collection of experimentally supported miRNA–lncRNA interactions. The interactions were identified by analyzing published PAR-CLIP and HITS-CLIP datasets with an in-house developed pipeline.

The analysis of CLIP-seq data is summarized in the following steps (Fig. 2):

Preprocessing of deep sequencing data. Tools such as FASTQC (www.bioinformatics.babraham.ac.uk/projects/fastqc), and cutadapt (https://code.google.com/p/cutadapt/) [47] were utilized for data quality control, contaminant detection and removal.

Alignment of reads. Alignment of CLIP-Seq reads against the reference genome was performed with GMAP/GSNAP [49], accordingly parameterized in order to identify reads in splice junctions.

Identification of CLIP-Seq enriched regions. Regions enriched in CLIP-Seq reads were formed by overlapping reads. In PAR-CLIP data, peaks were filtered to retain only regions with adequate T-to-C (sense strand) or A-to-G (antisense strand) incorporation in the same position (>5 % of the reads).

Annotation of peaks. A comprehensive reference set of transcripts including mRNAs, lncRNAs and pseudogenes was utilized for the annotation of enriched CLIP-Seq regions.

Guided MRE identification. A CLIP-peak-guided MRE search algorithm was utilized to compute interactions of expressed miRNAs taking into consideration the binding type, binding free energy, AU flanking content and MRE conservation.

3.4 miRNA–lncRNA In Silico Predicted Interactions

Most of the miRNA target prediction algorithms function in regions of mRNAs. miRNA–lncRNA interactions have not yet been characterized with in silico computational approaches.

DIANA-LncBase provides an extensive collection of in silico predicted miRNA–lncRNA targets. Predictions are provided for all the available miRNAs in miRBase v18 versus an integrated set of transcripts from different lncRNA resources. The interactions have been generated utilizing an appropriately modified version of the latest DIANA-microT [50] algorithm, applied in lncRNA spliced transcripts. Important features of microT algorithm for MRE scoring are the site accessibility (Sfold [51]), the binding free energy (RNA-Hybrid [52]), the AU flanking Content and the binding type. The miRNA binding site conservation feature has been removed from microT scoring model due to the observed variation of lncRNA sequence conservation. The algorithm combines individual MRE scores in order to predict a cumulative miRNA–lncRNA score.

3.5 LncBase Interface and Services: Utilization of the LncBase Repository

LncBase consists of two modules for experimentally and computationally derived miRNA–lncRNA interactions respectively.

The database allows advanced queries of multiple miRNA and lncRNA terms, specialized genomic locus search of predicted MREs, different filtering of the produced results, miRNA–lncRNA interaction visualization in the UCSC browser as well as a seamless interconnection with other DIANA tools. The interface is enhanced with lncRNA expression information, miRNA/target links to external databases, target conservation and miRNA related MeSH (Medical Subject Headings) terms.

3.5.1 LncBase Module of Experimentally Supported Interactions

This module is utilized in order to explore experimentally supported interactions. The user can enter miRNA and lncRNA names or identifiers separated by spaces. The results of a combined query with multiple miRNAs (hsa-miR-424-5p, hsa-let-7b-5p, hsa-miR-126-5p, hsa-miR-101-3p, hsa-miR-15b-5p, hsa-miR-34a-5p, hsa-miR-29b-3p) and lncRNAs (MALAT1, XIST, GAS5) are displayed in Figs. 3 and 4. miRNA–lncRNA displayed interactions include pairs of the queried terms.

"miRNA details" section in LncBase enables miRNA-guided analysis with other DIANA tools such as DIANA-microT-CDS [50] and DIANA-miRPath [14]. miRNA hsa-miR-126-5p that

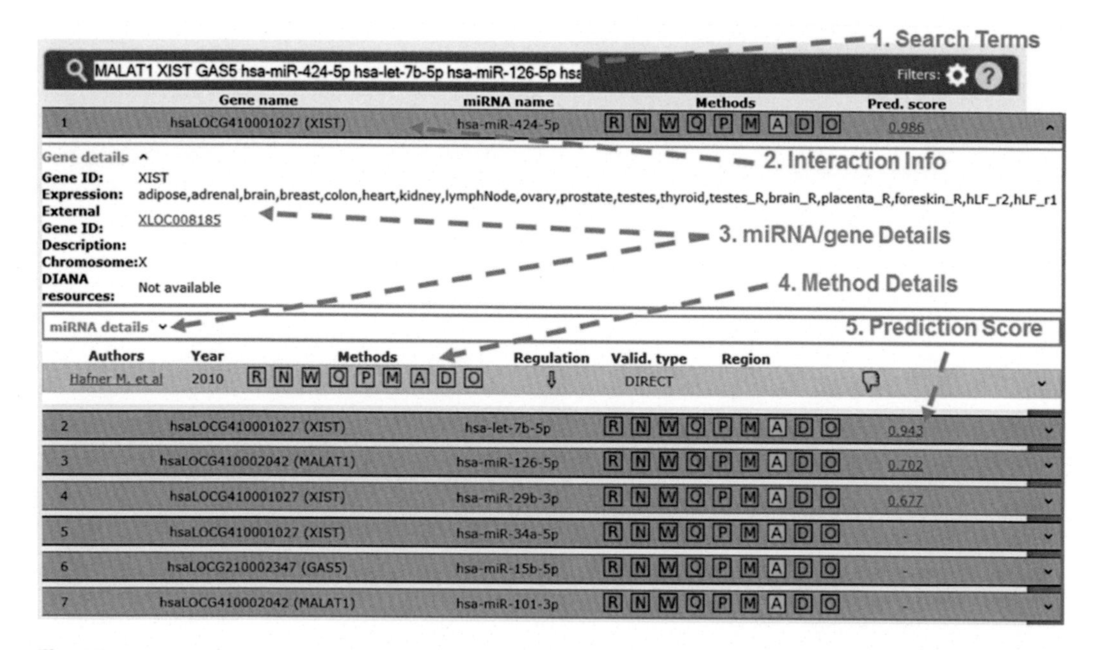

Fig. 3 A combined query using the LncBase module of experimentally supported interactions. miRNA–lncRNA displayed interactions include pairs of the queried terms. Further information, such as miRNA/gene details, external databases links, description of method validation, and related publications, is presented by expanding relevant result panels in each interaction. miRNA–lncRNA predicted interactions are escorted by microT scores, which constitute active links to corresponding entries in LncBase target prediction module

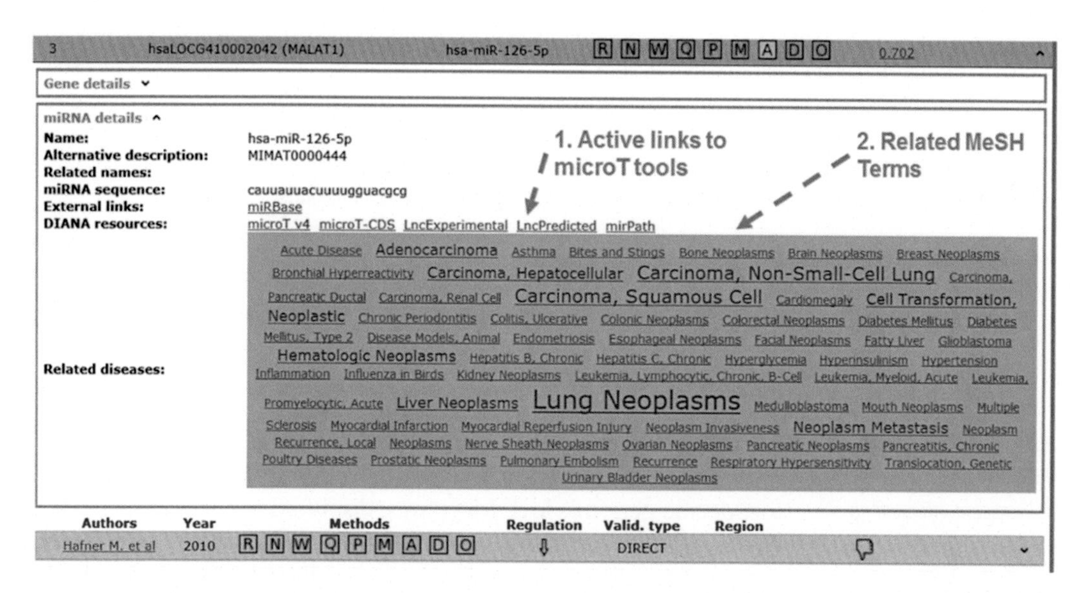

Fig. 4 The "*miRNA details*" section presents a weighted tag cloud with Medical Subject Headings (*MeSH*) terms derived from MedLine publications. Active links to corresponding miRNA entries in miRBase and other microT tools are also accessible in the miRNA expanded panel. miRNAs can be subjected to a pathway analysis or search of their mRNA targets following the provided links

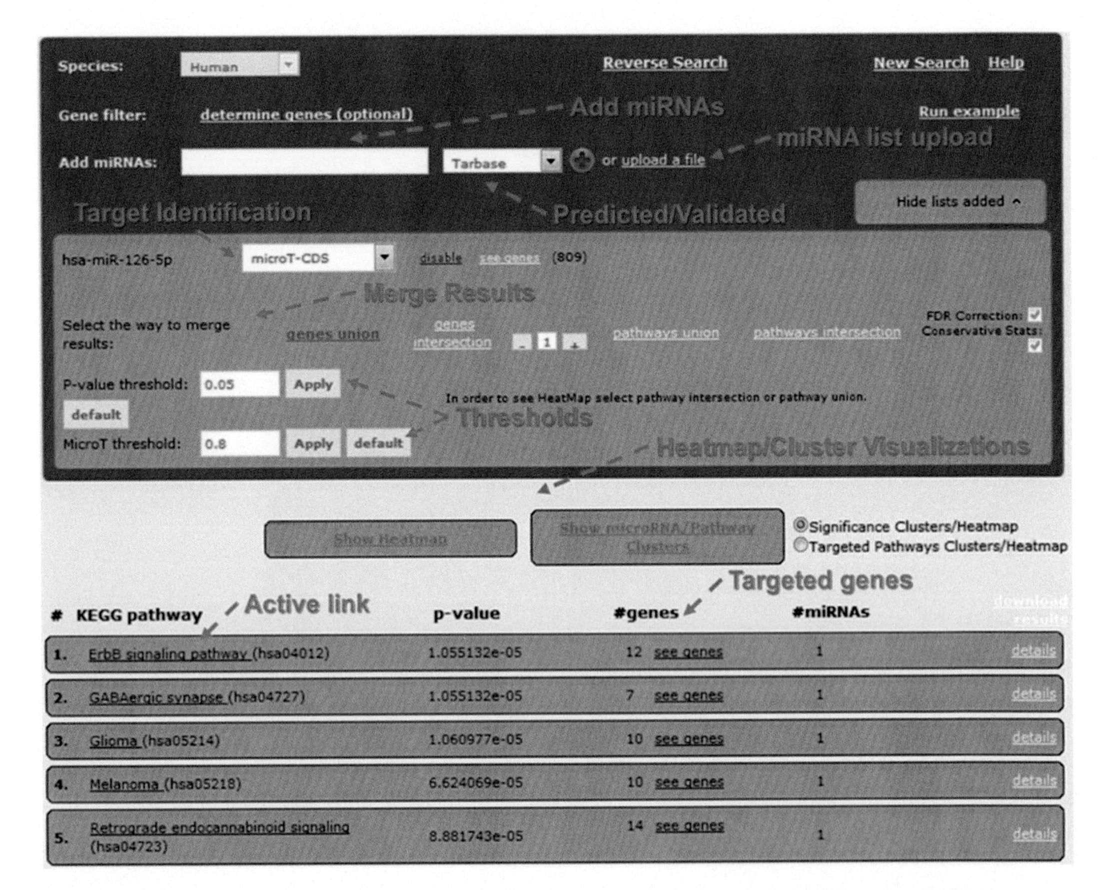

Fig. 5 miRNA hsa-miR-126-5p that targets MALAT1 based on LncBase experimentally supported interactions and in silico predictions is subjected to a pathway analysis using DIANA-miRPath. Optionally, the user can upload more miRNAs and select to either include their validated or predicted mRNA targets in the functional analysis. Several user-defined program options are provided, including: merging method selection, enrichment calculation methodologies as well as parameterization of microT score and *p*-values of targeted pathways. Sophisticated heatmap/cluster visualizations are available along with pathways merging methods selection. *Underlined pathway* descriptions are active links to enriched KEGG representations

targets MALAT1 based on LncBase experimentally supported interactions and in silico predictions is subjected to a pathway analysis with DIANA-miRPath, following the relevant link in LncBase interface (Fig. 5).

3.5.2 Querying the LncBase Module of Predicted Interactions

Additional search options are available in the LncBase predicted interactions module. Putative MREs on lncRNAs can be explored for specific genomic loci. Moreover, displayed results can be filtered based on lncRNA tissue expression data and microT interaction scores. The utilization of diverse user selected microT thresholds are highly recommended for fine-tuning prediction sensitivity and precision levels of the provided results.

LncBase supported predictions for a specific miRNA-lncRNA pair (hsa-miR-424-5p, XIST) and relevant visualization of MREs in a UCSC graphic are shown on Figs. 6 and 7.

XIST hsa-miR-424-5p — 1. Search Terms — Filters:

	Gene Id	miRNA name	miTG score	Experimentally Verified
1	hsaLOCG410001027 (XIST)	hsa-miR-424-5p	0.986	☐

2. Interaction Info

Gene details
miRNA details
pubMed links: miRNA | gene | both
UCSC graphic — 3. Visualization of MREs in UCSC graphic

Binding Type	Transcript position	Score	Conservation
7mer	5-33	0.0127679696477624	3

Position on chromosome: X:73072555-73072583
Conserved species: panTro2,rheMac2,canFam2
Binding area:

```
(Transcript)5'AGUUC          C  C    C        3'
                   UUAAAGCG UG AAUU GCUGCUG
                   I.III.I. II IIII IIIIIII
                   AGUUUUGU AC UUAA CGACGAC
(miRNA)       3'   A                          5'
```

4. MRE Info

Binding Type	Transcript position	Score	Conservation
6mer	501-529	0.00490171978280588	4
9mer	805-833	0.0217809948418491	1
6mer	3513-3541	0.00221694626390145	1
9mer	3940-3968	0.00621381155949468	1
6mer	4461-4489	0.0017446090225633	2
9mer	5562-5590	0.0133817377740668	1
7mer	10130-10158	0.0062804995238384	2
8mer	14883-14911	0.00790534996382945	0
6mer	16056-16084	0.00175891516001269	0
6mer	17956-17984	0.000785189744348558	0

5. Seed Conservation

Fig. 6 A combined query using LncBase computationally derived interactions module. miRNA/gene details are provided by expanding the result panel and by selecting the information links. Each interaction is described by a microT score, a color coded indication of experimental support and is escorted by description of the predicted MREs. Each MRE entry is enriched with additional information, such as the binding type, graphical representation of the binding area, MRE location on the transcript, the prediction score, and the number of species that the MRE was found to be conserved. Interactions can be further visualized and processed in the UCSC genome browser

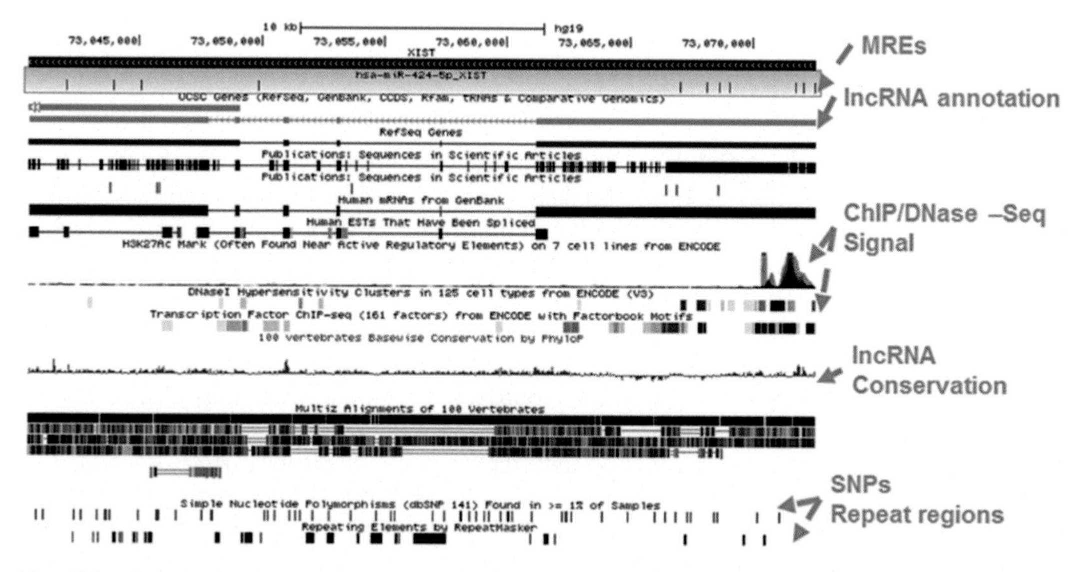

Fig. 7 Visualization of lncRNA–miRNA interaction in UCSC genome browser graphic upon user selection in the LncBase interface. MREs are shown along the annotated (un)spliced lncRNA transcript. Extra information tracks regarding ChIP/DNase-Seq signal, sequence conservation, SNPs and repeat regions are also provided. The *graphical representation* is an active link to the UCSC genome browser where the user is facilitated with all the available browser options

3.6 Interpretation of miRNA–lncRNA Interactions

lncRNA–miRNA interactions can be incorporated in miRNA functional analyses and are considered essential to the understanding of the ncRNA role in disease-, developmental stage-, and tissue-specific mechanisms. LncBase in silico predicted/experimentally supported data are freely available for further experimentation with in-house developed algorithms and custom pipelines. The precompiled data can be accessed from registered users in the DIANA webserver.

Two distinct formats are provided for the files that contain in silico and experimentally derived predictions.

miRNA–lncRNA experimentally supported entries in the downloaded file are described by a fixed number of comma separated fields. Every line consists of miRNA MIMAT, miRNA name, gene identifier, gene name as well as a positive indication of the experimental low/high-throughput method that supports the interaction.

A sample of the adopted miRNA–lncRNA experimentally supported interactions file format:

miRNA mimat, miRNA-name(miRBase-version), gene Identifier, gene Name, method1, method2, method3,…, methodn

MIMAT0000066,hsa-let-7e-5p(18),ENSG00000247556, RP11-46M12.1,\N,\N,\N,\N,\N,POSITIVE,\N,\N

MIMAT0000145, mmu-miR-133a-3p(18),linc-MD1,linc-MD1, POSITIVE,\N,\N,\N,\N,\N,\N,\N

MIMAT0000724,hsa-miR-372(18),ENSG00000251164,HULC,POSITIVE,\N,\N,\N,\N,\N,\N,\N

MIMAT0000732,hsa-miR-378a-3p(18),XLOC004784,CTD-2653M23.2,\N,\N,\N,\N,\N,POSITIVE,\N,\N

MIMAT0001341,hsa-miR-424-5p(18),XLOC008185,XIST,\N,\N,\N,\N,\N,POSITIVE,\N,\N

On the other hand, the downloaded miRNA–lncRNA interactions from the predicted lncBase module support details concerning the genomic location of the predicted MREs. Each interaction record is again a list of comma separated fields: Transcript Identifier, Gene Identifier followed by Gene Name, mature miRNA name, and microT interaction score, while the subsequent lines mark miRNA binding sites (MREs) within this lncRNA. MRE associated lines provide location (chromosome: coordinates) and the score information for the specific binding sites. The genomic coordinates of an MRE located in a splice junction are reported as *chromosome:MRE_start1;MRE_start2- MRE_stop1;MRE_stop2* as shown in the following example.

The adopted miRNA–lncRNA predicted interactions file notation is as follows:

TranscriptId,GeneId(name),Mirna-Name(miRBase-version), miTG-score

ENST00000554595,ENSG00000258827(RP11-537P22.2), hsa-miR-98(18),0.488

14:37567326-37567354,0.00992228797303395

ENST00000423737,ENSG00000224271(RP11-191 L9.4.1), hsa-let-7c(18),0.531

22:48134313;48250660-48134325;48250675, 0.00255419691016231

22:48251061-48251089,0.00502701330001512

22:48251074-48251102,0.00516245778717435

ENST00000426635,ENSG00000233237(LINC00472), hsa-let-7c(18),0.503

6:72130402-72130430,0.00430991581985486

Different ways of preprocessing and filtering miRNA–lncRNA interaction files can be applied.

An example of how to filter the predicted interactions file "lncPredicted_data.csv" is presented below. Interactions are initially filtered by specific miRNA/gene identifiers listed in the file: "ids_to_filter.txt". Remaining entries are subsequently checked (2nd filtering step) in order to report interactions with a microT score >0.7. The following command should be used in a terminal of a Linux-like operating system.

grep –f ids_to_filter.txt lncPredicted_data.csv|awk –F ',' '{if ($4>0.7) print}'>results.txt

The command will create a file named "results.txt" containing the refined interactions.

It is required that the downloaded data will be accordingly reformatted in order to be seamlessly included into multistep analyses.

3.7 LncBase Predictions into BED Format

BED format (http://genome.ucsc.edu/FAQ/FAQformat.html) is an efficient way to represent data described by genomic locations and is considered as a widely accepted input format for many bioinformatics tools and databases.

Registered users in DIANA tools server can change the notation of LncBase in silico predicted interactions into a BED format, by utilizing the appropriate perl script available in their download area. MREs in BED format can be subsequently color coded according to BED specifications and displayed as a custom track in a UCSC genome browser. The file can be also further processed with BEDtools (http://bedtools.readthedocs.org/en/latest/) in order to identify annotated regions, of certain regulation and function, overlapping miRNA binding sites on lncRNA.

3.8 Conservation of miRNA–lncRNA Interactions

The computationally predicted MREs in BED format can be examined for their conservation in the UCSC Genome Browser website following the next steps:

1. Open a UCSC Table browser (http://genome.ucsc.edu/cgi-bin/hgTables) and perform a species selection by specifying clade, genome, and assembly.

2. Select Comparative Genomic group and a relevant conservation table (e.g., phastCons or phyloP scores [53]).
3. In the "region" panel of UCSC table, select "define region" in order to upload the MREs BED file.
4. A wiggle file (http://genome.ucsc.edu/goldenpath/help/wiggle.html) with per base conservation is returned as an output.

4 Notes

1. DIANA tools webserver provides an extensive documentation of tool/database filtering options, functionalities, and detailed description of the allowed queried terms.
2. lncRNA/miRNA identifiers differ in-between databases and therefore the adopted naming nomenclature should be checked thoroughly to avoid inconsistencies in the results. At the moment, LncBase provides lncRNA–miRNA interactions for miRNAs included in miRBase v18, and lncRNAs retrieved from Cabili et al. [5], NONCODEv3 [54], and Ensembl v65 [55].
3. The genome assemblies utilized for in silico predictions in LncBase v1.0 are mm9 and hg19 for mouse and human species, respectively.
4. In silico predicted miRNA–lncRNA interactions are assigned in human and mouse species following "hsa" and "mmu" prefixes utilized in miRNA names.
5. The in silico predicted LncBase interactions cannot be easily processed and viewed with common file editors due to the large file size (>1 GB). The results file size can be reduced after appropriate filtering of the provided interactions.
6. microT threshold utilization filters predicted interactions with lower assigned scores. Sensitivity levels are increased with the use of low threshold values, while precision decreases. A recommended loose threshold is 0.7 corresponding in 28.3 % sensitivity and 50.8 % precision. On the other hand 0.9 is considered a more stringent microT threshold.

Acknowledgements

The authors would like to thank Ioannis S Vlachos, Dimitra Karagkouni, and Georgios Georgakilas for their helpful comments and suggestions.

This work has been supported from the project "TOM", "ARISTEIA" Action of the "OPERATIONAL PROGRAMME EDUCATION AND LIFELONG LEARNING" and is cofunded by the European Social Fund (ESF) and National Resources.

References

1. ENCODE Project Consortium (2012) An integrated encyclopedia of DNA elements in the human genome. Nature 489(7414):57–74
2. Huntzinger E, Izaurralde E (2011) Gene silencing by microRNAs: contributions of translational repression and mRNA decay. Nat Rev Genet 12(2):99–110
3. Griffiths-Jones S (2010) miRBase: microRNA sequences and annotation. Curr Protoc Bioinformatics Chapter 12:Unit12.9.1–Unit 12.9.10
4. Baker M (2011) Long noncoding RNAs: the search for function. Nat Methods 8(5):379–383
5. Cabili MN et al (2011) Integrative annotation of human large intergenic noncoding RNAs reveals global properties and specific subclasses. Genes Dev 25(18):1915–1927
6. Johnsson P et al (2014) Evolutionary conservation of long non-coding RNAs; sequence, structure, function. Biochim Biophys Acta 1840(3):1063–1071
7. Rinn JL, Chang HY (2012) Genome regulation by long noncoding RNAs. Annu Rev Biochem 81:145–166
8. Gutschner T, Diederichs S (2012) The hallmarks of cancer: a long non-coding RNA point of view. RNA Biol 9(6):703–719
9. Wang J et al (2010) CREB up-regulates long non-coding RNA, HULC expression through interaction with microRNA-372 in liver cancer. Nucleic Acids Res 38(16):5366–5383
10. Klein U et al (2010) The DLEU2/miR-15a/16-1 cluster controls B cell proliferation and its deletion leads to chronic lymphocytic leukemia. Cancer Cell 17(1):28–40
11. Cai X, Cullen BR (2007) The imprinted H19 noncoding RNA is a primary microRNA precursor. RNA 13(3):313–316
12. Salmena L et al (2011) A ceRNA hypothesis: the Rosetta Stone of a hidden RNA language? Cell 146(3):353–358
13. Paraskevopoulou MD et al (2013) DIANA-LncBase: experimentally verified and computationally predicted microRNA targets on long non-coding RNAs. Nucleic Acids Res 41 (Database issue):D239–D245
14. Vlachos IS et al (2012) DIANA miRPath v. 2.0: investigating the combinatorial effect of microRNAs in pathways. Nucleic Acids Res 40(Web Server issue):W498–W504
15. Georgakilas G et al (2014) microTSS: accurate microRNA transcription start site identification reveals a significant number of divergent pri-miRNAs. Nat Commun 5:5700
16. Hansen TB et al (2011) miRNA-dependent gene silencing involving Ago2-mediated cleavage of a circular antisense RNA. EMBO J 30(21): 4414–4422
17. Memczak S et al (2013) Circular RNAs are a large class of animal RNAs with regulatory potency. Nature 495(7441):333–338
18. Hansen TB et al (2013) Natural RNA circles function as efficient microRNA sponges. Nature 495(7441):384–388
19. Calin GA et al (2007) Ultraconserved regions encoding ncRNAs are altered in human leukemias and carcinomas. Cancer Cell 12(3):215–229
20. Faghihi MA et al (2010) Evidence for natural antisense transcript-mediated inhibition of microRNA function. Genome Biol 11(5):R56
21. Yoon JH et al (2012) LincRNA-p21 suppresses target mRNA translation. Mol Cell 47(4):648–655
22. Poliseno L et al (2010) A coding-independent function of gene and pseudogene mRNAs regulates tumour biology. Nature 465(7301): 1033–1038
23. Cesana M et al (2011) A long noncoding RNA controls muscle differentiation by functioning as a competing endogenous RNA. Cell 147(2): 358–369
24. Franco-Zorrilla JM et al (2007) Target mimicry provides a new mechanism for regulation of microRNA activity. Nat Genet 39(8):1033–1037
25. Imig J et al (2015) miR-CLIP capture of a miRNA targetome uncovers a lincRNA H19-miR-106a interaction. Nat Chem Biol 11(2): 107–114
26. Kallen AN et al (2013) The imprinted H19 lncRNA antagonizes let-7 microRNAs. Mol Cell 52(1):101–112
27. Zhou X et al (2014) Linc-RNA-RoR acts as a "sponge" against mediation of the differentiation of endometrial cancer stem cells by micro RNA-145. Gynecol Oncol 133(2):333–339
28. Wang K et al (2014) The long noncoding RNA CHRF regulates cardiac hypertrophy by targeting miR-489. Circ Res 114(9):1377–1388
29. Wang X et al (2015) Silencing of long noncoding RNA MALAT1 by miR-101 and miR-217inhibits proliferation, migration and invasion of esophageal squamous cell carcinoma cells. J Biol Chem 290(7):3925–3935
30. Leucci E et al (2013) MicroRNA-9 targets the long non-coding RNA MALAT1 for degradation in the nucleus. Sci Rep 3:2535
31. Han Y et al (2013) Hsa-miR-125b suppresses bladder cancer development by down-regulating

oncogene SIRT7 and oncogenic long noncoding RNA MALAT1. FEBS Lett
32. Wang T et al (2014) Hsa-miR-1 downregulates long non-coding RNA urothelial cancer associated 1 in bladder cancer. Tumour Biol 35(10):10075–10084
33. Prensner JR et al (2014) The long non-coding RNA PCAT-1 promotes prostate cancer cell proliferation through cMyc. Neoplasia 16(11): 900–908
34. Wang P et al (2015) Long non-coding RNA CASC2 suppresses malignancy in human gliomas by miR-21. Cell Signal 27(2):275–282
35. Wang K et al (2014) MDRL lncRNA regulates the processing of miR-484 primary transcript by targeting miR-361. PLoS Genet 10(7), e1004467
36. Gao Y et al (2015) LncRNA-HOST2 regulates cell biological behaviors in epithelial ovarian cancer through a mechanism involving microRNA let-7b. Hum Mol Genet 24(3):841–852
37. Chiyomaru T et al (2013) Genistein inhibits prostate cancer cell growth by targeting miR-34a and oncogenic HOTAIR. PLoS One 8(8), e70372
38. Ma MZ et al (2014) Long non-coding RNA HOTAIR, a c-Myc activated driver of malignancy, negatively regulates miRNA-130a in gallbladder cancer. Mol Cancer 13:156
39. Cao C et al (2015) The long intergenic noncoding RNA UFC1, a target of MicroRNA 34a, interacts with the mRNA stabilizing protein HuR to increase levels of beta-Catenin in HCC cells. Gastroenterology 148(2):415–26.e18
40. Zhang Z et al (2013) Negative regulation of lncRNA GAS5 by miR-21. Cell Death Differ 20(11):1558–1568
41. Braconi C et al (2011) microRNA-29 can regulate expression of the long non-coding RNA gene MEG3 in hepatocellular cancer. Oncogene 30(47):4750–4756
42. Fan M et al (2013) A long non-coding RNA, PTCSC3, as a tumor suppressor and a target of miRNAs in thyroid cancer cells. Exp Ther Med 5(4):1143–1146
43. Vlachos IS et al (2015) DIANA-TarBase v7.0: indexing more than half a million experimentally supported miRNA:mRNA interactions. Nucleic Acids Res 43(Database issue):D153–D159
44. Chi SW et al (2009) Argonaute HITS-CLIP decodes microRNA-mRNA interaction maps. Nature 460(7254):479–486
45. Hafner M et al (2010) Transcriptome-wide identification of RNA-binding protein and microRNA target sites by PAR-CLIP. Cell 141(1):129–141
46. Helwak A et al (2013) Mapping the human miRNA interactome by CLASH reveals frequent noncanonical binding. Cell 153(3):654–665
47. Bolger AM, Lohse M, Usadel B (2014) Trimmomatic: a flexible trimmer for Illumina sequence data. Bioinformatics 30(15):2114–2120
48. Trapnell C, Pachter L, Salzberg SL (2009) TopHat: discovering splice junctions with RNA-Seq. Bioinformatics 25(9):1105–1111
49. Wu TD, Nacu S (2010) Fast and SNP-tolerant detection of complex variants and splicing in short reads. Bioinformatics 26(7):873–881
50. Paraskevopoulou MD et al (2013) DIANA-microT web server v5.0: service integration into miRNA functional analysis workflows. Nucleic Acids Res 41(Web Server issue): W169–W173
51. Ding Y, Chan CY, Lawrence CE (2004) Sfold web server for statistical folding and rational design of nucleic acids. Nucleic Acids Res 32(Web Server issue):W135–W141
52. Rehmsmeier M et al (2004) Fast and effective prediction of microRNA/target duplexes. RNA 10(10):1507–1517
53. Pollard KS et al (2010) Detection of nonneutral substitution rates on mammalian phylogenies. Genome Res 20(1):110–121
54. Bu D et al (2012) NONCODE v3.0: integrative annotation of long noncoding RNAs. Nucleic Acids Res 40(Database issue): D210–D215
55. Flicek P et al (2012) Ensembl 2012. Nucleic Acids Res 40(Database issue):D84–D90

Chapter 22

Long Noncoding RNAs: An Overview

Dongmei Zhang, Minmin Xiong, Congjian Xu, Peng Xiang, and Xiaomin Zhong

Abstract

Recently an explosion in the discovery of long noncoding RNAs (lncRNAs) was obtained by high throughput sequencing. Genome-wide transcriptome analyses, in conjugation with research for epigenetic modifications of chromatins, identified a novel type of non-protein-coding transcripts longer than 200 nucleotides named lncRNAs. They are gradually emerging as functional and critical participants in many physiological processes. Here we give an overview of the characteristics, biological functions, and working mechanism for this new class of noncoding factors.

Key words Long noncoding RNAs, Epigenetic, Stem cell, Cancer

1 Introduction

The discovery of a large number of long noncoding RNAs (lncRNAs) characterized so far was ascribed to the technology of high throughput sequencing, which has enabled global analyses of transcriptome and epigenetic modifications of chromatins. Early studies identified that the vast majority of human genome gives rise to a huge number of transcripts without protein-coding potential [1, 2], among which lncRNA is a class arbitrarily defined as transcripts longer than 200 nucleotides lacking a canonical open reading frame (ORF) [3–8]. The biological functions of these lncRNAs were poorly investigated. They were believed to be either gene trash or transcriptional noise. However, subsequent functional studies indicated that lncRNAs are not by-products of gene transcription. On the contrary, more and more evidence showed that they are independent and active participants in multiple important biological processes, such as cell proliferation, differentiation, migration, and apoptosis [6–12]. Advanced studies implied that the working mechanisms of lncRNA mostly converge on their ability of regulating gene expression at almost every level, from epigenetic modifications of chromatins, transcriptional control,

Yi Feng and Lin Zhang (eds.), *Long Non-Coding RNAs: Methods and Protocols*, Methods in Molecular Biology, vol. 1402, DOI 10.1007/978-1-4939-3378-5_22,

mRNA stability, to translational control [13–19]. Recently, rapid progresses in lncRNA research have inspired overwhelming enthusiasm for this field. Here we overview the characteristics, biological functions, and working mechanism of these noncoding factors for a better understanding of lncRNAs.

2 The Characteristics of LncRNAs

Most lncRNAs were first discovered and characterized by large-scale sequencing of cDNA libraries, and identifying transcriptional signature from RNA pol II binding and epigenetic modifications of chromatin [3, 14]. Their full-length sequences were further confirmed by genome-wide tiling arrays and transcriptome sequencing data. As a result, lncRNA gene loci were found to be located in various genomic contexts. LncRNA genes can reside in genomic regions deprived of any known genes (e.g., HOTAIR, H19) [20, 21], or introns of coding genes (e.g., COLDAIR located in the first intron of the flowering repressor locus FLC) [22]. LncRNAs are also transcribed as antisense transcripts of known genes, most possibly near the promoter or the 3′ end regions (e.g., XIST/TSIX, KCNQ1/KCNQ1OT1) [23, 24]. Meanwhile, the machinery of transcriptional initiation produces many species of noncoding RNAs with unknown functions in both sense and antisense directions. They are called divergent transcripts, and promoter-associated or enhancer-associated RNAs [11, 25, 26]. The transcription of lncRNAs is mostly RNA pol II-dependent. LncRNA transcripts usually have a 5′ terminal methylguanosine cap and are polyadenylated. They are often spliced, although with fewer exons than coding genes. There were also evidences showing some lncRNAs lacking 3′ poly(A) tails, which are possibly generated from RNA pol III promoters [27, 28]. In addition, the splicing of nucleolar RNAs gives rise to a subset of lncRNAs deficient of 5′ caps and 3′ poly(A) tails [29]. In summary, there are few unique features in lncRNA structure that can help differentiate them from other mRNA molecules due to their common ways of biogenesis. Considering the absence of an ORF, lncRNAs probably perform their specific biological functions by forming secondary and tertiary structures of RNA molecules.

3 Physiological Functions of LncRNAs

The noncoding properties of lncRNAs developed many interests in their real physiological functions. With the effort to elucidate the biological roles of lncRNAs, researchers have found that lncRNAs are widely involved in developmental processes, differentiation, and reprogramming of stem cells, as well as a variety of human diseases.

3.1 Roles in Developmental Processes

Although only a small number of lncRNAs have been well studied for their biological functions, many lncRNAs were reported to be involved in multiple aspects of developmental process, including X chromosome inactivation (XCI), stem cell pluripotency, and somatic cell reprogramming, as well as cell lineage specification.

3.1.1 X Chromosome Inactivation

X chromosome inactivation is an important event to accomplish gene dosage compensation in female embryos with a redundant X chromosome. So far as we know, XIST/TSIX is a sense–antisense pair of lncRNAs responsible for controlling XCI during embryo development. XIST is highly expressed from the inactivated X (Xi) chromosome, but not from activated X (Xa). It coats the whole Xi and acts as a scaffold to recruit silencing complexes, such as PRC2, to inactivate gene expression from the Xi [30–32]. Meanwhile, TSIX is only expressed from the Xa in an antisense direction from a promoter downstream of XIST. TSIX downregulates XIST expression by a number of mechanisms [32–35]. Hence XIST and TSIX manifest a reciprocal expression pattern. The interaction between XIST and TSIX leads to the stable maintenance of Xi and Xa for a long term.

3.1.2 Stem Cell Pluripotency and Somatic Cell Reprogramming

The ability of maintaining pluripotent status while keeping differentiation potential is critical for the self-renewal of embryonic stem (ES) cells and induced pluripotent stem (iPS) cells, i.e., reprogrammed somatic cells. The pluripotent state was well known to be maintained or obtained through reprogramming technology by protein-coding stem cell factors, such as OCT-4, SOX2, and NANOG. As a novel class of gene expression regulators, lncRNAs were reported to be involved in the network controlling stem cell pluripotency and reprogramming.

Lipovich et al. found OCT-4 and NANOG directly regulated the expression of several conserved lncRNAs associated with pluripotency [36]. Using a customized lncRNA microarray, Stanton and colleagues identified lncRNAs specifically involved in human ES cells pluripotency and neurogenesis [37]. These lncRNAs physically interacted with SOX2/ PRC2 in hES cells, and with REST/ SUZ12 during neurogenesis, implicating a suppressive role in target gene expression associated with pluripotency and neurogenesis. Furthermore, a large scale loss-of-function study on most lncRNAs expressed in mouse ES cells integrated them into the well-characterized circuitry controlling ES cell pluripotent state. Lander and colleagues found dozens of lncRNAs were regulated by key transcription factors in ES cells, and responsible for maintaining the undifferentiated state by affecting gene expression together with chromatin regulatory proteins [10].

Similarly, in the process of somatic cell reprogramming, Rinn et al. defined lncRNAs whose expression was elevated in iPS cells, and whose transcription was under direct control of stem cells factors. One such lncRNA (lincRNA-RoR) was demonstrated to modulate reprogramming process [38]. Recently, the dynamic change in lncRNA types and expression levels during somatic reprogramming process was further revealed by single cell transcriptome analysis. Hundreds of lncRNAs were activated to repress lineage-specific genes and regulate metabolic gene expression [39]. These evidences implied the critical roles of lncRNAs in de novo establishment of pluripotency.

3.1.3 Lineage Specification

To illuminate the role of lncRNAs in early embryo development, Li et al. reported lncRNA expression profiles specific for each stage of mouse embryo cleavage based on the comprehensive analysis of single cell RNA-seq data [40]. Their results indicated that lncRNA might serve as a novel type of developmental regulators and stage-specific markers for early development studies, similar to protein-coding genes. Meanwhile, in later developmental stages such as lineage commitment of progenitor cells in adult tissues, the functions of lncRNAs were more widely explored. A number of lncRNA genes are encoded within the HOX gene clusters, which determine the anterior–posterior patterning in bilateral metazoans, for example HOTAIR from HOXC cluster, and HOTTIP and Mistral from HOXA cluster. They were demonstrated to regulate gene expression of either the host or a distant HOX gene [20, 41, 42]. In epidermis, the lncRNAs ANCR and TINCR played an opposite role in committing the undifferentiated or differentiated state to progenitor cells [43, 44]. During adipocyte development, numerous lncRNAs were transcriptionally activated by key transcription factors for adipogenesis, and were required to various degrees for proper differentiation of adipocytes [45]. In the immune system, the lncRNA lnc-DC directly bound to and activated STAT3 to promote dendritic cell differentiation [46]. In addition, lncRNAs were also reported to be associated with spermatogenesis and mammary stem cell expansion [47, 48], etc.

3.2 Functions in Human Diseases

Association with the initiation and progression of human diseases, especially cancers, is an important part of lncRNA functions. HOTAIR was reported to promote metastasis of cancers, probably dependent on PRC2-mediated epigenetic modification on target genes [11]. The expression level of HOTAIR was later found to be upregulated and correlated with poor prognosis of multiple types of cancers [9, 49, 50]. ANRIL is a lncRNA increased in prostate cancer and interacts with both CBX7/PRC1 [51] and SUZ12/PRC2 [52] to repress the transcription of the tumor suppressor

INK4A/INK4B. lincRNA-p21, a lncRNA encoded by the genomic region adjacent to p21 gene, is upregulated by p53 upon DNA damage. By associating with hnRNP-K, lincRNA-p21 mediates the suppressor function of p53 on target gene expression and modulates p53-dependent apoptosis [12]. A recently identified lncRNA FAL1 manifests amplification in gene copy number and increase in transcript level in multiple types of cancers. FAL1 serves as an oncogene by interacting with BMI1/PRC1 to suppress p21 expression and enhance the growth rate of cancer cells [53]. In addition, more and more lncRNAs were found to be associated with tumorigenesis, such as MALAT-1 [54], PCAT-1 [55], and PANDA [9].

3.3 LncRNAs and Imprinting

Genes that are expressed reciprocally and monoallelically, based on their parental origins, are called imprinting genes. LncRNAs are also involved in genomic imprinting. Some imprinted gene loci encode reciprocally expressed lncRNA genes and protein-coding genes, e.g., H19, the first identified lncRNA from mammalian cells, imprinted with IGF2 gene [56]. Other examples include KCNQ1/KCNQ1OT1 [24] and AIR/IGF2R [57]. Some of these lncRNAs play a major role in silencing not only the respective imprinted genes, but also flanking protein-coding genes. They may function as either the antisense transcript for silencing the imprinted counterpart, or the scaffold for the recruitment of G9a and PRC2 complexes to suppress the neighboring gene expression [58].

4 Mechanism for lncRNA Function

Since lncRNAs are involved in various physiological processes, it has elicited vast interest in exploring the mechanism of their functions. In most cases, lncRNAs accomplish their biological roles by participating in every stage of gene expression regulation, i.e., from epigenetic modifications of chromatins, transcriptional control, to posttranscriptional control.

To modulate gene expression at epigenetic level, lncRNA transcripts usually act as scaffolds to recruit epigenetic modification factors to chromatin regions under regulation. LncRNA HOTAIR displays affinity to both protein and DNA, which enables it to guide silencing complexes PRC2 and LSD1/CoREST/REST to DNA regions for H3K27 methylation and H3K4 demethylation [16]. Both of these epigenetic modifications are repressive markers for the expression of a large number of HOTAIR target genes spread across the genome, such as HOXD cluster genes. Similarly, ANRIL binds SUZ12/PRC2 and CBX7/PRC1 to suppress CDKN2A and CDKN2B expression [51, 52]. KCNQ1OT1 recruits both PRC2 and G9a to promote H3K4 trimethylation and

H3K9 methylation at KCNQ1 imprinting locus to repress flanking target genes [58].

At transcriptional level, lncRNAs play their roles in either promoting or repressing gene expression through multiple mechanisms. A specific subtype of lncRNA has been implicated in upregulating gene expression by serving as transcriptional enhancers, termed enhancer RNA (eRNA) [13, 17]. In addition, lncRNAs can act directly on transcription factors to modulate gene expression. The noncoding repressor of NFAT, lncRNA NRON, inhibits nucleocytoplasmic shuttling of the transcription factor NFAT to repress downstream gene transcription [19]. Sometimes lncRNAs act as decoys for transcription factors or competes with transcription factors for DNA binding sites, e.g., lncRNA PANDA sequesters NF-YA protein away from its proapoptotic target genes [9], and LncRNA GAS5 (growth arrest specific 5) competes with glucocorticoid receptor (GR) for binding sites of target DNA sequences [59]. Intriguingly, lncRNAs also activate or repress gene expression by directly regulating the organization of nuclear compartment, such as MALAT1 and TUG1 [60].

Besides the roles in epigenetic modification and transcriptional control, lncRNAs also demonstrated functions in posttranscriptional control, including mRNA processing, mRNA stability, and translational efficiency. One example for controlling mRNA processing is that MALAT1 associates with mRNA alternative splicing through interactions with Ser/Arg splicing factors. By modulating the nuclear localization and levels of phosphorylated Ser/ Arg proteins, MALAT1 directly affects the alternative splicing pattern of target gene pre-mRNAs [61]. Just like protein-coding genes, some lncRNAs and transcripts from pseudogenes harbor microRNA (miRNA) binding sites in their sequences. They function as miRNA sponges to sequester miRNAs away from target mRNAs [18, 62, 63], thus stabilizing the mRNA molecules. LncRNAs even exert their functions in regulating translation. Together with the general translation repressor Rck, lincRNA-p21 associates with and represses the translation of β-catenin and JUNB mRNAs [15]. It was also reported that a translational regulatory lncRNA (treRNA) downregulates E-Cadherin expression by inhibiting the translation of its mRNA [64].

Acknowledgments

This work was supported, in whole or in part, by the Recruitment Project of Hundred Person of Sun Yat-Sen University (X.Z.), National Natural Science Foundation 81302262 (X.Z.). We apologize to scientists whose work is not discussed here due to space constraints.

References

1. Derrien T, Johnson R, Bussotti G et al (2012) The GENCODE v7 catalog of human long noncoding RNAs: analysis of their gene structure, evolution, and expression. Genome Res 22(9):1775–1789
2. Djebali S, Davis CA, Merkel A et al (2012) Landscape of transcription in human cells. Nature 489(7414):101–108
3. Guttman M, Amit I, Garber M et al (2009) Chromatin signature reveals over a thousand highly conserved large non-coding RNAs in mammals. Nature 458(7235):223–227
4. Lee JT (2012) Epigenetic regulation by long noncoding RNAs. Science 338(6113):1435–1439
5. Lieberman J, Slack F, Pandolfi PP et al (2013) Noncoding RNAs and cancer. Cell 153(1):9–10
6. Orom UA, Shiekhattar R (2013) Long noncoding RNAs usher in a new era in the biology of enhancers. Cell 154(6):1190–1193
7. Prensner JR, Chinnaiyan AM (2011) The emergence of lncRNAs in cancer biology. Cancer Discov 1(5):391–407
8. Ulitsky I, Bartel DP (2013) lincRNAs: genomics, evolution, and mechanisms. Cell 154(1): 26–46
9. Hung T, Wang Y, Lin MF et al (2011) Extensive and coordinated transcription of noncoding RNAs within cell-cycle promoters. Nat Genet 43(7):621–629
10. Guttman M, Donaghey J, Carey BW et al (2011) lincRNAs act in the circuitry controlling pluripotency and differentiation. Nature 477(7364):295–300
11. Gupta RA, Shah N, Wang KC et al (2010) Long non-coding RNA HOTAIR reprograms chromatin state to promote cancer metastasis. Nature 464(7291):1071–1076
12. Huarte M, Guttman M, Feldser D et al (2010) A large intergenic noncoding RNA induced by p53 mediates global gene repression in the p53 response. Cell 142(3):409–419
13. Orom UA, Derrien T, Beringer M et al (2010) Long noncoding RNAs with enhancer-like function in human cells. Cell 143(1):46–58
14. Khalil AM, Guttman M, Huarte M et al (2009) Many human large intergenic noncoding RNAs associate with chromatin-modifying complexes and affect gene expression. Proc Natl Acad Sci U S A 106(28):11667–11672
15. Yoon JH, Abdelmohsen K, Srikantan S et al (2012) LincRNA-p21 suppresses target mRNA translation. Mol Cell 47(4):648–655
16. Tsai MC, Manor O, Wan Y et al (2010) Long noncoding RNA as modular scaffold of histone modification complexes. Science 329(5992):689–693
17. Wang D, Garcia-Bassets I, Benner C et al (2011) Reprogramming transcription by distinct classes of enhancers functionally defined by eRNA. Nature 474(7351):390–394
18. Wang Y, Xu Z, Jiang J et al (2013) Endogenous miRNA sponge lincRNA-RoR regulates Oct4, Nanog, and Sox2 in human embryonic stem cell self-renewal. Dev Cell 25(1):69–80
19. Willingham AT, Orth AP, Batalov S et al (2005) A strategy for probing the function of noncoding RNAs finds a repressor of NFAT. Science 309(5740):1570–1573
20. Rinn JL, Kertesz M, Wang JK et al (2007) Functional demarcation of active and silent chromatin domains in human HOX loci by noncoding RNAs. Cell 129(7):1311–1323
21. Brannan CI, Dees EC, Ingram RS et al (1990) The product of the H19 gene may function as an RNA. Mol Cell Biol 10(1):28–36
22. Heo JB, Sung S (2011) Vernalization-mediated epigenetic silencing by a long intronic noncoding RNA. Science 331(6013):76–79
23. Lee JT, Davidow LS, Warshawsky D (1999) Tsix, a gene antisense to Xist at the X-inactivation centre. Nat Genet 21(4):400–404
24. Kanduri C, Thakur N, Pandey RR (2006) The length of the transcript encoded from the Kcnq1ot1 antisense promoter determines the degree of silencing. EMBO J 25(10):2096–2106
25. Core LJ, Waterfall JJ, Lis JT (2008) Nascent RNA sequencing reveals widespread pausing and divergent initiation at human promoters. Science 322(5909):1845–1848
26. He Y, Vogelstein B, Velculescu VE et al (2008) The antisense transcriptomes of human cells. Science 322(5909):1855–1857
27. Dieci G, Fiorino G, Castelnuovo M et al (2007) The expanding RNA polymerase III transcriptome. Trends Genet 23(12):614–622
28. Kapranov P, Cheng J, Dike S et al (2007) RNA maps reveal new RNA classes and a possible function for pervasive transcription. Science 316(5830):1484–1488
29. Yin QF, Yang L, Zhang Y et al (2012) Long noncoding RNAs with snoRNA ends. Mol Cell 48(2):219–230
30. Brown CJ, Hendrich BD, Rupert JL et al (1992) The human XIST gene: analysis of a 17 kb inactive

X-specific RNA that contains conserved repeats and is highly localized within the nucleus. Cell 71(3):527–542

31. Clemson CM, McNeil JA, Willard HF et al (1996) XIST RNA paints the inactive X chromosome at interphase: evidence for a novel RNA involved in nuclear/chromosome structure. J Cell Biol 132(3):259–275
32. Zhao J, Sun BK, Erwin JA et al (2008) Polycomb proteins targeted by a short repeat RNA to the mouse X chromosome. Science 322(5902):750–756
33. Sado T, Hoki Y, Sasaki H (2005) Tsix silences Xist through modification of chromatin structure. Dev Cell 9(1):159–165
34. Sun BK, Deaton AM, Lee JT (2006) A transient heterochromatic state in Xist preempts X inactivation choice without RNA stabilization. Mol Cell 21(5):617–628
35. Ogawa Y, Sun BK, Lee JT (2008) Intersection of the RNA interference and X-inactivation pathways. Science 320(5881):1336–1341
36. Sheik Mohamed J, Gaughwin PM, Lim B et al (2010) Conserved long noncoding RNAs transcriptionally regulated by Oct4 and Nanog modulate pluripotency in mouse embryonic stem cells. RNA 16(2):324–337
37. Ng SY, Johnson R, Stanton LW (2012) Human long non-coding RNAs promote pluripotency and neuronal differentiation by association with chromatin modifiers and transcription factors. EMBO J 31(3):522–533
38. Loewer S, Cabili MN, Guttman M et al (2010) Large intergenic non-coding RNA-RoR modulates reprogramming of human induced pluripotent stem cells. Nat Genet 42(12):1113–1117
39. Kim DH, Marinov GK, Pepke S et al (2015) Single-cell transcriptome analysis reveals dynamic changes in lncRNA expression during reprogramming. Cell Stem Cell 16(1):88–101
40. Zhang K, Huang K, Luo Y et al (2014) Identification and functional analysis of long non-coding RNAs in mouse cleavage stage embryonic development based on single cell transcriptome data. BMC Genomics 15:845
41. Bertani S, Sauer S, Bolotin E et al (2011) The noncoding RNA Mistral activates Hoxa6 and Hoxa7 expression and stem cell differentiation by recruiting MLL1 to chromatin. Mol Cell 43(6):1040–1046
42. Wang KC, Yang YW, Liu B et al (2011) A long noncoding RNA maintains active chromatin to coordinate homeotic gene expression. Nature 472(7341):120–124
43. Kretz M, Webster DE, Flockhart RJ et al (2012) Suppression of progenitor differentiation requires thc long noncoding RNA ANCR. Genes Dev 26(4):338–343
44. Kretz M, Siprashvili Z, Chu C et al (2013) Control of somatic tissue differentiation by the long non-coding RNA TINCR. Nature 493(7431):231–235
45. Sun L, Goff LA, Trapnell C et al (2013) Long noncoding RNAs regulate adipogenesis. Proc Natl Acad Sci U S A 110(9):3387–3392
46. Wang P, Xue Y, Han Y et al (2014) The STAT3-binding long noncoding RNA lnc-DC controls human dendritic cell differentiation. Science 344(6181):310–313
47. Liang M, Li W, Tian H et al (2014) Sequential expression of long noncoding RNA as mRNA gene expression in specific stages of mouse spermatogenesis. Sci Rep 4:5966
48. Zhang Y, Xia J, Li Q et al (2014) NRF2/long noncoding RNA ROR signaling regulates mammary stem cell expansion and protects against estrogen genotoxicity. J Biol Chem 289(45):31310–31318
49. Kogo R, Shimamura T, Mimori K et al (2011) Long noncoding RNA HOTAIR regulates polycomb-dependent chromatin modification and is associated with poor prognosis in colorectal cancers. Cancer Res 71(20):6320–6326
50. Niinuma T, Suzuki H, Nojima M et al (2012) Upregulation of miR-196a and HOTAIR drive malignant character in gastrointestinal stromal tumors. Cancer Res 72(5):1126–1136
51. Yap KL, Li S, Munoz-Cabello AM et al (2010) Molecular interplay of the noncoding RNA ANRIL and methylated histone H3 lysine 27 by polycomb CBX7 in transcriptional silencing of INK4a. Mol Cell 38(5):662–674
52. Kotake Y, Nakagawa T, Kitagawa K et al (2011) Long non-coding RNA ANRIL is required for the PRC2 recruitment to and silencing of p15(INK4B) tumor suppressor gene. Oncogene 30(16):1956–1962
53. Hu X, Feng Y, Zhang D et al (2014) A functional genomic approach identifies FAL1 as an oncogenic long noncoding RNA that associates with BMI1 and represses p21 expression in cancer. Cancer Cell 26(3):344–357
54. Ji P, Diederichs S, Wang W et al (2003) MALAT-1, a novel noncoding RNA, and thymosin beta4 predict metastasis and survival in early-stage non-small cell lung cancer. Oncogene 22(39):8031–8041
55. Prensner JR, Iyer MK, Balbin OA et al (2011) Transcriptome sequencing across a prostate can-

cer cohort identifies PCAT-1, an unannotated lincRNA implicated in disease progression. Nat Biotechnol 29(8):742–749
56. Bartolomei MS, Zemel S, Tilghman SM (1991) Parental imprinting of the mouse H19 gene. Nature 351(6322):153–155
57. Lyle R, Watanabe D, te Vruchte D et al (2000) The imprinted antisense RNA at the Igf2r locus overlaps but does not imprint Mas1. Nat Genet 25(1):19–21
58. Pandey RR, Mondal T, Mohammad F et al (2008) Kcnq1ot1 antisense noncoding RNA mediates lineage-specific transcriptional silencing through chromatin-level regulation. Mol Cell 32(2):232–246
59. Kino T, Hurt DE, Ichijo T et al (2010) Noncoding RNA gas5 is a growth arrest- and starvation-associated repressor of the glucocorticoid receptor. Sci Signal 3(107):ra8
60. Yang L, Lin C, Liu W et al (2011) ncRNA- and Pc2 methylation-dependent gene relocation between nuclear structures mediates gene activation programs. Cell 147(4):773–788
61. Tripathi V, Ellis JD, Shen Z et al (2010) The nuclear-retained noncoding RNA MALAT1 regulates alternative splicing by modulating SR splicing factor phosphorylation. Mol Cell 39(6):925–938
62. Cesana M, Cacchiarelli D, Legnini I et al (2011) A long noncoding RNA controls muscle differentiation by functioning as a competing endogenous RNA. Cell 147(2):358–369
63. Poliseno L, Salmena L, Zhang J et al (2010) A coding-independent function of gene and pseudogene mRNAs regulates tumour biology. Nature 465(7301):1033–1038
64. Gumireddy K, Li A, Yan J et al (2013) Identification of a long non-coding RNA-associated RNP complex regulating metastasis at the translational step. EMBO J 32(20):2672–2684

INDEX

Yi Feng and Lin Zhang (eds.), *Long Non-Coding RNAs: Methods and Protocols*, Methods in Molecular Biology, vol. 1402, DOI 10.1007/978-1-4939-3378-5, © Springer Science+Business Media New York 2016

L

M

N

O

P

Q

R

S

T

V